Planets in Binary Star Systems

Astrophysics and Space Science Library

For other titles published in this series, go to
http://www.springer.com/series/5664

Nader Haghighipour
Editor

Planets in Binary Star Systems

Springer

Editor
Dr. Nader Haghighipour
Institute for Astronomy and
NASA Astrobiology Institute
University of Hawaii
2680 Woodlawn Drive
Honolulu, HI 96822
USA
nader@ifa.hawaii.edu

ISSN 0067-0057
ISBN 978-90-481-8686-0 e-ISBN 978-90-481-8687-7
DOI 10.1007/978-90-481-8687-7
Springer Dordrecht Heidelberg London New York

Library of Congress Control Number: 2010927320

Cover design: eStudio Calamar S.L.

Cover illustration: GG Tau Circumstellar Disk
Date: 08/17/04
Credit: Daniel Potter/University of Hawaii Adaptive Optics Group/Gemini Observatory/AURA/NSF

Printed on acid-free paper

Springer is part of Springer Science+Business Media (www.springer.com)

Preface

In 1988, in an article on the analysis of the measurements of the variations in the radial velocities of a number of stars, Campbell, Walker, and Yang reported an interesting phenomenon; the radial velocity variations of γ Cephei seemed to suggest the existence of a Jupiter-like planet around this star. This was a very exciting and, at the same time, very surprising discovery. It was exciting because if true, it would have marked the detection of the first planet outside of our solar system. It was surprising because the planet-hosting star is the primary of a binary system with a separation less than 19 AU, a distance comparable to the planetary distances in our solar system.

The moderately close orbit of the stellar companion of γ Cephei raised questions about the reality of its planet. The skepticism over the interpretation of the results (which was primarily based on the idea that binary star systems with small separations would not be favorable places for planet formation) became so strong that in a subsequent paper in 1992, Walker and his colleagues suggested that the planet in the γ Cephei binary might not be real, and the variations in the radial velocity of this star might have been due to its chromospheric activities.

Despite the 1992 article, the search for planets in binaries did not stop. Gamma Cephei was continuously monitored and more precise measurements of its radial velocity variations were obtained. In 2003, 15 years after the first announcement of the planet of this system, these efforts fruited, and in an article in *Astrophysical Journal*, Hatzes and his colleagues confirmed the existence of a Jupiter-like planet around the primary of γ Cephei. The planet became real, and so became many challenges that it introduced to the planetary science.

The 2003 confirmation of γ Cephei's planet, and the subsequent detection of giant planets in three other moderately close binary stars, GL 86, HD 41004 and HD 196885, marked the beginning of a new era on theoretical and observational research on planets in dual-star systems. During the past few years, much research has been carried out in this area, and a large number of excellent articles have been published on different aspects of observational and theoretical studies of planets in moderately close binaries. The depth of these articles, combined with their great diversity and the rich history of literature on the dynamical evolution of planets in dual-star systems has turned the field of planets in binaries into a well-established and an independent branch of exoplanetary science. This book is intended to introduce this field to the community. In doing so, this volume presents the reader with the current state of the research on the detection and formation of planets in binary stars, written by teams of experts on these topics. The first half of the book focuses on the observational evidence for the birthplace of planets in binary systems, and techniques of detecting planets in and around dual-stars. The second half discusses the status of theoretical research on the formation of planets in binaries, from planetesimals, to planetary embryos, and eventually to giant and terrestrial planets. The last chapter presents a complete review of the dynamics of planets in binary star systems and the possibility of habitable planet formation in these environments.

In making of this book, I had the privilege of collaborating with an outstanding team of authors and referees. I am grateful to the authors for their participation in this project and for their responsiveness during the editorial phase. I am also indebted to the referees, Richard Durisen, Eric Jensen, John Johnson, Greg Laughlin, Mercedes Lopez-Morales, Fred Rasio, Kevin Rauch, John Rayner, Steinn Sigurdsson, Gordon Walker, Russel White, and Jason Wright, for accepting to review chapters of this book, and for their constructive comments and suggestions. Each chapter in this book has been reviewed by at least one of these reviewers and myself. I am also thankful to the NASA Astrobiology Institute at the University of Hawaii for their continuous support during this project, and to the NASA Astrobiology Central for their financial support for the production of this book.

IfA/UH-NAI
University of Hawaii
September 2009

Nader Haghighipour

Contents

Chapter 1
Disks Around Young Binary Stars

Lisa Prato and Alycia J. Weinberger

1.1 Introduction

Multiple star systems provide a complicated mix of conditions for planet formation. Whereas circumstellar disks around single stars are likely routine sites for planet formation, binary systems can have circumprimary (around the more massive star), circumsecondary (around the less massive star), and circumbinary (around both stars) disks. These heterogeneous locations can provide opportunities as well as hazards.

The frequency and separation of young binary populations are perhaps most important when examined in light of the impact of companion stars on the potential for planet formation. Even for star-forming regions in which the binary frequency is similar to that of the local field population, roughly two thirds of all member stars form in multiple systems. For a certain range of stellar separations, the presence of a companion star will clearly impact the formation, structure, and evolution of circumstellar disks, and, hence any potential planet formation. Another aspect of planet formation in young binaries is that we can assume that the stars in a binary are relatively coeval. Differences in the planet forming properties between two stars of similar age in the same local environment provide key information for understanding what *stellar* properties are more or less favorable for planets. Thus, the two stars in a young binary provide a built-in control sample.

It is an observational fact that among young stars in many nearby star forming regions (SFRs) an excess binary population exists (Ghez et al. 1993; Leinert et al. 1993; Simon et al. 1993 and reviews in Mathieu et al. 2007; Duchêne et al. 2007). This overabundance of young doubles, in comparison to field stars in the solar

L. Prato (✉)
Lowell Observatory, 1400 West Mars Hill Road, Flagstaff, AZ 86001
e-mail: lprato@lowell.edu

A.J. Weinberger
Department of Terrestrial Magnetism and NASA Astrobiology Institute,
Carnegie Institution of Washington, 5241 Broad Branch Road NW, Washington, DC 20015
e-mail: weinberger@dtm.ciw.edu

N. Haghighipour (ed.), *Planets in Binary Star Systems*, Astrophysics and Space Science Library 366, DOI 10.1007/978-90-481-8687-7_1,

neighborhood (Duquennoy and Mayor 1991), anti-correlates with the property of stellar density (Prosser et al. 1994; Petr et al. 1998; Patience et al. 2002; Beck et al. 2003). Thus, the denser clusters, in which most stars form, contain a lower fraction of bound multiple systems, comparable to the fraction found among field stars. The maximum separation of bound systems is also related to the stellar density. Based on analysis of a two-point correlation function, the transition between the binary and large-scale clustering regimes, and hence the cutoff separation for the likelihood of a bound pair, increases from 400 AU (Orion Trapezium) to 5,000 AU (Ophiuchus) to 12,000 AU (Taurus), while the average stellar surface density decreases (Simon 1997). Studies of large samples of binaries in a wide variety of star-forming regions are key to unravelling the nature of binary formation mechanisms and the impact of environment on multiplicity fraction, distribution, and evolution.

An insoluble problem among main-sequence field stars is the possibility of prior dynamical evolution of the system (Portegies Zwart and McMillan 2005). The interactions between young stars and their associated circumstellar and circumbinary disks may set in motion such evolution. Examining systems while they are still young tells us about the initial *potential* for planet formation. Field star observations tell us if this potential was realized.

For very small-separation binaries, models indicate that planet formation should be possible in a circumbinary disk (Quintana and Lissauer 2006). Several examples of close young binaries with circumbinary disks are well known, such as DQ Tau (Mathieu et al. 1997), UZ Tau E (Prato et al. 2002; Martín et al. 2005) and HD 98800 B (Koerner et al. 2000; Prato et al. 2001). Figures 1.1 and 1.2 show the system of HD 98800. These pairs have separations of approximately 30 solar-radii to 1 AU (Basri et al. 1997; Prato et al. 2002; Boden et al. 2005). The GG Tau and UY Aur binaries, with stellar separations of tens of AU, are surrounded by angularly large and therefore well-studied circumbinary disks (McCabe et al. 2002; Close et al. 1998). *Spitzer* Space Telescope observations of main-sequence pairs revealed circumbinary debris disk material in 12 systems with stellar separations of several solar-radii to $\sim$5 AU (Trilling et al. 2007). However, in spite of these promising disk observations and model predictions, no planet has yet been detected orbiting a small separation, main-sequence binary (although a 2.4 M_{Jup} minimum-mass planet orbits the G6V star HD 202206 and its 0.83 AU substellar companion; Udry et al. 2002). This dearth of detections may simply reflect the difficulties inherent in radial velocity (RV) searches for planets around binaries and the fact that binaries are typically eliminated from RV samples (Eggenberger et al. 2004; Konacki 2005).

Models also indicate a favorable outcome for planet formation in the circumstellar disks of wide binaries (Quintana et al. 2007). Reservoirs for this process, the optically thick, circumstellar disks around component stars, are routinely observed in binary systems with separations as small as $\sim$14 AU (Hartigan and Kenyon 2003). A *Spitzer* study of dust evolution in the circumstellar disks of wide binaries shows no difference in the initial processing stages, such as grain growth and crystallization, between the binary and single stars (Pascucci et al. 2008). More than

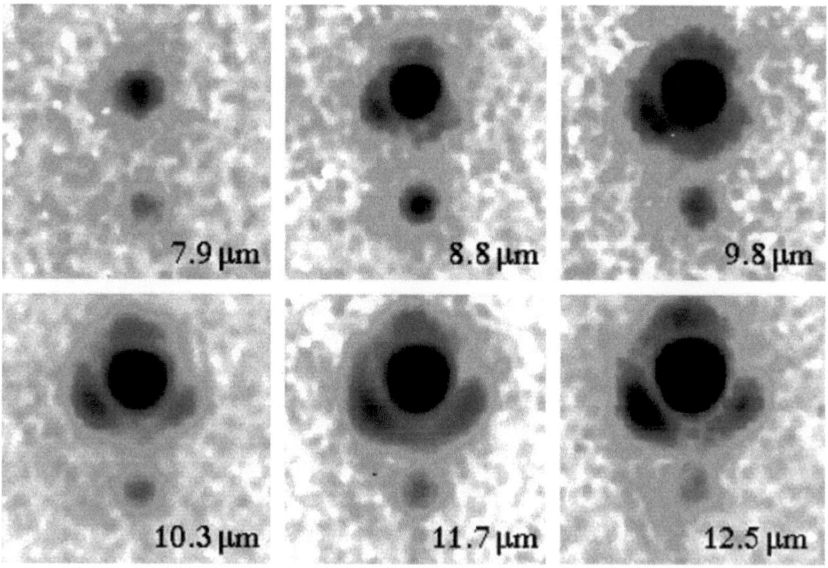

Fig. 1.1 Keck/MIRLIN imaging of the thermal infrared emission from the HD 98800 quadruple system oriented with up axis aligned due north. The spectroscopic binaries HD 98800A and HD 98800B are clearly resolved from each other and are identified, respectively, with northern and southern point sources separated by 0.8 arc sec (38 AU). Emission from HD 98800A steadily decreases with wavelength as λ^{-2} and is no longer detected in the 20 μm images. In contrast, radiation from the optical secondary, HD 98800B, increases dramatically out to 24.5 μm. Figure from Prato et al. (2001)

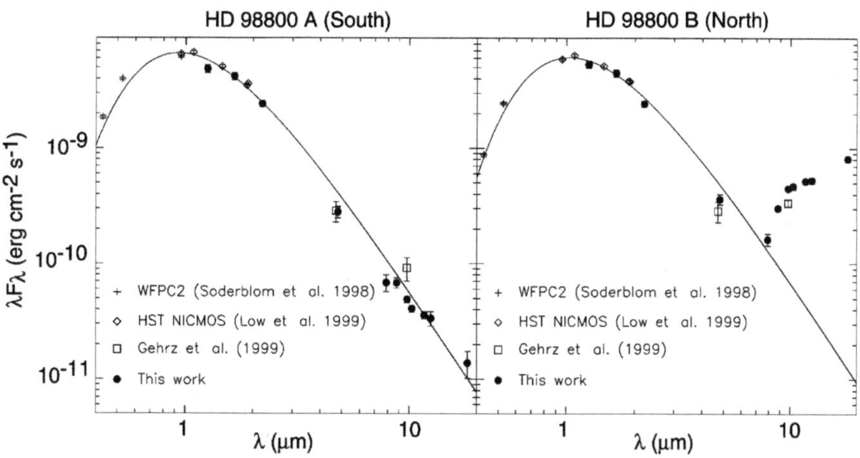

Fig. 1.2 Spectral energy distributions for HD 98800 A (*left*) and HD 98800 B (*right*). For HD 98800 A, no excess emission is evident out to the longest observed wavelength of 18.2 m; the 3932 K blackbody is plotted for reference. The 3562 K blackbody fit to HD 98800 B is also shown. Figure from Prato et al. (2001)

30 extrasolar planets (~20%) have been reported around one component in binaries with separations of tens of AU up to thousands of AU (Eggenberger et al. 2004; Raghavan et al. 2006) – circumstellar planet formation seems to be common in multiple systems.

Inevitable truncation of the outer portions of circumstellar disks in binaries with separations of a few to several tens of AU likely delineates a “planet-free” zone, at least for formation. Subsequent dynamical evolution in multiple systems could still bring planets into this region (Jang-Condell 2007). Interestingly, this fiducial separation is similar to that of the peak in the separation distribution for binaries in most SFRs (Patience et al. 2002). This planet-free regime of binary separation is also notably the least well-studied; components at such separations are too distant to be observed as spectroscopic binaries (orbital induced RV variations are on the order of star spot induced RV variations; Saar et al. 1998), yet too close to be easily angularly resolved. For example, two solar-mass stars in a circular orbit with a 10 AU semi-major axis would have a period of 22.4 years, and a maximum orbital velocity of ~1 km/s. At the 140 pc distance typical to the nearest SFRs, the maximum angular separation would be $0.07''$, slightly greater than the diffraction limit at a 10 m telescope in the near-infrared. Furthermore, angular resolution is more straightforward in a relatively face on orbit, but at the cost of modulation of the observed radial velocity by a $\sin i$ factor, where i is the angle between the plane of the sky and that of the orbit; a timescale of decades is required to observe a full cycle of radial velocity modulation in such a system.

A recent *Spitzer* study of the η Chamaeleontis cluster suggested that circumstellar disks were absent in 80% of the close binary systems while present around 80% of the single stars, although the study had a very small sample size and did not spatially resolve the binaries (Bouwman et al. 2006). At this point, such studies are suggestive, rather than definitive, of faster disk removal in close binaries.

It is not surprising that few data sets that go beyond initial binary identification exist, although there are some exceptions such as Hartigan and Kenyon (2003). We loosely define the binary separation regime most interesting, under-studied, and potentially treacherous to the formation and longevity of circumstellar disks, and therefore to the formation of planets, as spanning a few AU to 30 AU. This definition is naturally modulo eccentricity and mass-ratio, properties that could reinforce circumstellar disk destruction on short timescales.

In this chapter, we will discuss the current state of observations of disks in young multiple systems with an emphasis on circumstellar structures. Disks in solar analogue and low-mass stellar systems will be primarily considered. The topics covered in this review are (i) the evolution of inner disks in binaries (Section 1.2), (ii) the evolution of outer disks and the determination of disk masses as derived from submillimeter astronomy (Section 1.3), (iii) the orientation of disks in binary systems (Section 1.4), and (iv) the structure of debris disks in such environments (Section 1.5). We will present these topics through the lens of the potential for planet formation in these systems. In summary, Section 1.6 will present a discussion of future experiments and observations required to move knowledge in this field forward.

1.2 Inner Disks

Hydrogen emission line diagnostics (Hα or Brγ) and near-infrared colors are effective determinants of weak-lined (no or little inner disk material) and classical (optically thick inner disk) young stars (see Prato and Simon 1997; Martín 1998). Substantial line emission and near-infrared excesses attest to the presence of gas and warm dust located in the inner $\sim$1 AU of a circumstellar disk around a G–M spectral type young star. The inner few AU of a circumstellar disk delineate the likely site of terrestrial planet formation and giant planet migration, and thus are particularly important. Inner disks are thought to evolve quickly from optically thick to thin states; few systems have been found in the intermediate "transition" state. In an overview of about a dozen Taurus SFR transition objects, Najita et al. (2007) find that although their mass accretion rates are typically an order of magnitude lower than those of classical T Tauri stars, their median disk masses are about four times larger, consistent with a scenario in which an object with a massive disk is in transition because a Jovian-mass planet is opening a large gap and effectively starving the inner disk (D'Alessio et al. 2005). Several of Najita et al.'s transition systems are binaries (Najita et al. 2007), one with projected separation $\sim$120 AU (FQ Tau) and two with projected separations $\sim$30 AU (FO Tau and V773 Tau (Fig. 1.3); the A component of the latter is also a 51 day period spectroscopic binary; Welty 1995). These separations are relatively wide and easily allow for the presence of inner disks. If the interpretation of on-going planet formation is

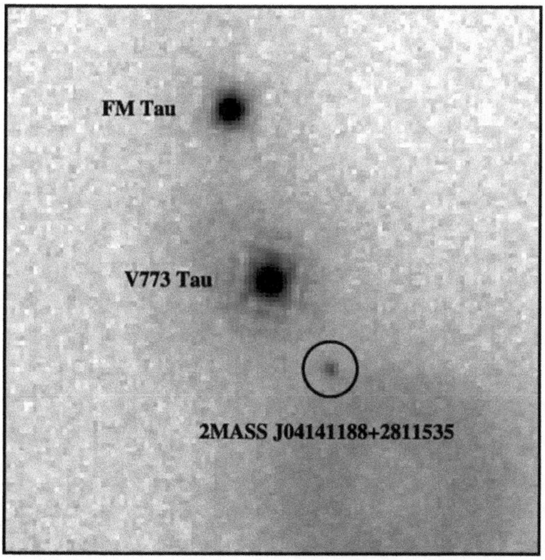

Fig. 1.3 X-ray image of V773 Tau in the 0.5–2 keV energy band obtained with XMM-Newton/UV Monitor image in the U-band of the young brown dwarf 2MASS J0414. This image is centered on the weak-line T Tauri star V773 Tau located only 24″ away from the brown dwarf. The Classical T Tauri star FM Tau is also visible. 2MASS J0414 is located on the PSF wings of V773 Tau, where no X-ray counterpart was found with the source detection algorithm. Figure from Grosso et al. (2007)

correct, these binaries illustrate the feasibility of this process in multiple systems. Confirmation of this hypothesis is, however, a challenge. One potential approach is to use high-resolution spectroscopy coupled with adaptive optics observations in order to angularly resolve these visual pairs and study the accretion signatures for the individual stars. A sufficiently massive planet could reveal its presence with radial velocity shifted hydrogen emission lines.

On the basis of the line emission and color diagnostics described above, Monin et al. (2007) classified a sample of young binaries with separations of ~15–1,500 AU. In an extensive search of the young star binary literature, only ~60 systems were found for which both component spectral types were known, and for which angularly resolved Hα, Brγ, K–L, or K–N (K = 2.2 μm, L = 3.4 μm, and N = 10 μm) color data were available. These few dozen systems are drawn from a variety of star-forming regions, and thus do not represent a homogeneous sample. This dramatically underscores the unavoidable small number statistics inherent in any analysis of this sample, and the pressing need for a substantial observational effort in this area.

In spite of the small sample size, Monin et al.'s analysis revealed intriguing results and trends (see Figure 4 of Monin et al. 2007). One surprising and relatively robust outcome is that mixed pairs, in which the components appear to be in different evolutionary stages, are not as rare as once thought (Prato and Simon 1997; Hartigan and Kenyon 2003, composing 38% of the sample, exclusive of pure weak-lined systems (Table 1, Monin et al. 2007)). Less statistically notable are the suggestions that mixed systems are more common among the larger separation pairs and that a slight majority of these systems are detected among the lower mass-ratio pairs. There is also a hint in the available data that the frequency of mixed pairs may vary between star forming regions; this was previously suggested by Prato and Monin (2001) (see their Table 1). Unfortunately, because of the sparse data, these results are all at the 2σ level at best. Because the young star binary distribution peaks at subarcsecond angular separations, producing statistically large (hundreds of stars) samples in the nearby SFRs will require years of work at relatively large facilities which supply adaptive optics capabilities.

The determination of the stellar properties associated with long-lived, inner circumstellar disks allows us to predict what kind of stars are most likely to host planets. For example, if disk-locking is the main mechanism for controlling stellar angular momentum, stellar rotation rates over time will depend on inner disk masses. A young, slowly rotating star locked to a massive inner disk is therefore a likely candidate for future planet formation. Angularly resolved high-resolution spectroscopy of close young binaries yields $v \sin i$ measurements, providing insight into either the alignment of stellar rotation axes, or component stellar rotation rates. This degeneracy can be resolved with time series observations designed to determine component rotation periods. If rotation axes are aligned, discrepant rotation periods suggest star-disk locking in only one component, or some other differential source of angular momentum loss. Binary formation with discrepant component rotation is difficult to explain but also cannot be ruled out. Figure 1.4 shows a young binary with component $v \sin i$'s discrepant by a

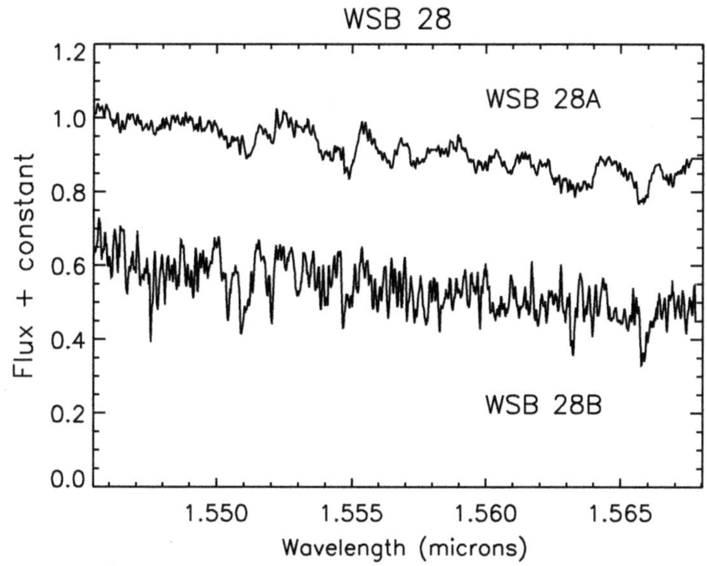

Fig. 1.4 R = 30,000 spectra of the components in the young binary WSB 28. The $v \sin i$'s are discrepant by a factor of 2–3, indicating either unaligned rotation axes or significantly different rotation periods. Veiling from circumstellar material cannot account for the shallow features in the M3 primary because this component is not associated with any circumstellar material, although the M7 secondary is (McCabe et al. 2006)

factor of 2–3. Intriguingly, this ~700 AU separation Ophiuchus binary, an M3 and an M7, is a mixed system (Prato et al. 2003). The rapidly rotating primary is not associated with dusty circumstellar material; however, the low mass companion is (McCabe et al. 2006), as we might expect from a disk-locking scenario. Similar discrepancies have also been observed in other systems, including the 30 AU separation young hierarchical triple, Elias 12, in Taurus (Schaefer 2004; Schaefer et al. 2006).

To determine the inclination of a stellar rotation axis, the system $v \sin i$ must first be measured. The sine of the inclination is proportional to $v \sin i$ multiplied by the period and divided by the stellar radius. The radius must be estimated based on models appropriate for the measured stellar parameters and is relatively uncertain but probably by a factor of less than two. For young binaries, the determination of the period is very challenging. Small separation systems require adaptive optics observations to resolve them, but most facilities offering this capability would be unlikely to schedule high cadence observations over a 10–20 day block conducive to rotation period determination. Furthermore, currently available adaptive optics systems function in the infrared regime, not ideal for measurements of flux modulation from star spots as the spot contrast and therefore signal amplitude is reduced at longer wavelengths. Larger separation systems (>1–2″) might be observed readily at 1–2 m telescopes which can be allocated for long term, multiple night programs. However, most stars in the nearest SFRs are relatively late type and are therefore faint, posing special challenges, particularly for large flux ratio binaries such as

WSB 28 (Fig. 1.4). The primary in this $\sim 5''$ pair has an I-band magnitude of ~ 13. The Gunn z filter, relatively similar to I-band, component flux ratio listed in Reipurth and Zinnecker (1993) is 0.06. Thus, the secondary's I magnitude is ~ 16. These observations are not impossible, but will require careful planning.

How much of an impact might selection effects have on the results presented here? Certainly small mass-ratio systems are more difficult to detect as well as to characterize, particularly in the most interesting small separation regime (Section 1.1). Systems classified as weak-lined T Tauris, unresolved, might also harbor truncated inner disks around the secondary stars. Such small structures could go undetected as the result of dilution from a relatively bright primary. Circumstellar disks with central holes that show excesses in the mid-infrared but not in the near-infrared, and which do not show signatures of accretion, may also be present but are effectively undetectable. Even if sensitive but low-angular resolution *Spitzer* observations could reveal the presence of such a structure, there is little recourse for ground-based mid-infrared follow up at sufficiently high sensitivity and angular resolution to determine structure and location. Only four of the circumstellar disks in the young binary sample of McCabe et al. (2006) (T Tau N and S, UZ Tau E, and RW Aur A) are brighter than the N = 4 mag limit of the VLTI mid-infrared instrument MIDI.

We must also take into account that the completeness of our knowledge of binary populations varies markedly between different star-forming regions, possibly leading to an inaccurate determination of differences in mixed pair fractions, etc., between regions. Although Taurus, given its small size and ready accessibility in the northern skies, is arguably the most thoroughly studied region, its faintest members are only now being surveyed for multiplicity (Konopacky et al. 2007). Ultimately, however, it is not enough to take a simple census of binary frequency and to characterize systems by their unresolved properties. Knowledge of the configuration of the circumstellar and circumbinary dust and gas is required to truly assess the planet-forming potential of young stars, and to determine if SFRs as a whole possess environments particularly conducive, or not, to planet formation. Global properties such as initial molecular cloud angular momentum, the presence of high mass, photoionizing sources, stellar density, etc., may all influence disk and thereby planet formation.

1.3 Outer Disks

Outer circumstellar disks, here taken to mean beyond about 10 AU, may also host planet formation. In addition, they provide an important reservoir of material that feeds the inner disk as well as a critical source of angular momentum transfer for interior material. The cool gas and dust in outer disks, including circumbinary disks, are best surveyed using far-infrared or submillimeter observations. Disks are usually optically thin at long wavelengths, so these observations have the additional benefit of providing total disk masses (Beckwith et al. 1990) in the region analogous to where giant planets formed in the Solar System.

Although estimates of the binary fraction were highly incomplete when the first submillimeter surveys were done, it was still clear immediately that binary stars with separations closer than 100 AU were deficient in disks (Beckwith et al. 1990; Jensen et al. 1994; Osterloh and Beckwith 1995). In a recent work, a survey of 150 young stars in Taurus (including 62 multiple stars) showed lower submillimeter fluxes, and hence disk masses, in binaries closer than $\sim$100 AU than in single stars, while wide binaries were similar to single stars in disk mass (Andrews and Williams 2005). Disks were present, albeit at these lower masses, in approximately the same fraction of multiple star as single star systems. Perhaps these disks can still form giant planets, but of lower average mass than the single stars.

The surveys described above were carried out with single dish telescopes and therefore have low spatial resolution incapable of distinguishing primary and secondary disks in the interesting separation range of $\leq$100 AU. A smaller number of objects have been surveyed with interferometers that can resolve the multiple systems. In one such survey, the primary stars of four binaries in Ophiuchus hosted higher mass disks, even when the secondaries were still accreting, while in four binaries in Taurus the circumsecondary disks were more massive (Patience et al. 2005). In these very young objects, the true "primary," i.e., more massive star of the pair, may have been misidentified in extinguished visual-wavelength data, or these trends may relate to the initial conditions. In another study of four wider systems, also in Taurus, the circumprimary disks were again the most massive (and again comparable to single stars in Taurus) (Jensen and Akeson 2003).

Models of disk dissipation generally show that the circumsecondary disk, which is expected to be truncated closer to the star than the circumprimary disk, should dissipate faster (Armitage and Clarke 1999). In single stars, however, outer disk mass is not correlated with stellar mass (Andrews and Williams 2005), therefore it is possible for circumsecondary disks to form with more mass than circumprimary disks. These initial conditions could overwhelm the difference in dissipation timescale.

Finally, it is worth noting that total disk masses for single stars or wide companions in Taurus or Ophiuchus, i.e., regions of low-mass star formation, are typically in the range 0.005–0.01 solar-masses although with wide dispersion and a substantial fraction (about 20%) of larger disk masses. For comparison, the mass of the "Minimum Mass Solar Nebula" necessary for forming our system's planets is about 0.01 $M_{\odot}$. Results for clusters with massive stars such as the Orion Nebula cluster (Bally et al. 1998) and NGC 2024 (Eisner and Carpenter 2003) suggest that there are fewer massive disks than in low-mass SFRs. It is possible that these disks dissipate more rapidly because of external radiation, for example. These regions of massive star formation are generally further away, so binary surveys are much less complete and single dish submillimeter measurements are less likely to resolve multiple stars (Fig. 1.5). However, just as early unresolved work in Taurus showed a general trend of lower disk mass with binarity, we expect this same trend to hold in regions of more massive star formation, as well.

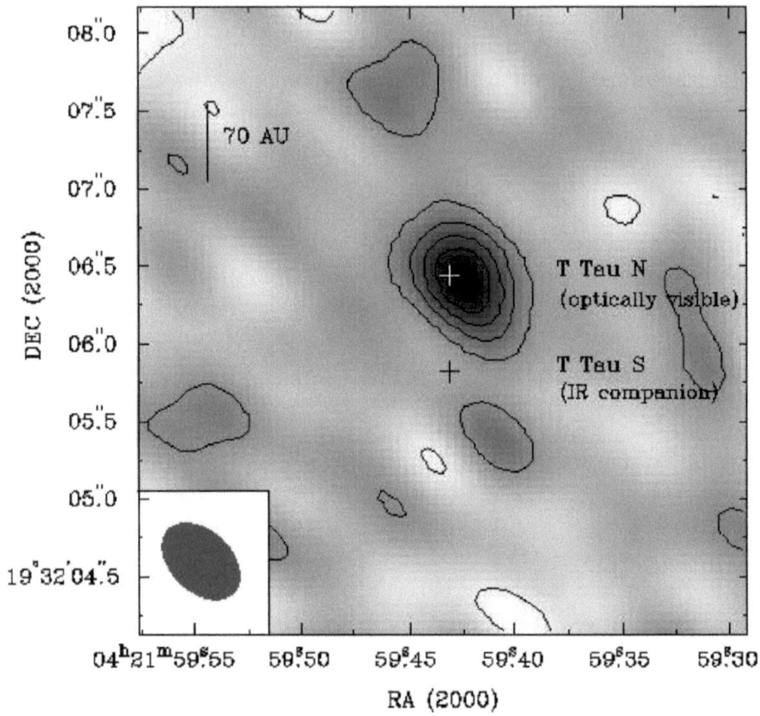

Fig. 1.5 Circumstellar disk around T Tau N. 108 GHz continuum emission using only the longest spacing array. The contour levels plotted are 2σ. The size of the error bars on the millimeter emission center position (*cross*) represents the absolute positional uncertainty for the millimeter image of 0.07 arc sec. Figure from Akeson et al. (1998)

1.4 Orientation of Disks in Young Binaries

A single star plus disk system contains a single plane: that of the disk. A binary system, however, is associated with four relevant planes: a circumstellar disk around each star, the plane of the binary orbit, and the plane of any circumbinary disk. Although circumbinary disks appear to be relatively rare in young systems (Jensen and Mathieu 1997), recent observations by Trilling et al. (2007) find evidence for circumbinary debris disks around 12 small separation main-sequence binaries of relatively early spectral type, A3–F8. Whether or not these disks are aligned with the binary orbit is not known. *Alignment* of circumstellar disks does not necessarily imply coplanarity of the binary orbital plane with that of the aligned disks (Fig. 1.6). The studies of Jensen et al. (2004) and Monin et al. (2006) trace circumstellar alignment, for relatively wide, angularly resolved young binaries, using the position angle of the integrated, linear polarization of the light scattered from a circumstellar disk. Because the position angle is parallel to the plane of the disk, it provides a proxy for disk orientation (but see additional discussion in Monin et al. (2007)). Jensen et al. (2004) and Monin et al. (2006) found that most simple binary systems

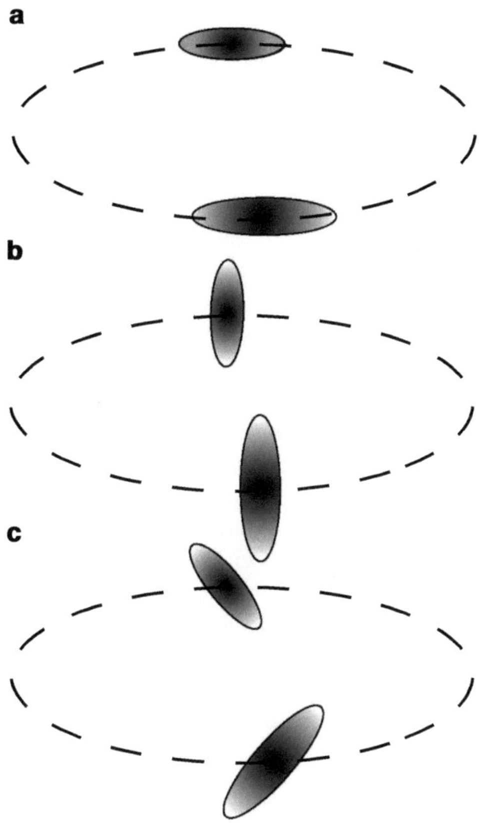

Fig. 1.6 Circumstellar disks in a simple binary orbit: (**a**) aligned and coplanar, (**b**) aligned but non-coplanar, (**c**) unaligned and non-coplanar

studied exhibit aligned disks with polarization position angles consistent to within <30 degrees, although higher order multiples show a large range of variation in polarization position angles.

The orientations of the highly collimated jets that emanate from many young star systems are also a proxy for determining disk orientations in unresolved binaries, as jets are thought to launch perpendicular to the inner circumstellar disks. Multiple misaligned jets are known to exist in a number of young systems (Monin et al. 2007 and references therein), suggesting that it is possible for small separation binaries to actually *form* with misaligned disks (Fig. 1.6, case c). Thus, formation models must account for this counterintuitive evidence.

The coplanarity of disks and binary orbits is readily studied for some well-separated pairs. Interestingly, it appears likely that circumbinary disks are aligned with close binary star orbits, e.g., for DQ Tau, UZ Tau E, and HD 98800 B (Mathieu et al. 1997; Prato et al. 2001; Prato et al. 2002). However, systems with a circumstellar disk around at least one component of a wider binary, e.g., HV Tau AB-C, HK Tau A-B, UZ Tau E-W, T Tau N-S, and HD 98800 N-S (Stapelfeldt et al. 2003, 1998; Prato et al. 2001, 2002; Akeson et al. 2002), *do not* appear to be coplanar. We note that with the possible exception of HK Tau, these systems are all higher order multiples, a condition which may well play a role in the non-coplanarity (Jensen et al. 2004). What are the implications of these observations for potential planet formation? An interesting case study is the quadruple system HD 98800. The wide pair has a separation of ~40 AU (Prato et al. 2001), a period of 300–430 years, and an inclination of ~88 degrees (Tokovinin 1999). Each component of the wide pair is a spectroscopic binary of similar properties, with K5 and K7 spectral type primaries, 262 and 315 day periods, and eccentricities of 0.484 and 0.781 for the A and B components, respectively. However, circumstellar material is only present around the HD 98800 B binary. The inclination of this stellar pair is 67 degrees (Boden et al. 2005). Recently, Akeson et al. (2007) showed that the associated circumbinary disk is likely warped by interactions with the distant A component. The primary mystery in this system, however, is the complete absence of circumbinary material surrounding HD 98800 A. Speculation by Prato et al. (2001) suggests that because of the relative orientations of the HD 98800 A and B circumbinary disks during repeated periastron passages of the wide pair over the ~10 Myr lifetime of the system, B's disk was perturbed and A's disk was completely disrupted. See the next section for more discussion of the planetesimal dynamics in this system.

The dynamics of circumbinary and even close circumstellar disks and the interrelationship between disks and orbits appears to be complex and is not yet well understood. We present these conclusions as a cautionary tale: even binaries with separations of a few tens of AU – or less – cannot be assumed to harbor aligned disks coplanar with binary orbits. In higher order multiples, misalignments may be the rule. It is possible that, in at least some cases, misalignment may have its origins in the formation dynamics of these systems.

1.5 Debris Disks and Binaries

Transitional and debris disks are generally older than the massive disks discussed earlier. Their primordial material, particularly gas, is partially or totally dissipated and remaining solids are large enough that their major destruction mechanism is collisions (either aggregation onto planets or disruptive). Giant planets must either have already formed, or will not form, in these systems, and terrestrial planets may be in their final stages of accumulation, perhaps eras akin to the late heavy bombardment in the Solar System.

To be detected in sensitivity-limited observations, debris disks must be closer to the Sun than the nearest sites of recent/ongoing star formation discussed earlier; this has the benefit that the effect of binarity on the disks can be observed in some detail. We will discuss two examples.

HD 141569 is a hierarchical triple system (Fig. 1.7) consisting of an A0-type primary star, which sports an extended disk containing small quantities of both gas and dust, and two M-type companion stars located about 1,000 AU away. The low mass stars, and presumably the whole system, are about 5 Myr old (Weinberger et al. 2000). Spiral structure at 200–500 AU in the primary's disk can be explained by either a highly eccentric ($e \geq 0.7$) binary A-BC orbit (Augereau and Papaloizou

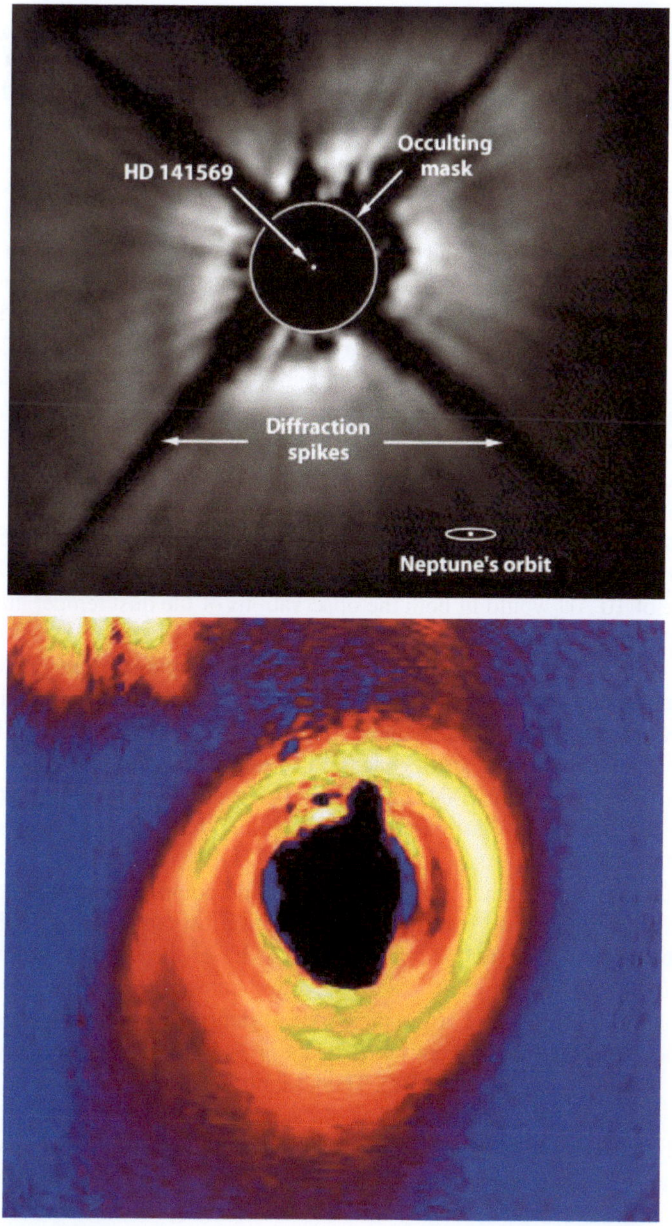

Fig. 1.7 *Top*: A near-infrared view of the disk surrounding HD 141569 recorded in 1998 by the Hubble Space Telescope. *Bottom*: Coronagraphic image of the protoplanetary disk around HD141569 taken with the Advanced Camera for Surveys (ACS) in the Hubble Space Telescope

2004; Quillen et al. 2005; Ardila et al. 2005) or a recent (~1,000 years ago) stellar flyby (Beust 2005). In both cases, the affected portion of the disk is at a radius of a few hundred AU, and structure in the disk at <150 AU must have another cause, perhaps a planet. Interestingly, the two M-type stars have no detectable disks down to the level (in $L_{\mathrm{IR,disk}}/L_*$; a measure of disk mass where $L_{\mathrm{IR,disk}}$ is the luminosity of the disk and L_* is that of its central star) of the primary's disk. This could be attributed to the small separation of their orbit, ~150 AU.

HD 98800 is a member of the ~8 Myr old TW Hya Association and the interesting arrangement of its fours stars and dust disk is described in the previous section and in Low et al. (1999), Koerner et al. (2000), and Prato et al. (2001). The system has characteristics of both a circumbinary and circumstellar disk. The HD 98800 B binary is eccentric (e = 0.78) with a semi-major axis of 1 AU (Boden et al. 2005). Based on its temperature, the inner edge of the dust disk sits at 1.2 – 2.1 AU (Prato et al. 2001). This is just barely consistent with estimates of the dynamical tidal truncation (Artymowicz and Lubow 1994). The A-B orbit is also significantly eccentric (0.3–0.6) with a periastron approach of perhaps 35 AU (Tokovinin 1999). The outer edge of the dust disk is less well constrained by the infrared/submillimeter observations, but is >5 AU and could be as large as 25 AU (Koerner et al. 2000). An outer size of 10 AU would fit both the observations of the dust temperature and the expected dynamical truncation from the A-B orbit.

While both of these systems provide interesting examples of the dynamical influence of multiplicity on the disk, they also illustrate that planet formation is possible under such complicated circumstances. The small dust grains in the HD 141569 A and HD 98800 B disks are regenerated in collisions (Weinberger et al. 1999; Augereau and Papaloizou 2004; Low et al. 1999) and indicate that planetesimals did form on timescales short enough that gas could have been present simultaneously with solid bodies.

Statistics of the incidence of debris disks around binaries are consistent with the idea that wide binaries do not affect disk evolution. A survey of 69 FGK stars including binaries of separations >500 AU finds 3/8 of the debris disks are around binary members (Bryden et al. 2006). A larger survey of old A and F-type stars (Trilling et al. 2007) looked at the incidence of debris disks as a function of separation and found that there were fewer disks in 3–50 AU separation systems than in closer or wider systems (Fig. 1.8). The total numbers of disks in this *Spitzer* survey were comparable to surveys of single stars and contained just as much dust. Although *Spitzer* did not resolve the individual components, the dust temperatures imply that both circumbinary and circumstellar debris disks are common, at least amongst stars somewhat more massive than the Sun (Fig. 1.8).

1.6 Future Tests and Observations

As noted throughout this review, binary star systems do indeed host circumstellar and circumbinary disks over their entire lifetimes from pre-main sequence to mature stars. At separations larger than a few hundred AU, disk evolution around a binary

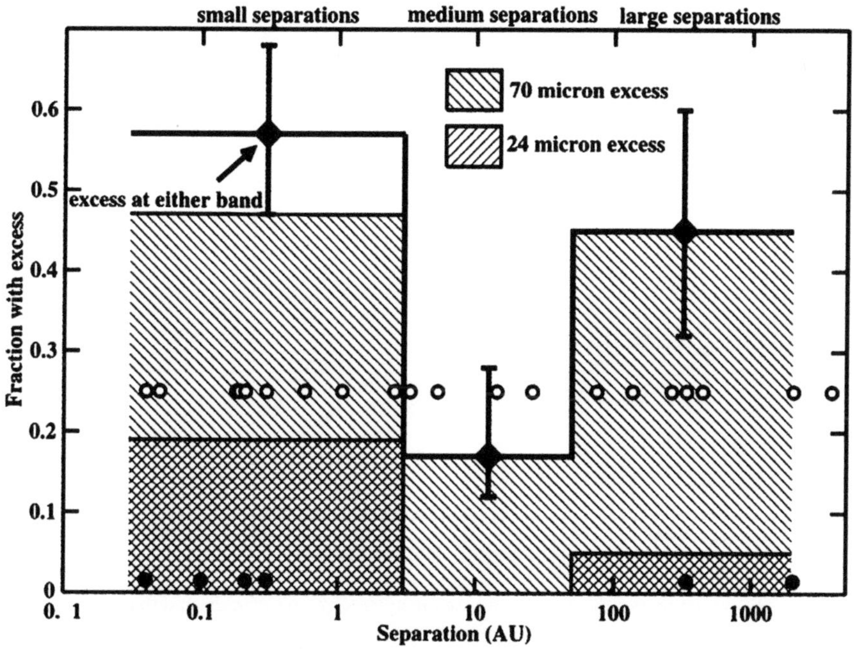

Fig. 1.8 Fraction of binary systems in each of three logarithmic bins (0–3 AU, 3–50 AU, 50–200 AU) that have 24 μm (diagonal pattern *lower left to upper right*), 70 μm (diagonal pattern *lower right to upper left*), or 24 or 70 μm excesses (clear). Binomial error bars are shown for the 24 or 70 μm excess category. Each category reads from the bottom of the plot (that is, the fraction of close binaries with 70 μm excesses is 47%). Some systems have excesses at both wavelengths, and the number of observed systems is not the same at 24 and 70 μm, so the combined fractions do not simply equal the sum of the two subcategories. The separations of the individual systems with excesses contained within each bin are indicated by the filled (24 μm) and open (70 μm) circles (with arbitrary y-axis values). Medium-separation systems have fewer excesses than small- or large-separation systems. Figure from Trilling et al. (2007)

component may proceed no differently than if the stars were not bound together. However, a tremendous observational effort is required to explore the most populous binary separation regime, and that of most scientific interest with respect to the impact of multiplicity on planet formation – a few to ~30 AU separations (Fig. 1.9).

With concerted observational attention, it seems a solvable problem to measure the dissipation timescales of primary and secondary disks. Ongoing spatially resolved spectroscopy with adaptive optics systems on large telescopes will assess the accretion parameters and optical depths of inner circumstellar disks in close binaries. Ground based interferometers will get detailed orbits for close binaries which can then be compared to disk sizes for empirical verification of dynamical estimates of tidal disruption and dissipation (Boden et al. 2005).

Progress on the determination of masses and orientations of circumstellar disks in young binaries, which cannot not be accomplished with the limited sensitivity and spatial resolution of the current generation of millimeter interferometers, will

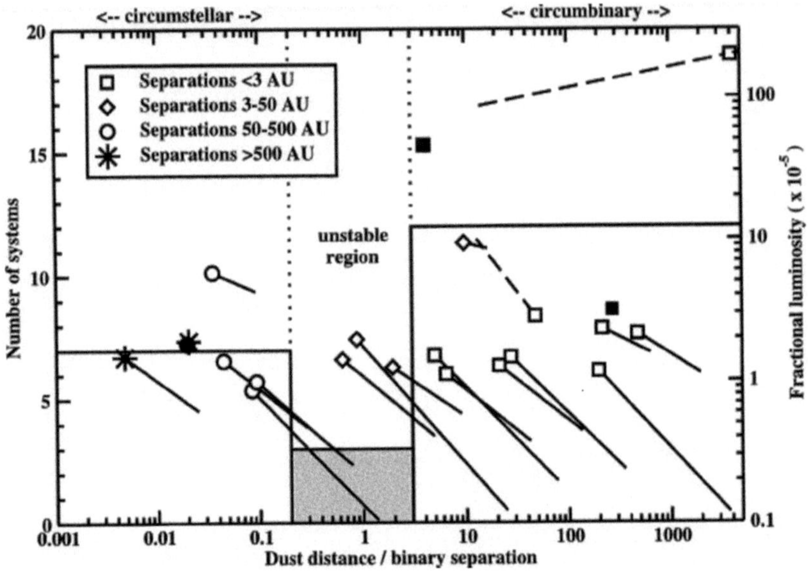

Fig. 1.9 Histogram of dust distance in units of binary separation (*left axis*) and fractional luminosity as a function of dust distance in units of binary separation (*right axis*). *Left axis: dashed vertical lines* show the approximate boundaries of the unstable zone (*histogram bar shaded gray*). Dust in three systems is found to reside within this dynamically unstable region. *Right axis:* there is no strong trend between fractional luminosity and dust location. Binary systems with small, medium, large, and very large physical separations are indicated. Not surprisingly, circumbinary disks are generally found in small-separation systems and circumstellar disks are found in large-separation systems. Dust in unstable regions is found only in medium separation systems. Fractional luminosities for the maximum-temperature cases are indicated by the symbols. "Tails" on the symbols indicate the locus of solutions, from maximum-temperature solutions (symbols) to 50 K (minimum reasonable) solutions at the other ends of the tails. Figure from Trilling et al. (2007)

advance markedly with the advent of the Atacama Large Millimeter Array (ALMA). Scheduled to begin full operation in 2012 and providing sub-arcsecond resolution and high sensitivity, ALMA will be able to probe component disk masses not only in nearby Taurus and Ophiuchus but also in clusters containing massive stars.

Imaging the dynamical effects of binarity on individual nearby disks will take advantage of ongoing, improving capabilities, such as more sensitive adaptive optics on 8–10 m ground-based telescopes. Furthermore, the James Webb Space Telescope (JWST), due to be launched in 2013, represents a leap in the quality of direct imaging in the near- and mid-infrared. A 6.5 m telescope, JWST will provide coronagraphic imaging at 1–23 μm for detailed disk studies.

The quantitative, detailed study of disks in young binaries has grown rapidly over the last decade, following quickly, and to some degree simultaneously, on the results of the initial discovery surveys. The upcoming decade promises to provide a number of powerful new and innovative tools and approaches for this field. Although many of the observations necessary to progress into the most interesting (and common) small binary separation regime require time at high-demand facilities, we emphasize that this work merits the investment: binary stars, and particularly young binary stars, dominate the stellar census. The particulars of planet formation in these systems determine the range of planetary system architectures present in the Galaxy.

The authors are grateful to the anonymous referee and to N. Haghighipour for comments that improved the presentation of this chapter. We thank S. Zoonematkermani for rendering Fig. 1.6.

References

R. L. Akeson, D. W. Koerner, & E. L. N. Jensen: ApJ, 505, 358 (1998)
R. L. Akeson, D. R. Ciardi, G. T. van Belle et al: ApJ, 566, 1124 (2002)
R. L. Akeson, W. M. K. Rice, A. F. Boden et al: ApJ, 670, 1240 (2007)
S. M. Andrews & J. P. Williams: ApJ, 631, 1134 (2005)
D. R. Ardila, S. H. Lubow, D. A. Golimowski et al: ApJ, 627, 986 (2005)
P. J. Armitage, C. J. Clarke, & C. A. Tout: MNRAS, 304, 425 (1999)
P. Artymowicz & S. H. Lubow: ApJ, 421, 651 (1994)
J. C. Augereau & J. C. B. Papaloizou: Astron. Astrophys., 414, 1153 (2004)
J. Bally, L. Testi, A. Sargent, & J. Carlstrom: AJ, 116, 854 (1998)
G. Basri, C. M. Johns-Krull, & R. D. Mathieu: AJ, 114, 781 (1997)
T. L. Beck, M. Simon, & L. M. Close: ApJ, 583, 358 (2003)
S. V. W. Beckwith, A. I. Sargent, R. S. Chini, & R. Guesten: AJ, 99, 924 (1990)
H. Beust: In *Space Telescoep Science Institute Mini-Workshop: Nearby Resolved Debris Disks*, http://www.stsci.edu/institute/itsd/information/streaming/archive/NRDD2005 (2005)
A. F. Boden, A. I. Sargent, R. L. Akeson et al: ApJ, 635, 442 (2005)
J. Bouwman, W. A. Lawson, C. Dominik, E. D. Feigelson, T. Henning, A. G. G. M. Tielens, & L. B. F. M. Waters: ApJ, 653, 47 (2006)
G. Bryden, C. A. Beichman, D. E. Trilling et al: ApJ, 636, 1098 (2006)
L. M. Close, A. Dutrey, F. Roddier et al: ApJ, 499, 883 (1998)
P. D'Alessio, L. Hartmann, N. Calvet et al: ApJ, 621, 461 (2005)
G. Duchêne, E. Delgado-Donate, K. E. Haisch Jr. et al: New Observational Frontiers in the Multiplicity of Young Stars. In *Protostars and Planets V*, ed by B. Reipurth, D. Jewitt, K. Keil (University of Arizona Press, Tucson 2007) pp 379–394
A. Duquennoy & M. Mayor: A&A, 248, 485 (1991)
J. A. Eisner & J. M. Carpenter: ApJ, 598, 1341 (2003)
A. Eggenberger, S. Udry, & M. Mayor: A&A, 417, 353 (2004)
A. M. Ghez, G. Neugebauer, & K. Matthews: AJ, 106, 2005 (1993)
N. Grosso, M. Audard, J. Bouvier, K. R. Briggs, & M. Güdel: A&A, 468, 557 (2007)
P. Hartigan & S. J. Kenyon: ApJ, 583, 334 (2003)
H. Jang-Condell: ApJ, 654, 641 (2007)
E. L. N. Jensen, R. D. Mathieu, & G. A. Fuller: ApJ, 429, L29 (1994)
E. L. N. Jensen & R. D. Mathieu: AJ, 114, 301 (1997)
E. L. N. Jensen & R. L. Akeson: ApJ, 584, 875 (2003)
E. L. N. Jensen, R. D. Mathieu, A. X. Donar et al: ApJ, 600, 789 (2004)
D. W. Koerner, E. L. N. Jensen, K. Cruz et al: ApJ, 533, L37 (2000)
M. Konacki: Nature, 436, 230 (2005)
Q. M. Konopacky, A. M. Ghez, E. L. Rice et al: ApJ, 663, 394 (2007)
Ch. Leinert, H. Zinnecker, N. Weitzel et al: A&A, 278, 129 (1993)
F. J. Low, D. C. Hines, & G. Schneider: ApJ, 520, L45 (1999)
E. L. Martín: AJ, 115, 351 (1998)
E. L. Martín, A. Magazzù, X. Delfosse et al: A&A, 429, 939 (2005)

R. D. Mathieu, K. Stassun, G. Basri et al: AJ, 113, 1841 (1997)
R. D. Mathieu, A. M. Ghez, E. L. N. Jensen et al: Young Binary Stars and Associated Disks. In *Protostars and Planets IV*, ed by V. Mannings, A. P. Boss, S. S. Russell (University of Arizona Press, Tucson 2007) pp 703–734
C. McCabe, G. Duchêne, & A. M. Ghez: ApJ, 575, 974 (2002)
C. McCabe, A. M. Ghez, L. Prato et al: ApJ, 636, 932 (2006)
J.-L. Monin, F. Ménard, & N. Peretto: A&A, 446, 201 (2006)
J.-L. Monin, C. Clarke, L. Prato et al: Disk Evolution in Young Binaries: From Observations to Theory. In *Protostars and Planets V*, ed by B. Reipurth, D. Jewitt, K. Keil (University of Arizona Press, Tucson 2007) pp 395–409
J. R. Najita, S. E. Strom, & J. Muzerolle: MNRAS, 378, 369 (2007)
M. Osterloh & S. V. W. Beckwith: ApJ, 439, 288 (1995)
I. Pascucci, D. Apai, E. E. Hardegree-Ullman, J. S. Kim, M. R. Meyer, & J. Bouwman: ApJ, 673, 477 (2008)
J. Patience, A. M. Ghez, I. N. Reid et al: AJ, 123, 1570 (2002)
J. Patience, R. L. Akeson, E. L. N. Jensen, & A. I. Sargent: Protostars and Planets V, 8603 (2005)
M. G. Petr, V. Coude Du Foresto, S. V. W. Beckwith et al: ApJ, 500, 825 (1998)
S. F. Portegies Zwart & S. L. W. McMillan: ApJ, 633, L141 (2005)
L. Prato & M. Simon: ApJ, 474, 455 (1997)
L. Prato, A. M. Ghez, R. K. Piña et al: ApJ, 549, 590 (2001)
L. Prato & J.-L. Monin: Correlations of Spectral Accretion Signatures in Young Binaries. In *The Formation of Binary Stars*, ed by H. Zinnecker & R. Mathieu (Astronomy Society of the Pacific, San Francisco 2001) pp 313–322
L. Prato, M. Simon, T. Mazeh et al: ApJ, 579, L99 (2002)
L. Prato, T. P. Greene, & M. Simon: ApJ, 584, 853 (2003)
C. F. Prosser, J. R. Stauffer, L. Hartmann et al: ApJ, 421, 517 (1994)
A. Quillen, P. Varniere, I. Minchev, & A. Frank: AJ, 129, 2481 (2005)
E. V. Quintana & J. J. Lissauer: Icarus, 185, 1 (2006)
E. V. Quintana, F. C. Adams, J. J. Lissauer et al: ApJ, 660, 807 (2007)
D. Raghavan, T. J. Henry, B. D. Mason et al: ApJ, 646, 523 (2006)
B. Reipurth & H. Zinnecker: A&A, 278, 81 (1993)
S. H. Saar, R. P. Butler, & G. W. Marcy: ApJ, 498, 153 (1998)
G. H. Schaefer: 2004, Ph.D. thesis, State Univ. New York (Stony Brook)
G. H. Schaefer, M. Simon, T. L. Beck, E. Nelan1, & L. Prato: AJ, 132, 2618
M. Simon, A. M. Ghez, Ch. Leinert et al: ApJ, 443, 625 (1995)
M. Simon: ApJ, 482, L81 (1997)
K. R. Stapelfeldt, J. E. Krist, F. Menard et al: ApJ, 502, L65 (1998)
K. R. Stapelfeldt, F. Ménard, A. M. Watson et al: ApJ, 589, 410 (2003)
A. A. Tokovinin: Astron. Lett., 25, 669 (1999)
D. E. Trilling, J. A. Stansberry, K. R. Stapelfeldt et al: ApJ, 658, 1289 (2007)
S. Udry, M. Mayor, D. Naef et al: A&A, 390, 267 (2002)
A. J. Weinberger, E. E. Becklin, G. Schneider, B. A. Smith, P. J. Lowrance, M. D. Silverstone, B. Zuckerman, & R. J. Terrile: ApJ, 535, L53 (1999)
A. J. Weinberger, R. M. Rich, E. E. Becklin, B. Zuckerman, & K. Matthews: ApJ, 544, 937 (2000)
A. D. Welty: AJ, 110, 776 (1995)

Chapter 2
Probing the Impact of Stellar Duplicity on Planet Occurrence with Spectroscopic and Imaging Observations

Anne Eggenberger and Stéphane Udry

2.1 Introduction

Over the past 14 years, Doppler spectroscopy has been very successful in detecting and characterizing extrasolar planets, providing us with a wealth of information on these distant worlds (e.g., Marcy et al. 2005a; Udry and Santos 2007b; Udry et al. 2007a). One important and considerably unexpected fact these new data have taught us is that diversity is the rule in the planetary world. Diversity is found not only in the characteristics and orbital properties of the ~340 planets detected thus far,[1] but also in the types of environments in which they reside and are able to form. This observation has prompted a serious revision of the theories of planet formation (e.g., Lissauer and Stevenson 2007; Durisen et al. 2007; Nagasawa et al. 2007), leading to the idea that planet formation may be a richer and more robust process than originally thought.

It is well known that nearby G, K, and M dwarfs are more likely found in pairs or in multiple systems. Specifically, 57% of the G dwarf primaries within 22 pc of the Sun have at least one stellar companion (Duquennoy and Mayor 1991). The multiplicity among K dwarfs is very similar (Halbwachs et al. 2003; Eggenberger et al. 2004b), and among nearby M dwarfs is close to 30% (Fischer and Marcy 1992; Delfosse et al. 2004). Altogether, these figures imply that more than half of the nearby F7–M4 dwarfs are in binaries or in higher order systems. Since these stars constitute the bulk of the targets searched for extrasolar planets via

[1] See the Extrasolar Planet Encyclopedia, http://exoplanet.eu/, for an up-to-date list.

A. Eggenberger (✉)
Observatoire de Genève, Université de Genève, 51 ch. des Maillettes, 1290 Sauverny, Switzerland
and
Laboratoire d'Astrophysique de Grenoble, UMR 5571 CNRS/Université J. Fourier,
BP 53, 38041 Grenoble Cedex 9, France
e-mail: anne.eggenberger@obs.ujf-grenoble.fr

S. Udry
Observatoire de Genève, Université de Genève, 51 ch. des Maillettes, 1290 Sauverny, Switzerland
e-mail: stephane.udry@obs.unige.ch

N. Haghighipour (ed.), *Planets in Binary Star Systems*, Astrophysics and Space Science Library 366, DOI 10.1007/978-90-481-8687-7_2,

Doppler spectroscopy, the question of the existence of planets in binary and multiple star systems is fundamental and cannot be avoided when one tries to assess the overall frequency of planets.

From the theoretical perspective, the existence of planets in binary and multiple star systems is not guaranteed a priori as the presence of a stellar companion may disrupt both planet formation and long-term stability. On the other hand, young binary systems often possess more than one protoplanetary disk (Monin et al. 2007 and references therein), meaning that planets may form around any of the two stellar components (circumstellar planets) and/or around the pair as a whole (circumbinary planets). Although theoretically both circumstellar and circumbinary planets should exist (Barbieri et al. 2002; Mayer et al. 2005; Boss 2006; Thébault et al. 2006; Quintana and Lissauer 2006; Haghighipour and Raymond 2007; Quintana et al. 2007; Pierens and Nelson 2007), our present planet search programs are essentially aimed at detecting circumstellar planets, and only these will be considered in this chapter. Our discussion will furthermore be focused on giant planets, which are less challenging to detect by means of the Doppler spectroscopy technique than lower mass objects.

Two different scenarios have been proposed to explain the formation of gaseous giant planets. According to the core accretion model, giant planets form in a protoplanetary disk through the accretion of solid planetesimals followed by gas capture (see, e.g., Lissauer and Stevenson (2007) for a review and references). Despite some remaining uncertainties, this scenario is commonly considered as the favored mechanism to explain the formation of giant planets. An important point in this model is that the protoplanetary cores that give rise to the giant planets may have to form beyond the snow line (i.e., beyond 1–4 AU for solar-type stars) to benefit from the presence of ices as catalysts.

An alternative way to view giant planet formation is to consider that these planets form by direct fragmentation of the protoplanetary disk. This is the so-called disk instability model (see Durisen et al. 2007 and chapter ... for a review and references). Since it is not clear yet whether real protoplanetary disks actually meet the requirements for fragmentation, and whether the fragments will live long enough to contract into permanent planets, the disk instability scenario has remained somewhat speculative. Observational tests that would help characterizing and quantifying the likelihood of forming giant planets by this method are thus desirable.

Regardless of the exact formation process, tidal perturbations from a stellar companion within $\sim$100 AU may affect planet formation by truncating, stirring, and heating a potential circumstellar protoplanetary disk (e.g., Artymowicz and Lubow 1994; Nelson 2000; Mayer et al. 2005; Pichardo et al. 2005; Boss 2006; Thébault et al. 2006). Disk truncation is a serious concern as it reduces the amount of material available for planet formation and it may cut the disk inside the snow line. This is a direct threat to planet formation in binary stars and explains why the naive outlook for planet formation in moderately close binaries is pessimistic.

The impact of disk stirring and heating on planet formation is not so easily understood and requires dedicated simulations. According to Nelson (2000), giant planet formation is inhibited in equal-mass binaries with a separation of 50 AU whatever the formation mechanism, whereas Boss (2006) claims that giant planets are able to form in binaries with periastrons as small as 25 AU. Other studies on the subject concluded that planetesimal accretion is perturbed but remains possible in various binary systems closer than 50 AU (Thébault et al. 2004, 2006), and that the two possible formation mechanisms may yield different predictions as to the occurrence of giant planets in binaries separated by 60–100 AU (Mayer et al. 2005). This last conclusion is particularly interesting since it implies that planets in 60–100 AU binaries might be used to identify the main formation mechanism for giant planets.

Assuming that planets can form in various types of binary systems, another important concern is their survival. The extensive body of literature on this subject can be summarized as follows. For low-inclination planetary orbits ($i \lesssim 39°$), the survival time is primarily determined by the binary periastron. A stellar companion with a periastron wider than approximately 5–7 times the planetary semimajor axis does not constitute a serious threat to the long-term ($\sim$5 Gyr) stability of Jovian-mass planets (e.g., Holman and Wiegert 1999; Fatuzzo et al. 2006). The survival time of planets on higher inclination orbits depends not only on the binary periastron, but also on the inclination angle (Innanen et al. 1997; Haghighipour 2006; Malmberg et al. 2007), meaning that planetary orbits become more easily unstable, even if the semimajor axis is quite large (several hundred of AU). This additional type of instability is due to the so-called Kozai mechanism, which causes synchronous oscillations of the planet eccentricity and inclination (e.g., Kozai 1962; Holman et al. 1997; Mazeh et al. 1997; Takeda and Rasio 2005).

To sum up, if giant planets are to form in binaries with a separation below $\sim$100 AU, then the most sensitive (but also less understood) issue regarding their occurrence in these systems seems to be whether or not these planets can form in the first place. This conclusion is quite appealing as it implies that quantifying the occurrence of planets in moderately close binaries may be a means of obtaining some observational constraints on the processes underlying planet formation. Yet, recent work made to explain the existence of a close-in Jovian planet around HD 188753 A emphasized the alternative possibility that moderately close double and multiple star systems originally void of giant planets may acquire one via dynamical interactions (stellar encounters or exchanges), in which case the present orbital configuration of the system would not be indicative of the planetary formation process (Pfahl 2005; Portegies Zwart and McMillan 2005). Pfahl and Muterspaugh (2006) have tried to quantify the likelihood that a binary system could acquire a giant planet in this way and concluded that dynamical processes could deposit Jovian planets in $\sim$0.1% of the binaries closer than 50 AU. Therefore, to test the possibility of forming giant planets in binaries closer than $\sim$100 AU, one needs not only to detect giant planets in these systems, but above all, to quantify their frequency.

From the observational perspective, the existence of planets in wide binaries and multiple star systems has been supported by observations almost since the first discoveries. In 1997 three planets were found to orbit the primary components of wide binaries HR 3522, HR 5185, and HR 458 (Butler et al. 1997), while another one was discovered around 16 Cyg B, the secondary component of a triple system (Cochran et al. 1997). Three years later, the detection of a giant planet around Gl 86 A (Queloz et al. 2000) brought a clear evidence that Jovian planets can also exist in the much closer spectroscopic binaries, as suggested previously by the possible detection of a giant planet around γ Cephei A (Campbell et al. 1988; Walker et al. 1992; Hatzes et al. 2003). These discoveries rapidly prompted a new interest in the study of planets in binaries, raising the possibility that planets may be common in double and multiple star systems.

When considering planets in binaries, it is important to note that most Doppler planet searches used to be, and still are, strongly biased against binaries closer than $\sim$200 AU. As a consequence, present data from these surveys provide incomplete information on the suitability of $\lesssim$200 AU binaries for planetary systems. Similarly, the actual frequency of planets in these systems remains unconstrained.

Recognizing early the importance and the interest of including binary stars in extrasolar planet studies, we have investigated the impact of stellar duplicity on planet occurrence for a few years. This investigation follows two different approaches. The first one uses Doppler spectroscopy to quantify the occurrence of giant planets in spectroscopic binaries (Eggenberger et al. 2003, 2008b). Combining the results from these surveys targeting moderately close binaries with the results from our "classical" planet searches with ELODIE (Perrier et al. 2003) and CORALIE (Queloz et al. 2000; Udry et al. 2000), we aim at quantifying the occurrence of giant planets in binaries with various separations. The second approach to our study makes use of direct imaging to probe the multiplicity status of nearby solar-type stars with and without planets. This work aims at tracing out the impact of stellar duplicity on planet occurrence and properties in binaries with typical separations between 35 and 250 AU (Udry et al. 2004; Eggenberger et al. 2004c, 2007b, 2008, 2008b).

The outline for this chapter is as follows. In Section 2.2 we present the results from classical Doppler planet searches, whose outcomes constitute the general framework within which lie more specific studies dedicated to binaries. In Section 2.3 we describe how direct imaging can be used to probe the impact of stellar duplicity on planet occurrence and to test whether the frequency of giant planets is reduced in binaries closer than $\sim$100 AU. In Section 2.4 we discuss some preliminary results from our Doppler surveys dedicated to the search for circumstellar planets in spectroscopic binaries. All these results are finally summarized in Section 2.5.

2.2 Results from Classical Doppler Planet Searches

Most of the information gathered to date on planets in binary and multiple star systems[2] has been obtained by "classical" Doppler surveys searching for planets around G and K dwarfs within 100 pc of the Sun (Udry et al. 2007a and references therein). Here, we present and discuss these observational results, together with the selection effects against binary systems that affect classical Doppler planets searches.

[2] For the sake of conciseness, we will henceforth call "planets in binaries" the planets residing either in true binaries or in hierarchical multiple systems.

2.2.1 Selection Effects Against Binaries in Doppler Planet Searches

In a general way, Doppler searches for planets around nearby G and K dwarfs avoid binaries closer than $2''$ to $6''$ (systems that we will call moderately close binaries in this chapter) (Udry et al. 2000; Perrier et al. 2003; Marcy et al. 2005b; Jones et al. 2006), meaning that these programs reject from their samples many star systems closer than ∼200 AU. The reason for this discrimination is twofold. First, the naive prospect of finding giant planets in moderately close binaries used to be quite poor and until recently moderately close binaries were not considered particularly interesting targets for planet searches. Secondly, double stars with an angular separation similar to, or smaller than, the size (projected onto the sky) of the spectrograph's fiber or slit present technical difficulties since they cannot be observed as two isolated stars. As explained in Section 2.4, this often complicates the extraction of the radial velocity, rendering classical cross-correlation techniques inadequate to search for planets in certain types of spectroscopic and visual binaries.

When designing our ELODIE and CORALIE planet search programs, we rejected from the main samples all the G and K dwarfs belonging either to "short-period" single-lined spectroscopic binaries ($\lesssim$10 years) or to double-lined spectroscopic binaries[3] (Udry et al. 2000; Perrier et al. 2003). This discrimination was performed in the first place on the basis of former radial-velocity measurements gathered with the two CORAVEL instruments,[4] but additional systems discovered later in the course of our planet programs met the same fate and were rejected as well. However, we kept in the samples single-lined spectroscopic binaries with long periods ($\gtrsim$10 years) since in those systems, not only was the prospect of finding giant planets higher than in double-lined spectroscopic binaries with more massive secondaries, but also the technical difficulties were thought to be minimal.

Our initial policy on wider binaries was less drastic and we kept all visual binaries in our ELODIE and CORALIE samples. However, the data accumulated in the early phases of the CORALIE program showed that radial velocity measurements of primary components of moderately close visual binaries were generally noisier and more variable than expected, suggesting that the secondaries in these systems often contribute to some extent to the recorded flux. Consequently, we flagged as second-priority targets, all the visual binaries closer than $\sim6''$ and with a magnitude difference $\Delta V \lesssim 4$. These targets are then observed less often than regular single stars.

---

[3] Single-lined spectroscopic binaries are systems for which only the spectrum of the primary component is detected, while double-lined spectroscopic binaries are systems for which the spectra of both components are detected. See Section 2.4 for further details on spectroscopic binaries.

[4] The two CORAVEL instruments (Baranne et al. 1979) were used extensively between 1977 and 1998 to monitor the radial velocity of more than 60,000 nearby stars at an intermediate precision (typically 300 m s^{-1}) in both hemispheres.

2.2.2 The Sample of Planets in Binaries

Thanks mostly to recent searches for common proper motion companions to planet-host stars (Section 2.3), the number of planets known to reside in binary and multiple star systems has been growing rapidly in the past few years (Patience et al. 2002; Eggenberger et al. 2004, 2007b; Mugrauer et al. 2005, 2006; Chauvin et al. 2006; Raghavan et al. 2006; Desidera and Barbieri 2007) and has now reached to 40 planets in 35 planetary systems. In terms of system architecture, these planets were found in binaries with projected separations between ~20 and ~12,000 AU. With few exceptions, all these planets orbit the primary components (Eggenberger et al. 2004; Raghavan et al. 2006; Desidera and Barbieri 2007). This last feature is partly a selection effect, the secondaries being often too faint to belong to the target samples used by Doppler planet searches. Not surprisingly, only a couple of planets were found in binary or multiple systems closer than ~100 AU. Although some theoretical models predict a shortage of giant planets in binaries closer than ~100 AU (Nelson 2000; Mayer et al. 2005; Thébault et al. 2006), current Doppler surveys are too severely biased against these particular systems to claim that observations meet theoretical predictions on this point. In particular, the fact that three of the few planets detected in binaries closer than 100 AU were found in systems with separations of about 20 AU likely reflects the selection effects just mentioned in Section 2.2.1. Indeed, for targets within 50 pc and for spectrographs like ELODIE or CORALIE, the separation range between ~10 and ~30 AU corresponds to both long-period spectroscopic binaries (that were kept in the samples) and to visual binaries that are compact enough for technical difficulties to remain acceptable if the secondary component is not too bright (see Section 2.4 for details). As a consequence, Doppler planet searches such as the ELODIE and the CORALIE surveys are presently more likely to detect planets in 10–30 AU systems than in $\lesssim$10 AU or in 30–100 AU systems.

The apparent lack of planets in binaries closer than ~20 AU is also worth noticing. According to theoretical models, the formation of giant planets in binaries closer than ~20 AU is possible only for low binary eccentricities, if at all (Nelson 2000; Thébault et al. 2004, 2006; Mayer et al. 2005; Boss 2006). Many short-period spectroscopic binaries may then be free from giant planets and the "limit" at ~20 AU might have a true meaning. Nonetheless, the present observational material does not allow us to rule out the alternative hypothesis that the lack of planetary detections in systems closer than ~20 AU actually reflects the discrimination against short-period spectroscopic binaries in classical Doppler surveys. On that basis, the question of the closest binaries susceptible of hosting circumstellar giant planets remains open.

To sum up, classical Doppler planet searches have brought observational evidence that circumstellar giant planets do exist in many types of binaries, including spectroscopic systems. Yet, this observational material is incomplete with regard to the closest binaries and we can derive from the present sample of planets in binaries only a minimum value for the fraction of planets residing in double and multiple star systems. This minimum fraction is 21%. Deriving the actual frequency of planets in binaries closer than ~200 AU and probing the existence of giant planets in binaries closer than ~20 AU both call for the need of planet search programs capable of dealing with spectroscopic and moderately close visual binaries. Two such programs are presently underway (Konacki 2005b; Eggenberger et al. 2003) and we discuss our own surveys in Section 2.4.

2.2.3 *Different Properties for Planets in Binaries?*

The first hint that planets found in binaries may possess some distinct properties and characteristics was brought by Zucker and Mazeh (2002). These authors pointed out that planets in binary systems seem to follow a different period-mass correlation than that of planets orbiting single stars. In a similar vein, in 2003, we performed a statistical study considering not only the period-mass but also the period-eccentricity relation (Eggenberger et al. 2004) (see also Mugrauer et al. 2005; Desidera and Barbieri 2007 for more recent studies). As shown in Fig. 2.1, our analysis confirms that the three planets with minimum masses[5] $M_2 \sin i \gtrsim 2$ M_{Jup} and periods $P \lesssim 40$ days all orbit the components of binaries or multiple stars. However, the inclusion, in our sample, of several newly discovered planets with periods longer than 100 days and minimum masses in the range 3–5 M_{Jup}, which were found in

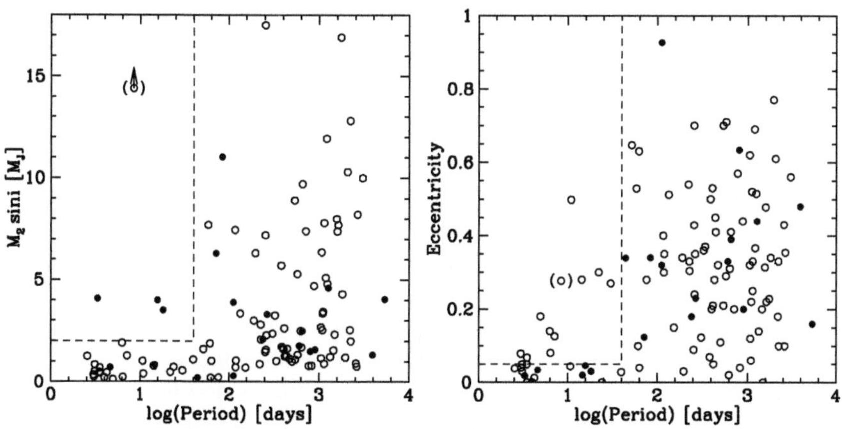

Fig. 2.1 *Left*: minimum mass versus orbital period for all the extrasolar planetary candidates known in 2003. Planets orbiting a single star are represented as *open circles*, while planets residing in binary or multiple star systems are represented as *dots*. The *dashed line* approximately delimits the zone where only extrasolar planets belonging to binaries are found. *Right*: eccentricity versus orbital period for the same planetary candidates as before. The *dashed line* approximately delimits the region where no planet-in-binary is found

[5] In the expression for the minimum mass, M_2 is the true mass of the planet and i is the inclination of the orbit with respect to the tangent plane of the sky.

binaries, decreases the significance of the negative period-mass correlation indicated by Zucker and Mazeh (2002). Yet, marginal signs of this correlation subsist in the form of a shortage of very massive planets ($M_2 \sin i \gtrsim 5\, \mathrm{M_{Jup}}$) on long-period orbits ($P \gtrsim 100$ days).

Our analysis also emphasizes that planets with periods $P \lesssim 40$ days which reside in binaries tend to have low eccentricities ($e \lesssim 0.05$) compared to their counterparts in orbits around single stars (Fig. 2.1). In other words, the minimum period for a significant eccentricity seems larger for planets in binaries ($P \sim 40$ days) than for planets around single stars ($P \sim 5$ days). The statistical significance of this finding is very modest, though, and calls for confirmation.

The two above-mentioned emerging trends are interesting because they might constitute the first observational evidence of two theoretical predictions. For instance, according to Kley (2000) the migration and mass growth rates of a Jovian protoplanet are enhanced when this object is embedded in a circumprimary disk in a 50–100 AU binary system. At the same time, the protoplanet's eccentricity decreases with time due to the damping action of the disk. Taken at face value, these theoretical predictions may provide a nice and self-consistent explanation for the observation that the most massive short-period planets are all found in binaries and have small eccentricities. Yet, the weak point in this reasoning is that the five circumprimary planets with periods shorter than 40 days, reside in systems with very different separations, from ~20 to ~1,000 AU. Kley's conclusions (Kley 2000) may thus apply to some of these systems, but not to all of them.

Another theoretical prediction that might find a first observational evidence in our results is the so-called Kozai migration. This migration process, specific to binaries, results from the coupling of the Kozai mechanism with tidal dissipation (Wu and Murray 2003; Fabrycky and Tremaine 2007). As shown by Takeda and Rasio (2005), if this mechanism has been at work in many planetary systems, it should have produced an excess of low-eccentricity planets. Again, this seems to provide a nice explanation to the observation that short-period planets found in binaries tend to have low eccentricities. The weak point here is that several requirements must be simultaneously met in order for the Kozai mechanism to operate (Holman et al. 1997; Wu and Murray 2003). Kozai migration may thus explain the low eccentricity of some of the five short-period planets found in binaries, but it is unlikely to explain the distinctive characteristics of all of them.

To summarize, the emerging trends seen in the period-mass and period-eccentricity diagrams are potentially interesting and might constitute a first indication that planetary migration can proceed differently in some binary systems than around single stars. To confirm and specify the present observational results, future investigations will have to improve on three points: (1) to increase the present sample of planets in binaries, (2) to systematically probe the presence of stellar companions to the known planet-hosting stars, and (3) to take into account the selection effects against moderately close binaries. We describe in the next two sections our efforts to tackle these issues, aiming at better understanding the impact of stellar duplicity, not only on planet occurrence, but also on planet properties and characteristics.

2.3 Results from Imaging Surveys

The problem of quantifying the impact of stellar duplicity on planet occurrence can be tackled in a somewhat indirect way by comparing the multiplicity among planet-bearing stars to the multiplicity among similar stars but without known planetary companions. Indeed, if the presence of a nearby stellar companion hinders planet formation, or drastically reduces the potential stability zones, the frequency of planets in binaries closer than a given separation (modulo eccentricity and mass-ratio) should be lower than the nominal frequency of planets around single stars. That is, the binary fraction among planet-hosting stars should be smaller than the binary fraction among single stars. Alternatively, if the presence of a nearby stellar companion stimulates planet formation one way or another, planets should be more common in binaries with a specific range of separations (again modulo eccentricity and mass-ratio) than around single stars. The binary fraction among planet-hosting stars should then be larger than the binary fraction among single stars. This indirect approach was first followed by Patience et al. (2002), who probed the multiplicity status of 11 planet-hosting stars and concluded that the companion star fraction among planet-bearing stars is not significantly different from that among field stars. The more than 300 planet-hosting stars known today, and the different conclusions of theoretical studies as to the impact of stellar duplicity on giant planet occurrence, both motivate a new analysis and a reconsideration of the multiplicity among planet-bearing stars.

To quantify the impact of stellar duplicity on planet occurrence and properties in binaries closer than $\sim$200 AU, we initiated in 2002 a large-scale adaptive optics search for stellar companions to $\sim$200 nearby solar-type stars with and without known planetary companions (Udry et al. 2004; Eggenberger et al. 2004c, 2007b, 2008, 2008b). To cover a substantial fraction of the sky, the main program was divided into two subprograms: a southern survey (130 stars) carried out with NAOS-CONICA (NACO) on the Very Large Telescope (VLT), and a northern survey (about 70 stars) carried out with PUEO on the Canada-France-Hawaii Telescope (CFHT). The southern survey has been completed, whereas, at the time of the writing of this chapter, the northern survey was still in progress. We present and discuss in this chapter observational and preliminary statistical results from our southern survey.

2.3.1 Our VLT/NACO Search for Stellar Companions to 130 Nearby Stars with and Without Planets

2.3.1.1 Sample and Observing Strategy

One major limitation that prevents all imaging surveys done to date (Luhman and Jayawardhana 2002; Patience et al. 2002; Mugrauer et al. 2005, 2006; Chauvin et al. 2006; Raghavan et al. 2006; Bonavita and Desidera 2007) to draw robust conclusions on the impact of stellar duplicity on planet occurrence is the absence of a well-defined control sample of non-planet-bearing stars. The use of a controlled sample is essential for two reasons. First, as explained in Section 2.2.1, Doppler planet searches suffer from noticeable selection effects against the closest binaries and these biases must be taken into account to obtain meaningful results. Second, to be rigorous, statistical studies must compare the multiplicity among planet-hosting stars with the multiplicity among similar stars but without planetary companions. To be as rigorous as possible, we included in our NACO survey both a subsample of planet-hosting stars and a controlled subsample of nearby field stars from our CORALIE planet search program showing no obvious evidence for planetary companions from radial-velocity measurements. Proceeding in this way, we had at hand high-precision radial-velocity data that place constraints on the potential giant-planet-bearing status of each comparison star. We matched the target selection criteria for Doppler planet searches, and minimized the corrections related to observational effects.

Our NACO survey therefore relies on a sample of 57 planet-host stars, together with 73 comparison stars (see Eggenberger et al. 2007b for further details on the definition of each subsample). Note that we purposely excluded from our observing list most planet-host stars observed by Patience et al. (2002) and by Chauvin et al. (2006) to avoid repeating existing observations. These stars will be included in our statistical analysis, though, balancing the two subsample sizes to about 70 stars in each subsample (Section 2.3.2). Since most of our targets are within 50 pc, the $13'' \times 13''$ field of view of NACO translates into a projected separation range of a few AU (diffraction limit) to about 325 AU. Recalling the theoretical predictions mentioned in Section 2.1, this means that our survey probes a large fraction of the separation range where the presence of a stellar companion should affect giant planet formation (hence giant planet occurrence) to some degree.

The survey observing strategy consisted of taking a first image of each of our targets (planet-hosting and controlled stars) to detect companion candidates. To distinguish true companions from unrelated background stars, we relied on two-epoch astrometry. Since most of our targets have a proper motion above $0.1''$ year^{-1}, astrometric parameters of bound companions are not expected to vary much over a few years, except for some orbital motion in the closest systems (Fig. 2.2). On the other hand, astrometric parameters of background objects without significant proper motion should vary according to the proper and parallactic motion of the primaries (Fig. 2.2). For relatively wide and bright companion candidates (projected separation $>10''$, magnitude in the K band <14), a pre-existing astrometric epoch could usually be found in the 2MASS catalog (Skrutskie et al. 2006), meaning that only one NACO observation was needed to identify true companions. However, due to the high angular resolution of NACO we could not rely on such preexisting data on a regular basis and we tried to re-observe the targets with companion candidates at a later epoch during the survey.

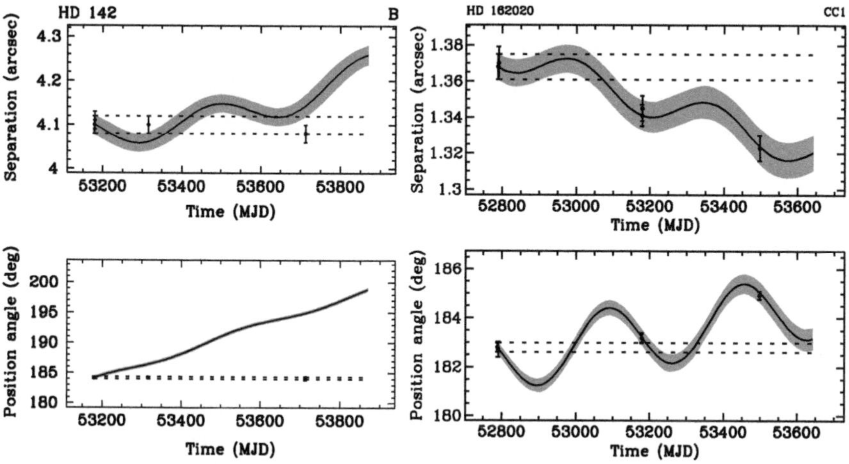

Fig. 2.2 Examples of multi-epoch astrometry from our NACO survey. *Solid lines* depict the evolution of angular separation and position angle for background objects with negligible proper motions. The *gray* zones are the related uncertainties. *Dots* represent our NACO observations and *dotted lines* depict the evolution expected for bound companions without significant orbital motion over the survey time span. The *left panels* show an example of true companion, while the *right panels* show an example of unrelated background star

2.3.1.2 Observational Results

Our NACO survey revealed 95 companion candidates in the vicinity of 33 targets. On the basis of two-epoch astrometry, we identified 19 true companions, 2 likely bound objects, and 34 background stars. The remaining 40 companion candidates (near 16 targets, most of them controlled stars) either lack second-epoch measurements (most of the objects), or have inconclusive astrometric results due to insufficiently sensitive images at one epoch (few objects). Follow-up observations have been carried out and will be used to complete the second-epoch observations.

The bound and likely bound systems identified in our NACO survey are listed in Table 2.1. Among planet-host stars, we discovered two very low mass companions to HD 65216, an early-M companion to HD 177830, and we resolved the previously known companion to HD 196050 into a close pair of M dwarfs. Besides these discoveries, our data confirm the bound nature of the companions to HD 142, HD 16141, and HD 46375. The remaining 11 true companions and the two likely bound objects all orbit control stars. These companions are late-K stars or M dwarfs, and have projected separations between 7 and 505 AU.

As illustrated on Fig. 2.3, the typical sensitivity of our survey enabled us to detect stellar companions down to ~M5 dwarfs at 0.2″, and down to the L-dwarf domain above 0.65″, providing us with a very complete census of the stellar multiplicity among our 130 targets.

Table 2.1 True (*upper portion*) and likely bound (*lower portion*) systems from our NACO survey. We refer the reader to Eggenberger et al. (2007b) for additional information on all these systems

Primary	Sample	Primary spec. type	Secondary	Secondary spec. type	Proj. sep. (AU)
HD 142 A	Planet	G1IV	HD 142 B	K8.5–M1.5	105.1 ± 1.8
HD 7895 A	Control	K1V	HD 7895 D	M2–M5	28.7 ± 0.8
HD 16141 A	Planet	G5IV	HD 16141 B	M1–M4	223 ± 11
HD 24331 A	Control	K2V	HD 24331 B	M4–M6	73.2 ± 1.7
HD 31412 Aa	Control	F8	HD 31412 Ab	M0–M3	7.1 ± 0.3
HD 40397 A	Control	G0	HD 40397 B	M0–M2	58.7 ± 1.7
HD 43834 A	Control	G5V	HD 43834 B	M3.5–M6.5	30.9 ± 0.3
HD 46375 A	Planet	K1IV	HD 46375 B	K9.5–M1.5	345 ± 12
HD 65216 A	Planet	G5V	HD 65216 Ba	M6.5–L0	255.2 ± 6.4
HD 65216 Ba		M6.5–L0	HD 65216 Bb	M7.5–L4	5.7 ± 1.1
HD 70923 A	Control	G0	HD 70923 B	M2–M5	36.9 ± 1.5
HD 78351 A	Control	G8/K0V	HD 78351 B	M1–M4	70.5 ± 2.6
HD 104263 A	Control	G5	HD 104263 B	M2.5–M4.5	68.6 ± 3.2
HD 129642 A	Control	K3V	HD 129642 B	M2.5–M5.5	157.4 ± 5.3
HD 154682 A	Control	G5V	HD 154682 B	M1.5–M4.5	45.3 ± 2.3
HD 177830 A	Planet	K0	HD 177830 B	M2–M5	97.1 ± 4.4
HD 196050 A	Planet	G3V	HD 196050 Ba	M1.5–M4.5	501 ± 22
HD 196050 Ba		M1.5–M4.5	HD 196050 Bb	M2.5–M5.5	19.7 ± 1.0
HD 223913 A	Control	G0V	HD 223913 B	K9.5–M2.5	314.0 ± 5.1
HD 82241	Control	F8V	CC1	M0–M3	16.3 ± 0.4
HD 134180	Control	K3V	CC2	M2.5–M5.5	505 ± 28

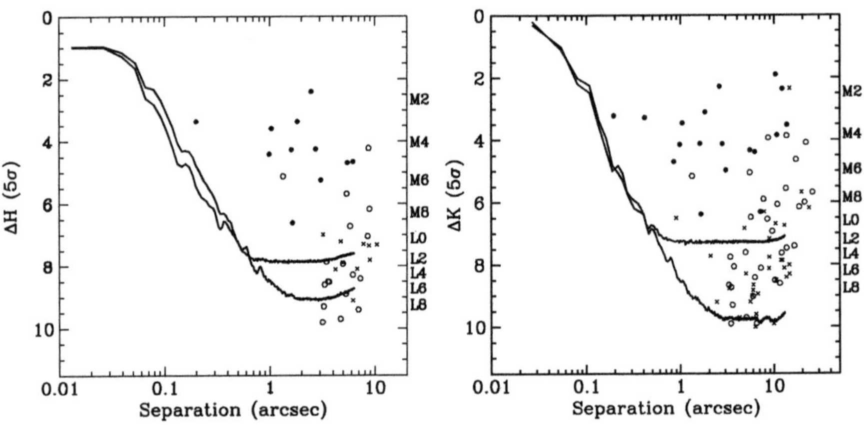

Fig. 2.3 Sensitivity limits and detections from our NACO survey in the H band (*left*) and in the K band (*right*). *Dots* represent bound and likely bound companions, *open circles* represent unbound objects, and *crosses* denote companion candidates with only one astrometric epoch. *Solid lines* are the median detection limits obtained with the two different detectors of NACO (a detector change occurred in the middle of our survey). Labels on the right-hand side of each plot show the relationship between magnitude (narrow-band photometry) and spectral type for companions to a typical old K0 dwarf

2.3.2 *The Impact of Stellar Duplicity on Planet Occurrence*

The observational results obtained in the context of our NACO survey form an unprecedented data set to study the impact of stellar duplicity on planet occurrence. Indeed, adding to our own results the targets surveyed by Patience et al. (2002) and Chauvin et al. (2006), we have a precise and homogeneous census of the multiplicity status of 73 planet-hosting stars and 66 comparison stars. We present here a preliminary statistical analysis aimed at obtaining a first quantification of the global impact of stellar duplicity on planet occurrence in binaries with mean semimajor axes between 35 and 250 AU.

2.3.2.1 Preliminary Statistical Analysis Based on the NACO Survey

A potentially sensitive issue in estimating the impact of stellar duplicity on planet occurrence is the exact definition of the controlled subsample, especially regarding the non giant-planet-bearing status of these stars. The main issue here is that a small amplitude radial velocity drift can just as well be the signature of a planetary companion as that of a more distant stellar companion. To test the sensitivity of our results to the exact definition of each subsample, we performed our first analysis based on two different sample redefinitions: (i) a loose re-definition where both subsamples were slightly modified except for a homogeneous cut-off at close separation ($\sim 0.7''$) to exclude the few stars with significant radial-velocity drifts; (ii) a more stringent redefinition where both subsamples were limited in distance to 50 pc, and where control stars showing any type of radial-velocity variation (small radial-velocity drifts, short-period variability, ...) were excluded. This additional selection was aimed at keeping in the controlled subsample as little potential planet-hosting stars as possible. Hereafter, the loosely redefined subsamples will be called "full" subsamples, while the more refined subsamples will be called "re-defined".

To quantify the global impact of stellar duplicity on giant planet occurrence, we computed the binary fraction for the four subsamples described above. According to our data, the binary fraction among planet-hosting stars is $5.5 \pm 2.7\%$ (4/73) for the full subsample and $4.9 \pm 2.7\%$ (3/62) for the redefined subsample. For control stars, we obtain binary fractions of $13.7 \pm 4.2\%$ (9/66) and $17.4 \pm 5.2\%$ (9/52) for the full and redefined subsamples, respectively. These results translate into a difference in binary fraction (controlled – planet-hosting) of $8.2 \pm 5.0\%$ for the full subsample and of $12.5 \pm 5.9\%$ for the redefined one. Although the relative errors on these results are quite large due to the small number of available companions, both sample definitions yield a positive difference with a statistical significance of 1.6-2.1σ. In physical terms, this positive difference implies that planets (mainly giant ones) are less frequent in binaries with mean semimajor axes between 35 and 250 AU than around single stars. In other words, stellar duplicity seems to negatively impact the occurrence of giant planets in such binary systems.

To extend the investigation one step further and to seek for a possible trend with mean semimajor axis, we computed the difference in binary fraction for a few bins

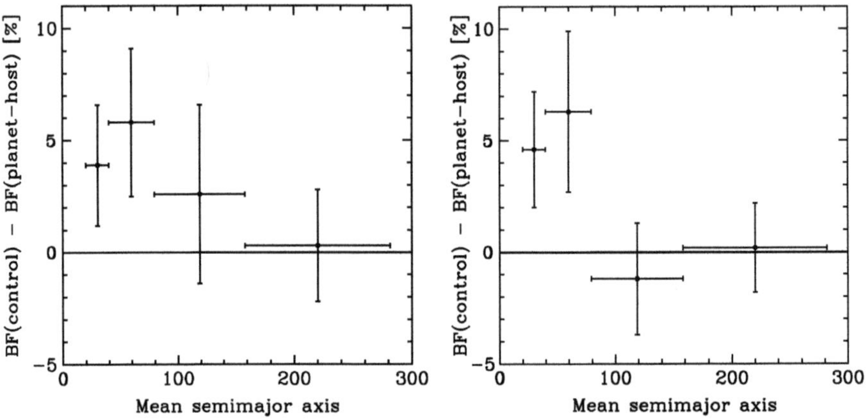

Fig. 2.4 Difference (in per cent) between the binary fraction among control stars and the binary fraction among planet-hosting stars as a function of binary mean semimajor axis. The *left plot* is based on the redefined subsamples, while the *right plot* is based on the full subsamples

in separation between 20 and 280 AU. The results for both the full and redefined subsamples are shown in Fig. 2.4. These two plots show that the difference in binary fraction does not seem to spread uniformly over the range of the semimajor axis studied here. But it seems rather concentrated below ~100 AU. This result is appealing since it might corroborate the theoretical studies that predict a negative impact of stellar duplicity on planet formation in binaries closer than ~100 AU. Nonetheless, as partly visible on Fig. 2.4, the small number of true companions available for the statistics still limits our analysis.

Given the range of semimajor axes considered in our analysis, the apparent lower frequency of planets in binaries closer than ~100 AU is likely to be related to the actual formation of these planets rather than their long-term survival. Recalling the conclusions from theoretical studies, one possible explanation to our observations would be that the disk instability scenario may be a more viable mechanism for the formation of giant planets, and as suggested by Mayer et al. (2005), this mechanism is inhibited in binaries closer than ~100 AU. However, a weak point in this argument is that Mayer et al. (2005) did not actually study planet formation via core-accretion. The prediction by these authors that the formation of giant planets via core-accretion model proceeds undisturbed in binaries with separations down to ~60 AU is solely based on the temperature profiles of their simulated disks, whereas additional effects (especially those affecting the relative velocities among planetesimals) may come into play to inhibit planet formation. Our observational results can, however, confirm that core-accretion may in fact be the only formation mechanism for planets, and that its efficiency is reduced in binaries closer than ~100 AU. This point of view may be consistent with the conclusion by Thébault et al. (2004) who state that planetesimal accretion is possible in the γ Cephei system (semimajor axis of 19 AU), but requires a delicate balance between gas-drag and secular perturbations by the secondary star.

2.3.2.2 Concluding Remarks on the Results from Imaging Surveys

The preliminary statistical results presented above are quite encouraging and already extend beyond what has been done before since the analyses by Patience et al. (2002), Raghavan et al. (2006), and Bonavita and Desidera (2007) could not correct the results for the selection effects against moderately close binaries. By adding about 70 stars to the statistics, the future results from our northern survey will make a valuable contribution to the analysis and will improve the statistical significance of the present results. The completion of the second-epoch measurements from our NACO survey will also strengthen our conclusions.

One point on which all observational studies agree is that if stellar duplicity impacts the formation and/or survival of circumstellar giant planets only in some types of binaries, it will not be easy to identify and quantify this effect in practice. This conclusion may result from practical limitations in the surveys (small sample sizes, difficulty to correct for selection effects, the need for radial-velocity data to ensure that controlled stars are free from giant planets, . . .), but it may alternatively have a more physical origin (e.g., not only binary semimajor axis, but also eccentricity and mass-ratio will likely play key roles in determining the impact of stellar duplicity on planet formation and evolution. Dynamical evolution may also significantly alter the initial distributions of planet-forming material and destroy the imprints of the formation process.). Further advances on both the theoretical and the observational fronts will be needed to specify this point. From the observational perspective, as we will see in the next section, Doppler searches for planets in spectroscopic binaries constitute another avenue to study the impact of stellar duplicity on the occurrence of giant planets in $\lesssim$200 AU binaries.

2.4 Results from Doppler Planet Searches in Spectroscopic Binaries

Nearby binary systems closer than 2–6″ can be classified into two categories: true spectroscopic binaries and moderately close visual binaries. True spectroscopic binaries are unresolved systems whose binary nature is known through the periodic translation of their spectra or, more pragmatically, through their periodic variations in radial velocity. Moderately close visual binaries are generally long-period spectroscopic systems as well, but they possess the additional property of being spatially resolved. Strictly speaking, this makes a small difference in terms of data analysis. We will also ignore this here and only consider the spectroscopic nature of all these systems. However, we will distinguish single-lined spectroscopic binaries (SB1s), for which only the spectrum of the primary star is detected, from double-lined spectroscopic binaries (SB2s), for which the spectra of both components are detected. The corollary of this distinction is that for SB1s we can only measure the radial velocity of the primary star, whereas for SB2s we can measure the individual velocities of both components. It should be noted that the classification of spectroscopic binaries into single- and double-lined systems is not absolute and depends on the instrument used for the observations and on the procedure used to analyze the data.

Until 2000–2002, planet searches in binaries closer than 2–6″ ($\lesssim$200 AU) were only of marginal interest. The discovery of two giant planets in the single-lined spectroscopic binaries Gl 86 and γ Cephei (projected separations of 20 AU and 19 AU, respectively Queloz et al. 2000; Hatzes et al. 2003) and the observation that the few most massive short-period planets all orbit the components of double or multiple star systems (Udry et al. 2002; Zucker and Mazeh 2002), changed this point of view and led to an ever-increasing interest for planet searches in moderately close binaries. Despite such an interest, classical Doppler surveys still avoid most 2–6″ binaries. The main issue with these systems is that each stellar component cannot be observed individually. That is, Doppler data of systems closer than 2–6″ consist generally of a composite spectrum made of two (or possibly more) stellar spectra, not of a single stellar spectrum. Obviously, this introduces some complications into the extraction of the radial velocity, rendering classical one-dimensional cross-correlation techniques not well adapted to the search for circumstellar planets in moderately close binaries. The inclusion of spectroscopic binaries into Doppler planet searches thus necessitated the development of data reduction techniques specially designed to extract precise radial velocities from composite spectra.

A rather natural way to analyze composite spectra and to extract precise radial velocities for the individual components of double-lined spectroscopic binaries is to generalize the concept of one-dimensional cross-correlation to that of two-dimensional correlation. This approach was followed some time ago by S. Zucker and T. Mazeh, who developed a two-dimensional correlation algorithm named TODCOR (Zucker and Mazeh 1994). Because we are interested in including spectroscopic binaries in our radial-velocity planet searches, these authors modified their TODCOR algorithm to allow it to work with our ELODIE and CORALIE echelle spectra. This resulted in a new multi-order TODCOR algorithm (Zucker et al. 2003), which has already produced some very interesting results (Zucker et al. 2003, 2004; Eggenberger et al. 2007a, 2008b). We are now using this algorithm extensively to search for planets in spectroscopic and moderately close visual binaries.

We present in this section some results from our ongoing searches for planets in spectroscopic binaries. Our presentation will follow an increasing order of difficulty in terms of radial-velocity extraction. We start with the easiest systems that are single-lined spectroscopic binaries (SB1s) and end with more complicated double-lined spectroscopic ones (SB2s).

2.4.1 Planet Searches in Single-Lined Spectroscopic Binaries

In order to obtain the first quantification of the occurrence of planets in the closest binaries capable of hosting circumstellar planets, we initiated in 2001, a systematic Doppler search for short-period circumprimary planets in single-lined spectroscopic binaries (Eggenberger et al. 2003, 2008b). The prime motivation for this program was the observation that the few most massive short-period planets are all in binaries or multiple systems (Section 2.2.3). The restriction of our survey to SB1s was motivated by two considerations. First, the faintness of the secondary components in these systems gave us good hopes that we could use our standard cross-correlation technique to extract precise radial velocities for the primary components. Second, the prospect of finding giant planets is higher in SB1s than in SB2s with similar separations but more massive secondaries. Our survey for giant planets in SB1s was thus designed as a first exploratory investigation that may be complemented later, in the case of positive results, by an additional survey targeting SB2s.

2.4.1.1 Sample and Observations

Our sample of binaries was selected on the basis of former CORAVEL surveys carried out to study the multiplicity among G and K dwarfs of the solar neighborhood (Duquennoy et al. 1991b; Halbwachs et al. 2003). Basically, we retained all the 140 SB1 candidates with periods longer than $\sim$1.5 years (some of them with well-characterized orbits, others with long-period drifts). Note that CORAVEL velocities have a typical precision of 300 $m\,s^{-1}$ and thus cannot be used to search for planets. To search for planets in our sample of 140 SB1s we took 10–15 additional high-precision radial-velocity measurements of each system, either with the ELODIE spectrograph (Observatoire de Haute-Provence, France; Baranne et al. 1996; Perrier et al. 2003) or with the CORALIE spectrograph (La Silla Observatory, Chile; Queloz et al. 2000; Udry et al. 2000). Given our initial aim to analyze these high-precision data with standard cross-correlation techniques, we rejected during the observations, the systems that turned out to be SB2s at the higher resolution of ELODIE and CORALIE, as well as the binaries that were resolved within the guiding field of the telescope. After this additional selection, we ended up with 101 SB1s that form the core of our survey.

2.4.1.2 First Analysis Based on One-Dimensional Cross-Correlation

As a first step in the analysis, the spectra obtained with ELODIE and CORALIE were reduced online, and the radial velocities were extracted using our standard cross-correlation pipeline. When searching for planets in binaries, what we are interested in are not the radial velocities themselves but instead the residual (radial) velocities around the binary orbits. The planet search was thus carried out by searching for short-period variations in these residual velocities.

Figure 2.5 shows the distribution of the residual-velocity variations for our 101 targets. These variations are quantified by a normalized root-mean-square (rms), which is the ratio of the external error (i.e., the standard deviation around the orbit or around the drift) to the mean internal error (i.e., the mean of individual photon-noise errors). As shown by Fig. 2.5, most of our targets (74%) have normalized rms close to 1, indicating that no source of radial-velocity variation other than the orbital

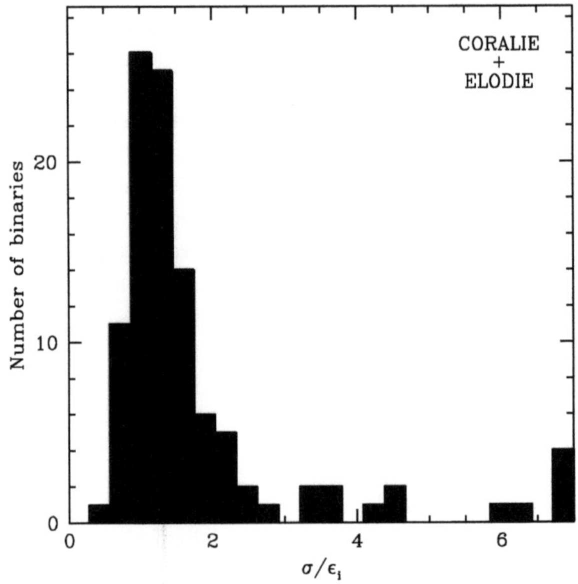

Fig. 2.5 Normalized residual-velocity rms for all our SB1s. σ is the standard deviation around a Keplerian orbit or around a drift, and ϵ_i is the mean measurement uncertainty. Systems with an rms larger than 7 are all gathered together in the last bin

motion is present (see Fig. 2.6 for an example). In contrast, 12.5% of our targets are clearly variable and exhibit normalized rms greater than 3 (see Fig. 2.6 for an example). The remaining systems (13.5%) are marginally variable, with normalized rms between 2 and 3.

In terms of planetary prospects, the most interesting systems are the variable and marginally variable binaries. Nonetheless, the presence of a planetary companion in orbit around the primary star is not the only way to produce residual-velocity variations like those observed. Alternative possibilities include: (i) the primary star is intrinsically variable, (ii) the system is an unrecognized SB2 (i.e., an SB1 when analyzed via one-dimensional cross-correlation, but an SB2 when analyzed via two-dimensional correlation), and (iii) the system is in fact triple and the secondary is itself a short-period spectroscopic binary. Assuming that planets are as common in close binaries as around single stars, we expect to find only one or two planets more massive than 0.5 M_{Jup} and with periods shorter than ~40 days in our sample. This rough estimation shows that most of the observed residual-velocity variations are probably not related to the presence of planetary companions, but likely stemmed from the binary or multiple nature of our targets. Therefore, to identify the few potential planet-bearing stars among the several variable and marginally variable systems, we must find a way to precisely characterize the cause of the residual-velocity variations.

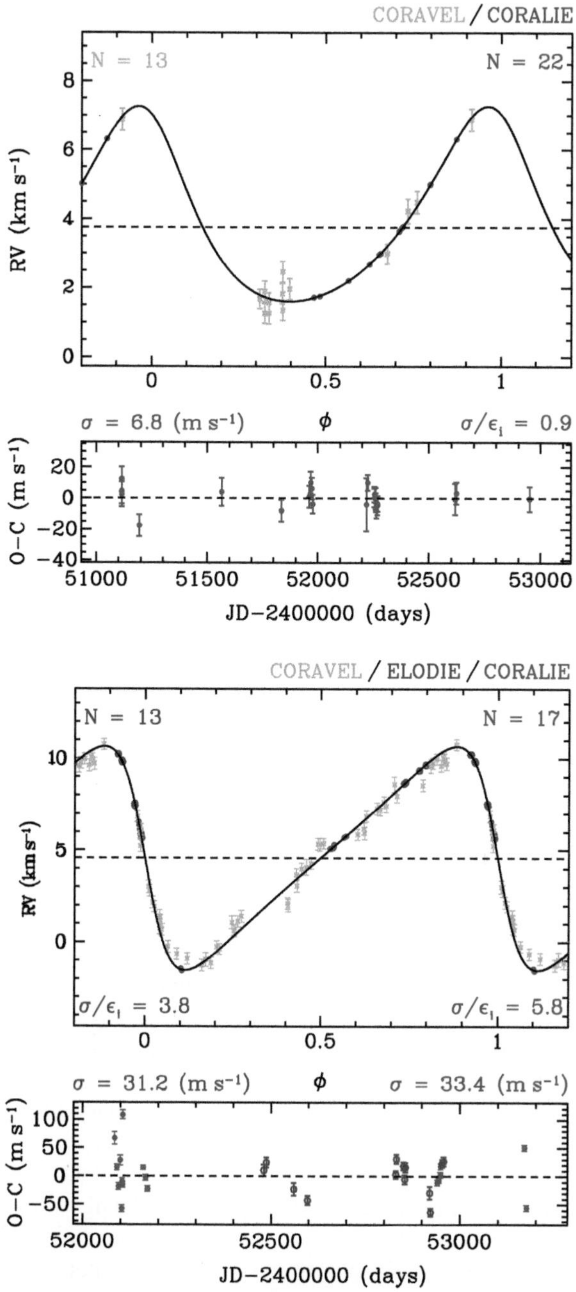

Fig. 2.6 *Top*: example of a binary exhibiting no residual-velocity variation. CORAVEL data are depicted as stars (*large error bars*) while CORALIE data are depicted as *dots*. The *bottom panel* shows the residual velocities (CORALIE data only). *Bottom*: example of a binary with variable residual velocities. This system was exceptionally observed with both ELODIE and CORALIE. Figures on the *top* refer to the ELODIE velocities (represented as *circles*), while figures on the *bottom* refer to the CORALIE velocities (represented as *dots*)

2.4.1.3 Identifying the Origin of Residual-Velocity Variations

Binaries with intrinsically variable primaries can be identified similar to single active stars by considering the chromospheric emission flux in the Ca II H and K lines. Using one-dimensional cross-correlation techniques, identifying triple systems and unrecognized SB2s is feasible in some instances (Santos et al. 2002; Eggenberger et al. 2003, 2008b), but two-dimensional correlation is a much more efficient tool for this purpose. We are thus presently analyzing all the variable and marginally variable systems with the two-dimensional algorithm TODCOR. This work is in progress and only four variable systems have been studied in some detail so far. Of these four systems, two turned out to be triple star systems (see Fig. 2.7 or Eggenberger et al. 2003, 2008b for an example), while the two others turned out to be unrecognized SB2s (see Fig. 2.8 for an example). None of these four systems shows hints of the presence of a circumprimary planet.

2.4.1.4 Preliminary General Results on Planet Searches in SB1s

The present results from our search for circumprimary short-period planets in SB1s show that in most of these systems (74%) the secondary component is so faint (magnitude difference $\Delta V \gtrsim 6$) that it does not contribute significantly to the recorded flux. Doppler data of such systems can be analyzed similar to the Doppler data of single stars and the precision achieved on the measurements of the radial velocities of their primary stars are as good as those of single stars.

In contrast, analyzing the Doppler data of the 26 SB1s that exhibit residual-velocity variations is not straightforward. In many of these systems the secondary component (and also possibly the tertiary component) significantly contributes to the recorded flux ($\Delta V \in [\sim 3, \sim 6]$), rendering the use of two-dimensional correlation mandatory to unambiguously identify the origin of the variations observed, and hence to search for circumprimary planets. Our current results do not enable us to precisely characterize our detection capabilities in terms of circumprimary planet searches, but we estimate that typical precisions on the radial velocity of the primary star range between 10 and 20 $\mathrm{m\,s^{-1}}$. Although these precisions are not as good as for single stars, they remain good enough to search for giant planets.

The preliminary results from our search for circumprimary giant planets in SB1s thus confirm that such a program has grounds for existence. So far, our survey has unveiled no promising planetary candidate, but the data of 22 variable and marginally variable systems remain to be analyzed in detail with two-dimensional correlation. Since contamination effects stemming from the stellar companions are likely to prevail over potential planetary signals, two-dimensional analyses must be completed before concluding on the existence or absence of planets in our sample. All we can say at present is that less than 22% of the SB1s from our sample have a short-period ($P \lesssim 40$ days) giant (minimum mass $\gtrsim 0.5\ \mathrm{M_{Jup}}$) planetary companion. Definitive results from our program will enable us to obtain a much stronger constraint.

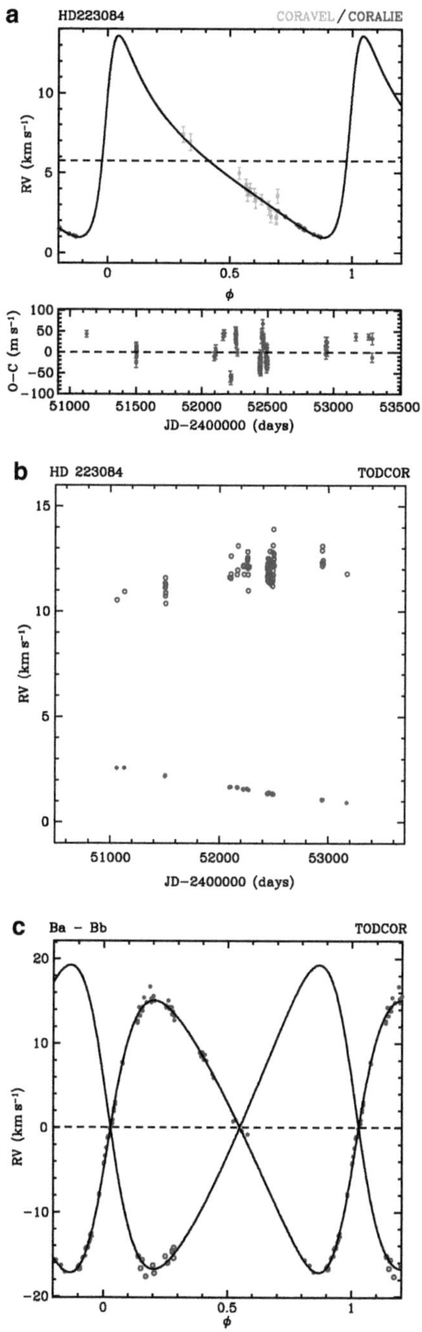

Fig. 2.7 An example of triple system: HD 223084. (**a**) CORAVEL (*crosses, large error bars*) and CORALIE (*dots*) velocities for HD 223084. The binary orbit is tentative and is used only as a proxy to compute residual velocities. The *bottom panel* of Fig. (**a**) shows the residual velocities (CORALIE data only). (**b**) TODCOR velocities for HD 223084 A (*dots*) and HD 223084 Ba (*open circles*) after having removed the 202-day modulation of the Ba–Bb inner pair. (**c**) SB2 orbit for HD 223084 Ba (*dots*) and HD 223084 Bb (*open circles*). This orbit is characterized by a period of 202 days and velocity semiamplitudes of 16.1 and 18 km s^{-1} for components Ba and Bb, respectively

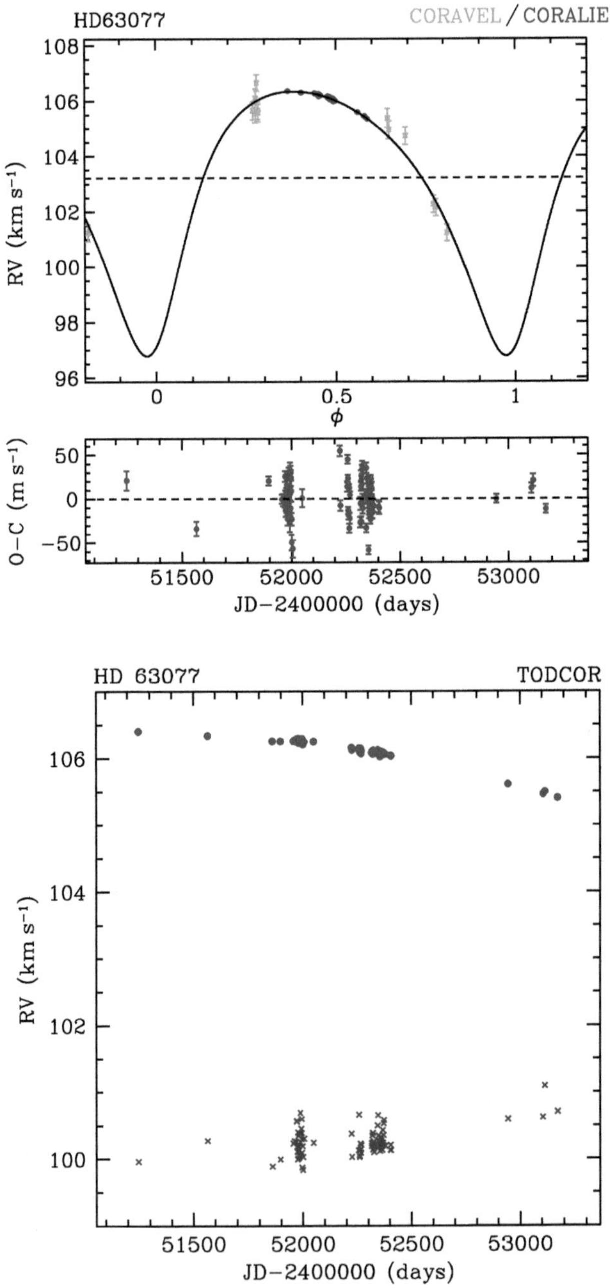

Fig. 2.8 An example of unrecognized SB2: HD 63077. *Top*: CORAVEL (*crosses, large error bars*) and CORALIE (*dots*) velocities for HD 63077. The binary orbit is tentative and is used only as a proxy to compute residual velocities. The *bottom panel* shows the residual velocities (CORALIE data only). *Bottom*: TODCOR velocities for HD 63077 A (dots) and HD 63077 B (*crosses*)

2.4.2 Planet Searches in Double-Lined Spectroscopic Binaries

Double-lined spectroscopic binaries have not been systematically included in any of our observing programs yet, but several of the visual binaries closer than $\sim 6''$ and with magnitude differences of $\Delta V \lesssim 4$ are in fact unrecognized SB2s (i.e., SB1s when analyzed via one-dimensional cross-correlation, but SB2s when analyzed via two-dimensional correlation). To properly analyze the data of these systems and to characterize the feasibility of Doppler searches for circumstellar planets in SB2s, we are presently conducting a series of observational tests and simulations on SB2 systems with various characteristics. To illustrate both the interest in including SB2s in planet searches and the challenges faced by Doppler planet searches in such systems, we present here the results we have obtained for our best-studied case, the triple system HD 188753 (Eggenberger et al. 2007a).

2.4.2.1 The Example of HD 188753

HD 188753 has attracted much attention since July 2005 when Konacki (2005) reported the discovery of a 1.14-M_{Jup} planet on a 3.35-day orbit around the primary component of this triple star system. Aside from the planet, HD 188753 consists of a primary star (HD 188753 A) orbited by a visual companion, HD 188753 B, which itself is a spectroscopic binary with two components HD 188753 Ba and HD 188753 Bb. The visual orbit of the AB pair is characterized by a period of 25.7 years, a semimajor axis of 12.3 AU (0.27$''$ separation) and an eccentricity of 0.5 (Söderhjelm 1999). The spectroscopic orbit of HD 188753 B has a period of 155 days (Griffin 1977; Konacki 2005). What renders this discovery particularly important and interesting is that the periastron distance of the AB pair may be small enough to preclude giant planet formation around HD 188753 A through the canonical planet-formation models (Nelson 2000; Mayer et al. 2005; Boss 2006; Jang-Condell 2007). The discovery of a close-in giant planet around this star has thus been perceived as a serious challenge to planet-formation theories, though the alternative possibility that HD 188753 A might have acquired its planet through dynamical interactions was also pointed out (Pfahl 2005; Portegies Zwart and McMillan 2005).

When observed with ELODIE, HD 188753 reveals itself as an SB2, the spectrum of the faintest component (Bb) being undetectable in most of our observations. Our TODCOR radial velocities for HD 188753 A and HD 188753 Ba are displayed in Fig. 2.9. These velocities confirm that HD 188753 Ba is a spectroscopic binary with a period of 155 days. However, our velocities for HD 188753 A show a steady decrease consistent with the 25.7-year orbital motion of the AB pair, but no sign of the 3.35-day planetary signal as reported by Konacki (2005). Instead, our results indicate that the residuals around the long-period drift are basically noise and the rms of 60 m s^{-1} can be interpreted as the precision we achieve on the measurement of the radial velocity of this star. Monte Carlo simulations run to check our ability to detect the potential planet around HD 188753 A showed that we had both the precision and

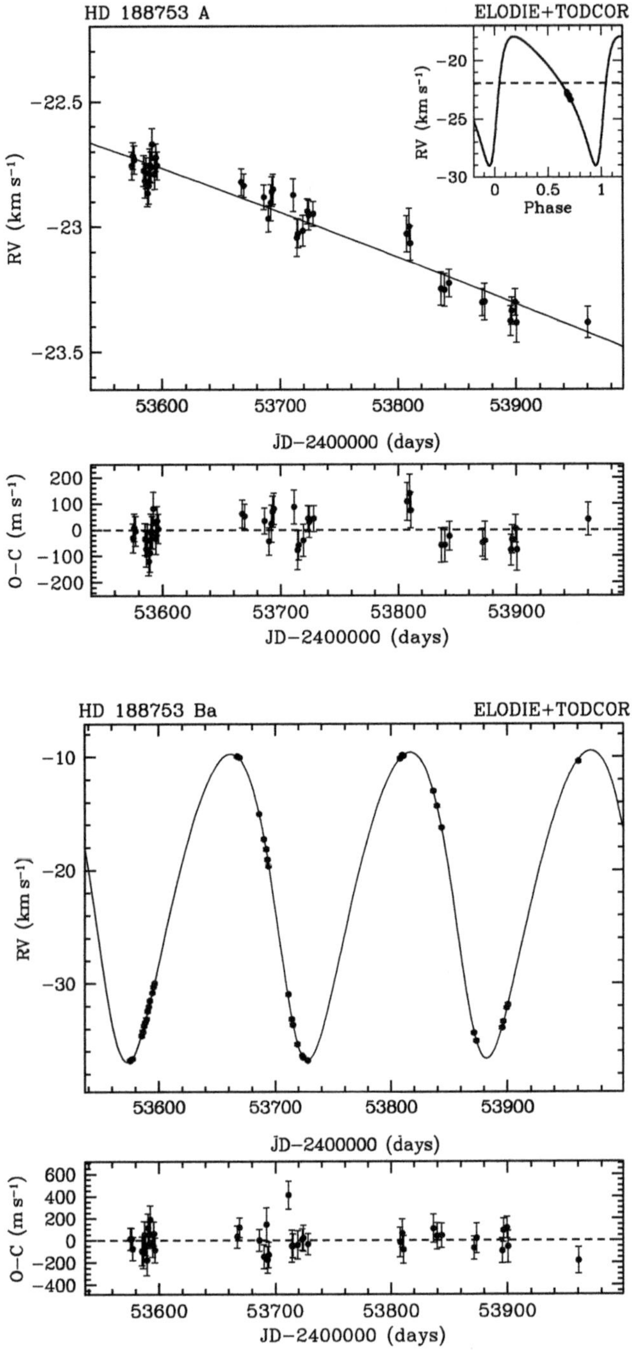

Fig. 2.9 Radial velocities and orbital solutions for HD 188753 A (*top*) and HD 188753 Ba (*bottom*). For component A, the *solid line* represents the 25.7-year orbital motion of the visual pair shown in full in the inset. For component Ba, the orbital solution corresponds to the 155-day modulation and it includes a linear drift to take the 25.7-year orbital motion into account

the temporal sampling required to detect a planetary signal like the one reported by Konacki (2005). On that basis, we conclude that our data show no evidence of a 1.14 M_{Jup} on a 3.35-day orbit around HD 188753 A.

In addition to the question of whether there is or is not a hot Jupiter around HD 188753 A, our analysis of HD 188753 raises several more questions. In particular, the precision of 60 m s^{-1} obtained on the radial velocity of HD 188753 A looks abnormally poor compared to the results presented in Section 2.4.1. The triple nature of HD 188753 may partly explain this result, but it is probably not the primary reason. Rather, the search for circumprimary planets in SB2s seems to require higher quality data (mainly a better spectral resolution) than the search for circumprimary planets in SB1s. Investigations are currently underway to specify this point.

2.4.2.2 Concluding Remarks on Planet Searches in SB2s

Based on our current experience, Doppler searches for circumprimary planets look more challenging in SB2s than in SB1s, even when using two-dimensional correlation. Clearly, considerable work remains to be done to precisely characterize our detection capabilities in spectroscopic binaries with various characteristics and to identify the main factor that limits our precision for each type of system. Nonetheless, including SB2s in planet searches is desirable since these systems are the most susceptible of providing us with interesting constraints on planet-formation mechanisms. Furthermore, SB2s are the potential targets for circumbinary planet searches, which offer a still unexplored research field worth of interest.

2.5 Conclusion and Perspectives

Over the past 5 years, binaries have become increasingly interesting targets of planet searches. On one hand, Doppler surveys have shown that giant planets exist even in spectroscopic binaries (Queloz et al. 2000; Hatzes et al. 2003; Zucker et al. 2004), raising the possibility that planets may be quite common in binary and multiple star systems. On the other hand, theoretical studies have shown that the presence of a stellar companion within ~100 AU likely affects the formation and subsequent evolution of circumstellar giant planets (Kley 2000; Nelson 2000; Mayer et al. 2005; Boss 2006; Thébault et al. 2006), leaving potential imprints in the occurrence, characteristics, and properties of the planets residing in these systems. The study of circumstellar planets found in binaries closer than ~100 AU might thus provide a unique means to probe the formation and evolution processes at work in planetary systems.

Imaging surveys searching for stellar companions to the known planet-bearing stars have been very successful, revealing several new binary planet-hosting systems and yielding a precise characterization of the multiplicity status of more than 70 planet-hosting stars (Luhman and Jayawardhana 2002; Patience et al. 2002;

Mugrauer et al. 2005, 2006; Chauvin et al. 2006; Raghavan et al. 2006; Eggenberger et al. 2007b). Additionally, our NACO survey has provided us with the multiplicity among a control sample of about 70 nearby stars showing no evidence for giant planetary companions, and affected by the same selection effects than planet-hosting stars. A preliminary statistical analysis based on our NACO data brings the first observational evidence that the occurrence of giant planets is reduced in binaries closer than ~100 AU (Eggenberger et al. 2008). Given our present knowledge of planet-formation mechanisms, two possible explanations can be put forward to explain this result: either disk instability is a viable formation mechanism that accounts for the existence of a significant number of the planets known presently, or core accretion is the only formation channel but its efficiency is reduced in binaries closer than ~100 AU. Differentiating between these two possibilities will require additional work, both on the theoretical and on the observational sides. Yet, the important point to notice is that observations have caught up with theoretical studies on the investigation of the impact of stellar duplicity on giant planet formation, meaning that some theoretical predictions can now be confronted with observational results.

The recent discoveries from imaging surveys have somewhat decreased the statistical significance of the emerging trends suggesting that short-period planets found in binary and multiple star systems possess distinctive characteristics and properties compared to their counterparts orbiting single stars (Zucker and Mazeh 2002; Eggenberger et al. 2004; Mugrauer et al. 2005; Desidera and Barbieri 2007). The most robust feature in this respect is still the observation that the few most massive short-period planets all orbit the components of binaries or triple stars. Nonetheless, such planets are still sparse and even the most recent statistical studies remain affected by the selection effects against moderately close binaries and by the uncertain multiplicity status of many planet-hosting stars. The combined results from our NACO and PUEO surveys will remove these two last uncertainties to a large extent, allowing for a major reinvestigation of possible differences in the eccentricity distributions of planets found in binaries and around single stars.

Over the past few years, significant effort has been put into extending radial-velocity planet searches to spectroscopic and moderately close visual binaries (Zucker et al. 2003, 2004; Konacki 2005a, 2005b; Eggenberger et al. 2007a, 2008b). In a general way, planet searches in moderately close binaries are still in their early phases and only partial results are available. Current results demonstrate that Doppler searches for giant planets are technically feasible in single-lined and in some types of double-lined spectroscopic binaries. However, the feasibility of planet searches in double-lined spectroscopic binaries with small magnitude differences remains to be characterized and confirmed.

Final results from the presently ongoing planet searches in spectroscopic binaries are awaited with great interest for several reasons. First, Doppler planets searches are the best tool to expand the size of the still limited sample of planets residing in binary and multiple star systems. Second, these surveys constitute the only current possibility to directly probe the occurrence of giant planets in binaries closer than ~200 AU and to characterize the closest systems potentially capable of hosting circumstellar giant planets. In particular, Doppler searches for planets in spectroscopic binaries will provide us with stronger constraints on the reality of the 20-AU "limit" and on its possible interpretation as a minimum separation for considering that a binary possibly harbors a giant planet. Finally, by probing the occurrence of giant planets in binaries closer than ~35 AU, planet searches in spectroscopic binaries will nicely complement the results from our NACO and PUEO surveys. Gathering together the observational results from our imaging and spectroscopic programs, we might then obtain some constraints as to whether most giant planets found in binaries closer than ~100 AU actually formed in these systems or were deposited at their present location through dynamical interactions.

As planet searches progress, the conviction that planets are common objects in the universe continually strengthen. The discovery of planets in environments previously considered as relatively hostile to their existence (spectroscopic binaries, pulsars, ...) has contributed to this development, showing that planet formation is not as easily inhibited as originally thought. In addition to the encouraging results obtained thus far for giant planets, the expectation that terrestrial planets form alongside their Jovian counterparts suggests that discoveries are limited by instrumental sensitivity rather than the availability of planets. Even if the presence of a nearby stellar companion lowers the efficiency of planet formation, theoretical studies support the existence of circumstellar terrestrial planets in many types of binaries (Barbieri et al. 2002; Haghighipour and Raymond 2007; Quintana et al. 2007). Circumbinary planets are also expected to exist around various types of binary systems (Quintana and Lissauer 2006; Pierens and Nelson 2007) and searches for circumbinary planets offer a still unexplored field of investigation for planet hunters. In view of the potential information they can yield on the overall frequency of planets and on the processes underlying planet formation, planet searches in and around binaries are thus not only meaningful but also desirable.

Acknowledgements We thank A.-M. Lagrange for helpful comments on our manuscript. A. E. acknowledges support from the Swiss National Science Foundation through a fellowship for prospective researcher.

References

P. Artymowicz and S. H. Lubow. Dynamics of binary-disk interaction. 1: Resonances and disk gap sizes. *ApJ*, 421:651–667, February 1994.

A. Baranne, M. Mayor, and J. L. Poncet. CORAVEL – A new tool for radial velocity measurements. *Vistas in Astronomy*, 23:279–316, 1979.

A. Baranne, D. Queloz, M. Mayor, G. Adrianzyk, G. Knispel, D. Kohler, D. Lacroix, J.-P. Meunier, G. Rimbaud, and A. Vin. ELODIE: A spectrograph for accurate radial velocity measurements. *A&AS*, 119:373–390, October 1996.

M. Barbieri, F. Marzari, and H. Scholl. Formation of terrestrial planets in close binary systems: The case of alpha Centauri A. *A&A*, 396:219–224, December 2002.

M. Bonavita and S. Desidera. The frequency of planets in multiple systems. *A&A*, 468:721–729, June 2007.

A. P. Boss. Gas giant protoplanets formed by disk instability in binary star systems. *ApJ*, 641:1148–1161, April 2006.

R. P. Butler, G. W. Marcy, E. Williams, H. Hauser, and P. Shirts. Three new "51 Pegasi–Type" planets. *ApJL*, 474:L115, January 1997.

B. Campbell, G. A. H. Walker, and S. Yang. A search for substellar companions to solar-type stars. *ApJ*, 331:902–921, August 1988.

G. Chauvin, A.-M. Lagrange, S. Udry, T. Fusco, F. Galland, D. Naef, J.-L. Beuzit, and M. Mayor. Probing long-period companions to planetary hosts. VLT and CFHT near infrared coronographic imaging surveys. *A&A*, 456:1165–1172, September 2006.

W. D. Cochran, A. P. Hatzes, R. P. Butler, and G. W. Marcy. The discovery of a planetary companion to 16 Cygni B. *ApJ*, 483:457, July 1997.

X. Delfosse, J.-L. Beuzit, L. Marchal, X. Bonfils, C. Perrier, D. Ségransan, S. Udry, M. Mayor, and T. Forveille. M dwarfs binaries: Results from accurate radial velocities and high angular resolution observations. In R. W. Hilditch, H. Hensberge, and K. Pavlovski, editors, *ASP Conf. Ser. 318: Spectroscopically and Spatially Resolving the Components of the Close Binary Stars*, pages 166–174, December 2004.

S. Desidera and M. Barbieri. Properties of planets in binary systems. The role of binary separation. *A&A*, 462:345–353, January 2007.

A. Duquennoy and M. Mayor. Multiplicity among solar-type stars in the solar neighbourhood. II - Distribution of the orbital elements in an unbiased sample. *A&A*, 248:485–524, August 1991.

A. Duquennoy, M. Mayor, and J.-L. Halbwachs. Multiplicity among solar type stars in the solar neighbourhood. I – CORAVEL radial velocity observations of 291 stars. *A&AS*, 88:281–324, May 1991b.

R. H. Durisen, A. P. Boss, L. Mayer, A. F. Nelson, T. Quinn, and W. K. M. Rice. Gravitational instabilities in gaseous protoplanetary disks and implications for giant planet formation. In B. Reipurth, D. Jewitt, and K. Keil, editors, *Protostars and Planets V*, pages 607–622, 2007.

A. Eggenberger, S. Udry, and M. Mayor. Planets in binaries. In *ASP Conf. Ser. 294: Scientific Frontiers in Research on Extrasolar Planets*, pages 43–46, 2003.

A. Eggenberger, S. Udry, and M. Mayor. Statistical properties of exoplanets. III. Planet properties and stellar multiplicity. *A&A*, 417:353–360, April 2004.

A. Eggenberger, J.-L. Halbwachs, S. Udry, and M. Mayor. Statistical properties of an unbiased sample of F7-K binaries: Towards the long-period systems. In C. Allen and C. Scarfe, editors, *Revista Mexicana de Astronomia y Astrofisica Conference Series*, pages 28–32, August 2004b.

A. Eggenberger, S. Udry, M. Mayor, J.-L. Beuzit, A. M. Lagrange, and G. Chauvin. Detection and properties of extrasolar planets in double and multiple star systems. In J. Beaulieu, A. Lecavelier Des Etangs, and C. Terquem, editors, *ASP Conf. Ser. 321: Extrasolar Planets: Today and Tomorrow*, page 93, December 2004c.

A. Eggenberger, S. Udry, T. Mazeh, Y. Segal, and M. Mayor. No evidence of a hot Jupiter around HD 188753 A. *A&A*, 466:1179–1183, May 2007.

A. Eggenberger, S. Udry, G. Chauvin, J.-L. Beuzit, A.-M. Lagrange, D. Ségransan, and M. Mayor. The impact of stellar duplicity on planet occurrence and properties. I. Observational results of a VLT/NACO search for stellar companions to 130 nearby stars with and without planets. *A&A*, 474:273–291, October 2007b.

A. Eggenberger, S. Udry, G. Chauvin, J.-L. Beuzit, A.-M. Lagrange, and M. a Mayor. Probing the impact of stellar duplicity on planet occurrence. In D. Fischer, F. Rasio, S. Thorsett, and A. Wolszczan, editors, *Proceedings of the Conference on Extreme Solar Systems, ASP Conf. Ser.*, 398:179, 2008.

A. Eggenberger, S. Udry, M. Mayor, G. Chauvin, B. Markus, J.-L. Beuzit, A. M. Lagrange, T. Mazeh, S. Zucker, and D. Ségransan. Extrasolar planets in double and multiple stellar systems. In S. Hubrig, M. Petr-Gotzens, and A. Tokovinin, editors, *Multiple Stars Across the H-R Diagram, ESO Astrophysic Symposia*, p. 169, 2008b.

D. Fabrycky and S. Tremaine. Shrinking binary and planetary orbits by Kozai cycles with tidal friction. *ApJ*, 669:1298–1315, November 2007.

M. Fatuzzo, F. C. Adams, R. Gauvin, and E. M. Proszkow. A statistical stability analysis of earth-like planetary orbits in binary systems. *PASP*, 118:1510–1527, November 2006.

D. A. Fischer and G. W. Marcy. Multiplicity among M dwarfs. *ApJ*, 396:178–194, September 1992.

R. F. Griffin. The multiple star HD 188753 (ADS 13125). *The Observatory*, 97:15–18, February 1977.

N. Haghighipour. Dynamical stability and habitability of the γ Cephei binary-planetary system. *ApJ*, 644:543–550, June 2006.

N. Haghighipour and S. N. Raymond. Habitable planet formation in binary planetary systems. *ApJ*, 666:436–446, September 2007.

J. L. Halbwachs, M. Mayor, S. Udry, and F. Arenou. Multiplicity among solar-type stars. III. Statistical properties of the F7-K binaries with periods up to 10 years. *A&A*, 397:159–175, January 2003.

A. P. Hatzes, W. D. Cochran, M. Endl, B. McArthur, D. B. Paulson, G. A. H. Walker, B. Campbell, and S. Yang. A planetary companion to γ Cephei A. *ApJ*, 599:1383–1394, December 2003.

M. Holman, J. Touma, and S. Tremaine. Chaotic variations in the eccentricity of the planet orbiting 16 Cygni B. *Nature*, 386:254–256, March 1997.

M. J. Holman and P. A. Wiegert. Long-term stability of planets in binary systems. *AJ*, 117:621–628, January 1999.

K. A. Innanen, J. Q. Zheng, S. Mikkola, and M. J. Valtonen. The Kozai mechanism and the stability of planetary orbits in binary star systems. *AJ*, 113:1915, May 1997.

H. Jang-Condell. Constraints on the formation of the planet in HD 188753. *ApJ*, 654:641–649, January 2007.

H. R. A. Jones, R. P. Butler, C. G. Tinney, G. W. Marcy, B. D. Carter, A. J. Penny, C. McCarthy, and J. Bailey. High-eccentricity planets from the Anglo-Australian planet search. *MNRAS*, 369:249–256, June 2006.

W. Kley. Evolution of an embedded planet in a binary system. In *IAU Symposium*, page 211, 2000.

M. Konacki. An extrasolar giant planet in a close triple-star system. *Nature*, 436:230–233, July 2005.

M. Konacki. Precision radial velocities of double-lined spectroscopic binaries with an iodine absorption cell. *ApJ*, 626:431–438, June 2005.

Y. Kozai. Secular perturbations of asteroids with high inclination and eccentricity. *AJ*, 67:591, November 1962.

J. J. Lissauer and D. J. Stevenson. Formation of giant planets. In B. Reipurth, D. Jewitt, and K. Keil, editors, *Protostars and Planets V*, pages 591–606, 2007.

K. L. Luhman and R. Jayawardhana. An adaptive optics search for companions to stars with planets. *ApJ*, 566:1132–1146, February 2002.

D. Malmberg, M. B. Davies, and J. E. Chambers. The instability of planetary systems in binaries: How the Kozai mechanism leads to strong planet–planet interactions. *MNRAS*, 377:L1–L4, April 2007.

G. W. Marcy, R. P. Butler, S. S. Vogt, D. A. Fischer, G. W. Henry, G. Laughlin, J. T. Wright, and J. A. Johnson. Five new extrasolar planets. *ApJ*, 619:570–584, January 2005a.

G. Marcy, R. P. Butler, D. Fischer, S. Vogt, J. T. Wright, C. G. Tinney, and H. R. A. Jones. Observed properties of exoplanets: Masses, orbits, and metallicities. *Progress of Theoretical Physics Supplement*, 158:24–42, 2005b.

L. Mayer, J. Wadsley, T. Quinn, and J. Stadel. Gravitational instability in binary protoplanetary discs: new constraints on giant planet formation. *MNRAS*, 363:641–648, October 2005.

T. Mazeh, Y. Krymolowski, and G. Rosenfeld. The high eccentricity of the planet orbiting 16 Cygni B. *ApJl*, 477:L103, March 1997.

J.-L. Monin, C. J. Clarke, L. Prato, and C. McCabe. Disk evolution in young binaries: From observations to theory. In B. Reipurth, D. Jewitt, and K. Keil, editors, *Protostars and Planets V*, pages 395–409, 2007.

M. Mugrauer, R. Neuhäuser, A. Seifahrt, T. Mazeh, and E. Guenther. Four new wide binaries among exoplanet host stars. *A&A*, 440:1051–1060, September 2005.

M. Mugrauer, R. Neuhäuser, T. Mazeh, E. Guenther, M. Fernández, and C. Broeg. A search for wide visual companions of exoplanet host stars: The Calar Alto Survey. *Astronomische Nachrichten*, 327:321, May 2006.

M. Nagasawa, E. W. Thommes, S. J. Kenyon, B. C. Bromley, and D. N. C. Lin. The diverse origins of terrestrial-planet systems. In B. Reipurth, D. Jewitt, and K. Keil, editors, *Protostars and Planets V*, pages 639–654, 2007.

A. F. Nelson. Planet formation is unlikely in equal-mass binary systems with A ˜ 50 AU. *ApJL*, 537:L65–L68, July 2000.

J. Patience, R. J. White, A. M. Ghez, C. McCabe, I. S. McLean, J. E. Larkin, L. Prato, S. S. Kim, J. P. Lloyd, M. C. Liu, J. R. Graham, B. A. Macintosh, D. T. Gavel, C. E. Max, B. J. Bauman, S. S. Olivier, P. Wizinowich, and D. S. Acton. Stellar companions to stars with planets. *ApJ*, 581:654–665, December 2002.

C. Perrier, J.-P. Sivan, D. Naef, J. L. Beuzit, M. Mayor, D. Queloz, and S. Udry. The ELODIE survey for northern extra-solar planets. I. Six new extra-solar planet candidates. *A&A*, 410:1039–1049, November 2003.

E. Pfahl. Cluster origin of the triple star HD 188753 and its planet. *ApJL*, 635:L89–L92, December 2005.

E. Pfahl and M. Muterspaugh. Impact of stellar dynamics on the frequency of giant planets in close binaries. *ApJ*, 652:1694–1697, December 2006.

B. Pichardo, L. S. Sparke, and L. A. Aguilar. Circumstellar and circumbinary discs in eccentric stellar binaries. *MNRAS*, 359:521–530, May 2005.

A. Pierens and R. P. Nelson. On the migration of protoplanets embedded in circumbinary disks. *A&A*, 472:993–1001, September 2007.

S. F. Portegies Zwart and S. L. W. McMillan. Planets in triple star systems: The case of HD 188753. *ApJL*, 633:L141–L144, November 2005.

D. Queloz, M. Mayor, L. Weber, A. Blécha, M. Burnet, B. Confino, D. Naef, F. Pepe, N. Santos, and S. Udry. The CORALIE survey for southern extra-solar planets. I. A planet orbiting the star Gliese 86. *A&A*, 354:99–102, February 2000.

E. V. Quintana and J. J. Lissauer. Terrestrial planet formation surrounding close binary stars. *Icarus*, 185:1–20, November 2006.

E. V. Quintana, F. C. Adams, J. J. Lissauer, and J. E. Chambers. Terrestrial planet formation around individual stars within binary star systems. *ApJ*, 660:807–822, May 2007.

D. Raghavan, T. J. Henry, B. D. Mason, J. P. Subasavage, W.-C. Jao, T. D. Beaulieu, and N. C. Hambly. Two suns in the sky: Stellar multiplicity in exoplanet systems. *ApJ*, 646:523–542, July 2006.

N. C. Santos, M. Mayor, D. Naef, F. Pepe, D. Queloz, S. Udry, M. Burnet, J. V. Clausen, B. E. Helt, E. H. Olsen, and J. D. Pritchard. The CORALIE survey for southern extra-solar planets. IX. A 1.3-day period brown dwarf disguised as a planet. *A&A*, 392:215–229, September 2002.

M. F. Skrutskie, R. M. Cutri, R. Stiening, M. D. Weinberg, S. Schneider, J. M. Carpenter, C. Beichman, R. Capps, T. Chester, J. Elias, J. Huchra, J. Liebert, C. Lonsdale, D. G. Monet, S. Price, P. Seitzer, T. Jarrett, J. D. Kirkpatrick, J. E. Gizis, E. Howard, T. Evans, J. Fowler, L. Fullmer, R. Hurt, R. Light, E. L. Kopan, K. A. Marsh, H. L. McCallon, R. Tam, S. Van Dyk, and S. Wheelock. The two micron all sky survey (2MASS). *AJ*, 131:1163–1183, February 2006.

S. Söderhjelm. Visual binary orbits and masses post Hipparcos. *A&A*, 341:121–140, January 1999.

G. Takeda and F. A. Rasio. High orbital eccentricities of extrasolar planets induced by the Kozai mechanism. *ApJ*, 627:1001–1010, July 2005.

P. Thébault, F. Marzari, H. Scholl, D. Turrini, and M. Barbieri. Planetary formation in the γ Cephei system. *A&A*, 427:1097–1104, December 2004.

P. Thébault, F. Marzari, and H. Scholl. Relative velocities among accreting planetesimals in binary systems: The circumprimary case. *Icarus*, 183:193–206, July 2006.

S. Udry, M. Mayor, D. Naef, F. Pepe, D. Queloz, N. C. Santos, M. Burnet, B. Confino, and C. Melo. The CORALIE survey for southern extra-solar planets. II. The short-period planetary companions to HD 75289 and HD 130322. *A&A*, 356:590–598, April 2000.

S. Udry, M. Mayor, D. Naef, F. Pepe, D. Queloz, N. C. Santos, and M. Burnet. The CORALIE survey for southern extra-solar planets. VIII. The very low-mass companions of HD 141937, HD 162020, HD 168443 and HD 202206: Brown dwarfs or "superplanets"? *A&A*, 390:267–279, July 2002.

S. Udry, A. Eggenberger, J.-L. Beuzit, A.-M. Lagrange, M. Mayor, and G. Chauvin. The binarity status of stars with and without planets probed with VLT/NACO. In C. Allen and C. Scarfe, editors, *Revista Mexicana de Astronomia y Astrofisica Conference Series*, pages 215–216, August 2004.

S. Udry, D. Fischer, and D. Queloz. A decade of radial-velocity discoveries in the exoplanet domain. In B. Reipurth, D. Jewitt, and K. Keil, editors, *Protostars and Planets V*, pages 685–699, 2007a.

S. Udry and N. C. Santos. Statistical properties of exoplanets. *ARAA*, 45:397–439, September 2007b.

G. A. H. Walker, D. A. Bohlender, A. R. Walker, A. W. Irwin, S. L. S. Yang, and A. Larson. Gamma Cephei - Rotation or planetary companion? *ApJ*, 396:L91–L94, September 1992.

Y. Wu and N. Murray. Planet migration and binary companions: The case of HD 80606b. *ApJ*, 589:605–614, May 2003.

S. Zucker and T. Mazeh. Study of spectroscopic binaries with TODCOR. 1: A new two-dimensional correlation algorithm to derive the radial velocities of the two components. *ApJ*, 420:806–810, January 1994.

S. Zucker and T. Mazeh. On the mass-period correlation of the extrasolar planets. *ApJL*, 568: L113–L116, April 2002.

S. Zucker, T. Mazeh, N. C. Santos, S. Udry, and M. Mayor. Multi-order TODCOR: Application to observations taken with the CORALIE echelle spectrograph. I. The system HD 41004. *A&A*, 404:775–781, June 2003.

S. Zucker, T. Mazeh, N. C. Santos, S. Udry, and M. Mayor. Multi-order TODCOR: Application to observations taken with the CORALIE echelle spectrograph. II. A planet in the system HD 41004. *A&A*, 426:695–698, November 2004.

Chapter 3
The Detection of Extrasolar Planets Using Precise Stellar Radial Velocities

Artie P. Hatzes, William D. Cochran, and Michael Endl

3.1 Introduction

The discovery of extrasolar planets (Latham et al. 1989; Wolszczan et al. 1992; Mayor and Queloz 1995) is arguably one of the most exciting developments in astronomical observations in the past two decades. At the time of this writing over 400 planets were known to orbit around stars other than the Sun. Majority of these planets have been discovered using the so-called "Doppler wobble" method: the measurement of the subtle Doppler shifts of the stellar absorption lines due to the motion of the star about the star-planet barycenter.

The measurement of stellar radial velocities is a technique that dates back over 100 years, starting with the first measurements by Huggins (1868) and Vogel (1872). With such an old technique, why did it take a century to discover extrasolar planets? The answer is simple: precision. Classical techniques could rarely achieve standard errors of better than a few tenths of kilometer per second. By comparison, the best modern stellar radial velocity measurements can achieve a radial velocity (RV) precision of $\approx 1\ \mathrm{m\,s^{-1}}$.

Figure 3.1 shows the velocity amplitude of a solar-mass star about the barycenter as a function of orbital period due to the presence of stellar companions in the mass range 0.05–0.5 $M_\odot$. Also shown is the velocity amplitude as a function of period for substellar companions with masses in the range of 0.35–10 M_{Jupiter} ($0.01 M_\odot$). The horizontal line shows the nominal precision of traditional methods for the measurement of a star's radial velocity (here we take $\sigma \approx 0.5\ \mathrm{km\,s^{-1}}$). The small points show the RV amplitude and periods for known extrasolar planets around solar-type stars. The large symbols show the location of the first two extrasolar planets discovered: HD 114762 b (square) and 51 Peg b (triangle). The large dot marks the location of Jupiter.

A.P. Hatzes (✉)
Thüringer Landessternwarte, D-07778 Tautenburg, Germany
e-mail: artie@tls-tautenburg.de

W.D. Cochran and M. Endl
McDonald Observatory, The University of Texas at Austin, Austin, 78712 TX
e-mail: wdc@astro.as.utexas.edu; mike@astro.as.utexas.edu

N. Haghighipour (ed.), *Planets in Binary Star Systems*, Astrophysics and Space Science Library 366, DOI 10.1007/978-90-481-8687-7_3,

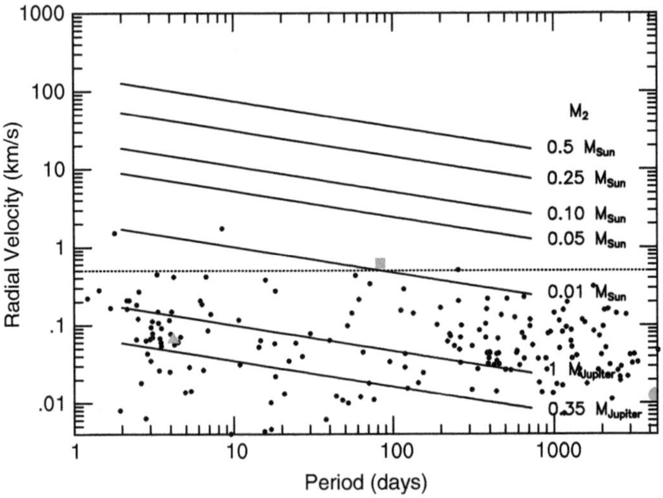

Fig. 3.1 The velocity of a solar-type star about the barycenter due to various mass companions. Companion masses are in solar masses except for the bottom two which are in Jupiter masses. Small points are the known extrasolar planets. Large points are 51 Peg b (*triangle*), HD 114762 b (*square*), and Jupiter (*dot*)

From this figure it is clear that classical techniques could have only discovered those extrasolar planets for which the host star had the highest amplitude reflex motion. HD 114762 was discovered using a more traditional method for measurement of the stellar radial velocity, but only because of the relatively high RV amplitude due to the mass of the planet (11 M_{Jupiter}) and its short period (90 days). The companion to 51 Peg or even a Jupiter analog would have been impossible to discover using traditional RV measurement methods.

This book is devoted to planets in binary stars. Even though most planet searches have focused on single stars, the basic technique for the detection of planets in binary systems is the same – the measurement of precise stellar radial velocities. In this chapter we will review the various methods for achieving a very precise stellar radial velocity measurement required for the detection of extrasolar planets. We end this contribution by presenting a few examples of planets in binary stars.

3.2 Traditional Methods of Stellar RV Measurements

Most astronomical detectors such as charge-coupled devices (CCDs) do not record any wavelength information; they merely record the intensity of light as a function of location on the detector. This spectral intensity, as a function of detector position (pixel) must be converted into intensity versus wavelength by observing a calibration source, typically a Thorium-Argon hollow cathode lamp. The problem for precise stellar RV measurements is that the wavelength comparison source is taken at a different time (either before or after the stellar spectrum), and usually the light from the star traverses a different optical path through the telescope+spectrograph. Most modern spectrographs do not have the mechanical and thermal stability to achieve an intrinsic velocity precision at the level of $10\,\mathrm{m\,s^{-1}}$. These uncontrolled or unmeasured instrumental shifts ultimately limit one's ability to detect extrasolar planets with the Doppler method.

Figure 3.2 shows the instrumental shifts in meters per second from two spectrographs, the CES spectrograph of the European Southern Observatory at La Silla, and the coude spectrograph of the 2.7 m telescope at McDonald Observatory. These shifts were calculated from a long time series of observations of an iodine absorption cell (see below). Both show long-term trends, where the spectrograph shifts by at least $100\,\mathrm{m\,s^{-1}}$ over 1–2 h. There are also short-term shifts up to $100\,\mathrm{m\,s^{-1}}$ on

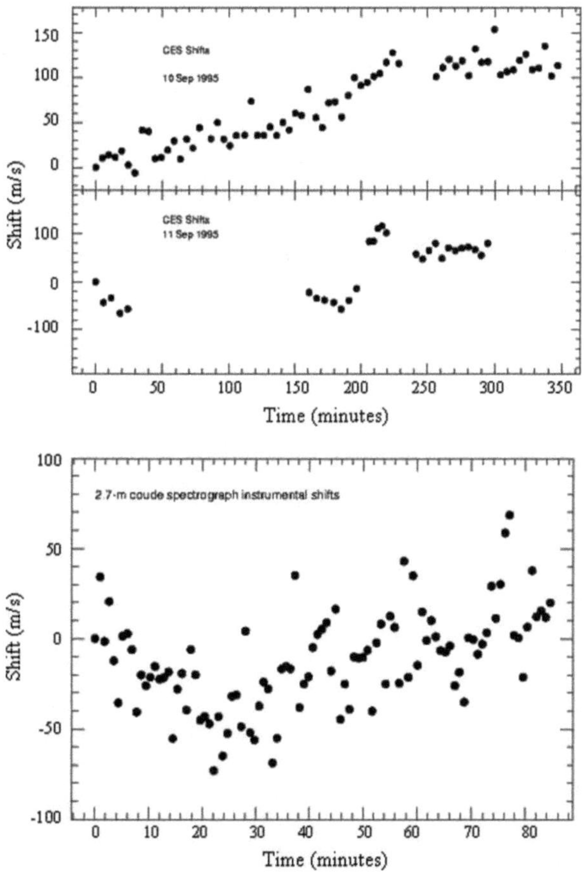

Fig. 3.2 Instrumental shifts for two spectrographs. *Top*: The CES Spectrograph of the European Southern Observatory. *Bottom*: The coude spectrograph of the 2.7 m telescope of McDonald Observatory

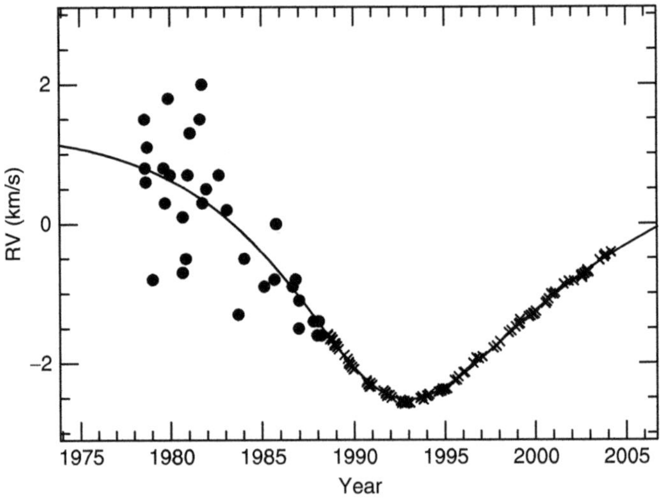

Fig. 3.3 RV measurements for the binary star γ Cep. *Dots* represent measurements made using traditional RV measurements. *Crosses* are those taken with precise RV measurements, in this case with an iodine absorption cell. The *solid line* represents the binary orbital solution

time scales of several minutes. Clearly, if one had to depend on the intrinsic stability of these spectrographs, it would be impossible to detect extrasolar planets around stars with an expected reflex motion around $10\,\mathrm{m\,s^{-1}}$.

Figure 3.3 shows RV measurements of γ Cep, a binary star whose primary star hosts a giant planet (Hatzes et al. 2003). The dots represent measurements made with more traditional techniques by Griffin et al. (2002). The crosses are modern precise RV measurements taken at McDonald Observatory with an iodine absorption cell. The improvement is dramatic. One can see slight variations on top of the binary orbit (solid line) with the precise RVs that are due to the planetary companion. These are completely hidden in the older measurements due to the much poorer RV precision. These can only detect the high amplitude orbital motion due to the stellar companion.

In the following sections we describe in detail how astronomers were able to achieve a factor of 100–1,000 improvement in the RV precision in the last two decades. These improvements allowed for the phenomenal progress in the field of extrasolar planets.

Modern precise stellar radial velocity measurements correct for the instrumental shifts like the ones shown in Fig. 3.2 by recording the wavelength reference simultaneously to that of the stellar spectrum. We will discuss three approaches for doing this: (1) using telluric features produced by the Earth's atmosphere, (2) gas absorption cells, and (3) simultaneous Thorium-Argon calibration.

3.3 The Telluric Technique

Griffin and Griffin (1973) were the first to point out that one factor limiting the precision of stellar radial velocity measurements is that the light from the wavelength comparison source is taken at a different time and usually traverses a different optical path. They proposed using telluric O_2 lines at 6,300 Å as a wavelength reference that is recorded simultaneously with the stellar spectrum. As starlight passes through the Earth's atmosphere, molecular oxygen produces sharp absorption features. If one measures radial velocity shifts with respect to these telluric lines, instrumental shifts are minimized because both the reference and stellar light illuminate the spectrograph in the same manner, and are recorded at the same time.

Figure 3.4 shows the spectral region covered by the telluric O_2 features. The top panel is a spectrum of Vega, an A0-type star that has no spectral features in this wavelength region. The narrow lines appearing as doublets are caused by atmospheric O_2. The bottom panel shows a spectrum of a solar-type star, μ Her where the O_2 features are now superimposed on the stellar spectrum. Doppler shifts of the star are computed with respect to the telluric O_2 lines.

Griffin and Griffin (1973) suggested that a velocity precision of 15–20 m s^{-1} was possible using the telluric method. During the initial stages of the McDonald Observatory Planet Search Program, we employed the telluric O_2 technique to search for extrasolar planets and our experience indeed confirms that a velocity precision of $\approx$20 m s^{-1} is indeed possible. Figure 3.5 shows radial velocity measurements for HD 114762 using the telluric method (Cochran et al. 1991). These are shown by

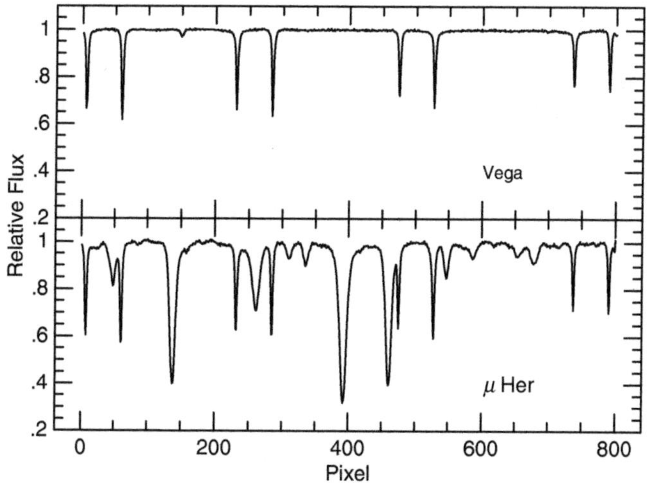

Fig. 3.4 The telluric method for RV measurements. (*Top*) A spectrum of Vega. The absorption features shown are due only to telluric O_2. (*Bottom*) A spectrum of the solar-type star μ Her. Spectral features are from both the star and telluric O_2

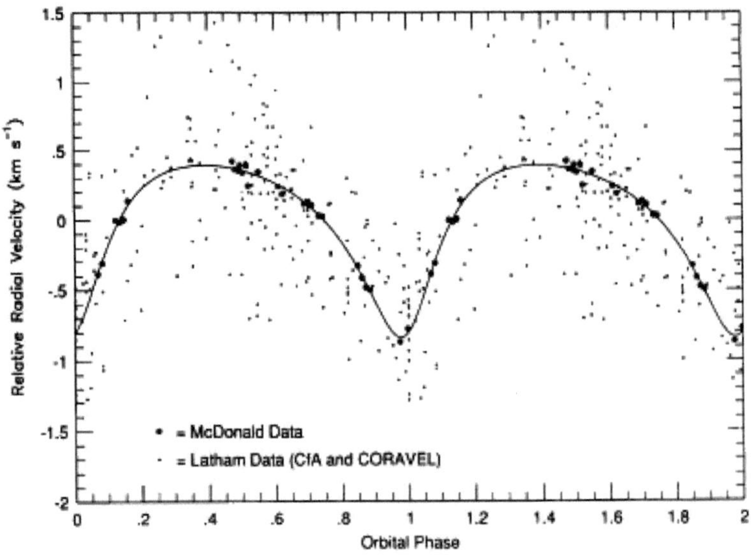

Fig. 3.5 Radial velocity measurements of HD 114762 taken with the telluric method (*large dots*) and with traditional techniques (*small dots*). The *solid line* represents the orbital solution. Figure from Cochran et al. (1991)

the large dots. Also shown as small dots are the measurements from the discovery paper of Latham et al. (1989). These were taken with a more traditional technique for the RV measurement with a precision of $\approx$500 m s^{-1}. The improvement is substantial. The traditional method was also able to detect this sub-stellar companion, but it required far more measurements than the telluric method.

The advantages of the telluric method is that it is simple, inexpensive, and easy to use. Almost any high resolution spectrograph can be used to achieve an RV precision of a few tens of meters per second. This method has also been extended to the near infrared using the telluric A and B bands between 6,860 and 6,930 Å, and 7,600 and 7,700 Å, respectively (Guenther and Wuchterl 2003). There are, however, significant disadvantages. Suitable O_2 lines are found only over a very narrow wavelength range and this translates into a poorer RV precision. The biggest disadvantage is that one is unable to control the Earth's atmosphere. Pressure and temperature changes, as well as winds in the earth's atmosphere are ultimately what limit the RV precision.

3.4 Absorption Cells

The telluric method could be improved if we could somehow control the absorbing gas that creates the wavelength reference. This is the principle behind the use of a gas absorption cell. One chooses a gas that creates a characteristic absorption spectrum (preferably with lines not found in the stellar or Earth's atmospheres) and inserts it into a sealed glass cell that is then placed in the optical path of the telescope. This cell is temperature stabilized so that the absorption spectrum of the gas is constant. Like the telluric method, the wavelength reference will be superimposed on the stellar spectrum, only in this case there are no shifts of the wavelength reference due to temperature or pressure changes, or wind! The use of a gas absorption cell as a wavelength metric is not a new idea and has been employed by solar astronomers in the past (Beckers 1976; Koch and Wöhl 1984). Marcy and Butler (1992) were the first to apply this technique to stellar radial velocities.

3.4.1 The Hydrogen-Fluoride Absorption Cell

The first use of a gas cell for planet detection via the RV method was the pioneering work of Campbell and Walker (1979). This program used the 3–0 band R-branch absorption lines of Hydrogen-Fluoride (HF) at 8,670–8,770 Å as the velocity metric. Figure 3.6 shows the spectral region of the HF at 8,700 Å along with spectrum of a star taken with and without the HF cell.

Campbell and Walker (1979) were able to demonstrate that the HF cell could achieve an RV precision of $13\,\mathrm{m\,s^{-1}}$. From Fig. 3.1 one can see that these authors had sufficient RV precision to have detected most of the currently known extrasolar planets. Indeed, their measurements were the first to detect the possible perturbations of the binary orbit of γ Cep, and they hypothesized that these could result from a planetary companion (Campbell et al. 1988).

Later in the chapter we will show that their RV measurements did show evidence for an extrasolar planet. Why did Campbell et al. (1988) not discover more extrasolar planets? Simply because the sample size was too small. The CHFT sample consisted only of 26 solar-type stars. We currently believe that approximately 5% of solar-type stars have giant planets. With this frequency, Campbell et al. (1988) should have found at most one extrasolar planet. Indeed, they did detect the planetary companion to γ Cep; however, Walker et al. (1992) later cast doubt on this by suggesting that this additional RV variations might be due to stellar rotation. If they had a significantly larger sample size, or a bit of luck, the field of extrasolar planets would be entering its third decade.

Although the HF cell is capable of achieving good RV precision, there were several disadvantages to using HF as the absorbing medium:

- It provides absorption features over only a limited wavelength range in the optical (about 100 Å).
- HF is sensitive to pressure shifts.
- HF requires a rather large path length (about 1 m) to produce suitable absorption lines. This could be a problem if space was limited in your spectrograph.
- HF is a highly corrosive and dangerous gas. The HF cell had to be filled for each observing run. So safety concerns are a real issue.

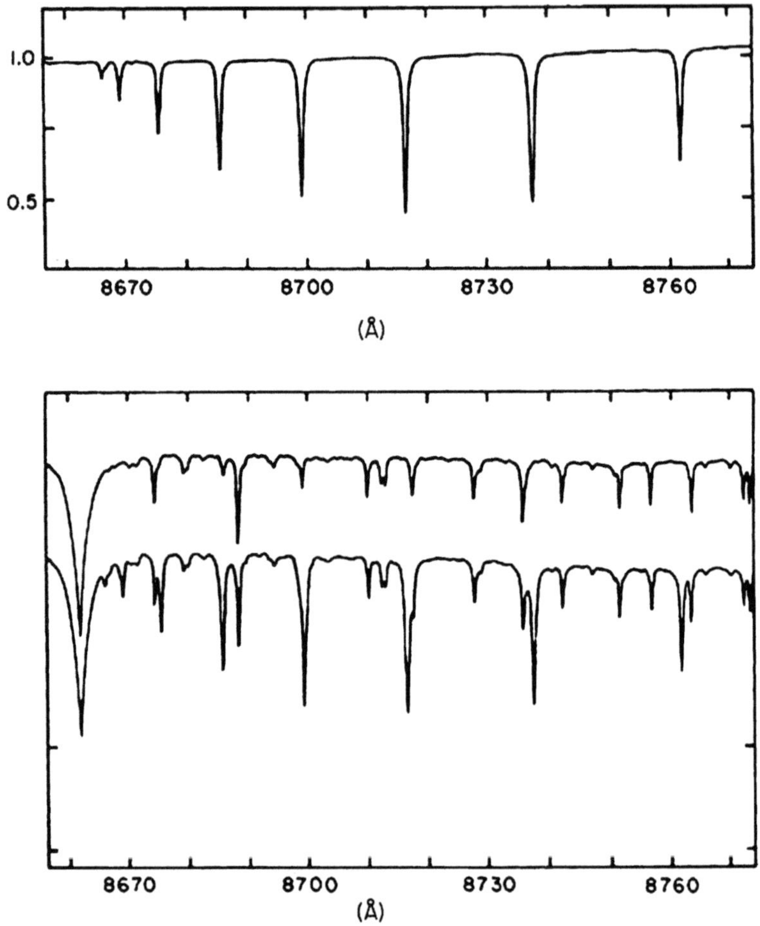

Fig. 3.6 *Top*: Spectrum of the HF cell at 8,720 Å. *Bottom*: Spectrum of a star taken without top spectrum and with HF. Figure from Campbell and Walker (1979)

3.4.2 Iodine Absorption Cells

An alternative to the HF cell is the use of iodine absorption cell. Molecular iodine has several advantages over HF for absorption cells: (1) It is a relatively benign gas that can be permanently sealed in a glass cell. The amount of iodine that is used is so small that there is no health hazard should the cell break. (2) I_2 has useful absorption lines over the interval 5,000–6,000 Å. This increases the number of spectral lines used for the Doppler measure by about a factor of 10 over the telluric and HF techniques. (3) A typical pathlength for an I_2 cell is about 10 cm, which can easily fit in front of the entrance slit of most spectrographs. (4) The I_2 cell can be stabilized at relatively modest temperatures (50–70°C). (5) Molecular iodine is less sensitive to pressure shifts than HF. (6) The rich density of narrow I_2 absorption lines enables one to model the instrumental profile of the spectrograph (more on this later).

Figure 3.7 shows a typical iodine gas absorption cell. Cells are typically about 10–15 cm in length and a have a width of a few cm. The iodine is permanently sealed in a glass that is surrounded by a heater foil and a temperature sensor that ensures that the cell is thermally stabilized during its use. Figure 3.8 shows the spectrum of a white light source taken through an iodine cell with an echelle spectrograph. Note the rich forest of spectral lines.

Iodine cells have become a popular method for the measurement of precise stellar radial velocities. Such cells are currently in use at the McDonald Observatory 2.7 m (Wittenmyer et al. 2006) and 2.1 m telescopes (Cochran et al. 2004), The Hobby-Eberly Telescope (Cochran et al. 2004), Lick Observatory (Butler et al. 1996), ESO's VLT+UVES (Kürster et al. 2000), Keck Observatory (Vogt et al. 2000), Tautenburg Observatory (Hatzes et al. 2005), the Anglo Australian Telescope

Fig. 3.7 A typical iodine gas absorption cell. The cell is about 15 cm long and 5 cm in diameter. When in use, the cell is surrounded by a heater, temperature sensor, and insulation

Fig. 3.8 A spectrum of molecular iodine taken with an echelle spectrograph. Each band represents a different spectral order and redder wavelengths are to the right and bottom. The iodine spectral lines begin to become visible at about 5,000 Å

(Tinney et al. 2001), Bohyunsan Optical Astronomical Observatory (Kim et al. 2006), to name a few. The majority of known extrasolar planets discovered by the Doppler method were made using iodine absorption cells.

3.4.2.1 Details on the Use of the Iodine Absorption Cell

After constructing the cell, a very high resolution ($R = \lambda/\delta\lambda = 500{,}000$), high signal-to-noise spectrum of the cell is made with a Fourier Transform Spectrometer (FTS). This spectrum must be taken at the same temperature that the cell will be operated at the telescope. This high resolution spectrum is required not only to provide the wavelength reference to for the RV measurement, but also to model the instrumental profile (IP) of the spectrograph, or equivalently the δ-function response of the spectrograph primarily caused by the optics. This is important because changes in the IP, particularly asymmetric ones, can introduce shifts in the spectral lines of order 1–10 m s^{-1}. Since the FTS spectrum is taken at much higher resolution than the spectrograph used for the RV measurements (these typically have R = 60,000 – 100,000), when this is re-binned to the observed I_2 spectrum used for the stellar observations, it is a suitable approximation of the δ-function response of the spectrograph. Comparing this to the observed I_2 taken with the spectrograph optics enables one to reconstruct the IP (Valenti et al. 1995).

The observational procedure consists of taking a "template" spectrum, I_S (known as the intrinsic stellar spectrum), of each target star made without the I_2 cell. This requires a high signal-to-noise spectrum. For the best precision this template should be deconvolved by the instrumental profile as determined using the iodine spectral lines. To measure the IP, a spectrum of the I_2 cell must be made with the spectrograph optics illuminated in the same manner as the stellar observations. This is commonly done by taking a spectrum of a hot, rapidly rotating early-type star through the cell. The velocity computation consists of computing the model spectrum, I_m which best matches the observed spectrum:

$$I_m = k[T_{I_2}(\lambda) I_S(\lambda + \delta\lambda)] * IP \qquad (3.1)$$

where T_{I_2} is the transmission function of the I_2 cell, k is a normalization factor, $\delta\lambda$ is the wavelength (Doppler) shift, IP is the spectrograph instrumental profile, and $*$ represents the convolution. Use of the IP modeling in the calculation of the Doppler shift was pioneered by Butler et al. (1996), and with modifications by Endl et al. (2000).

The results of the modeling process is shown in Fig. 3.9. The observed spectrum (solid line) of a star taken through the I_2 cell has absorption lines from both the star and the gas cell. The "pure" stellar template spectrum (without I_2) is shown as the dashed line, and the high resolution FTS I_2 spectrum re-binned to the dispersion of the stellar observation is shown as the dotted line. The model spectrum computed according to Eq. (3.1) is shown as the dot-dashed line, which is coincident with the solid observed star spectrum. The difference of the observed and model spectrum is shown as the long-dashed line at the top.

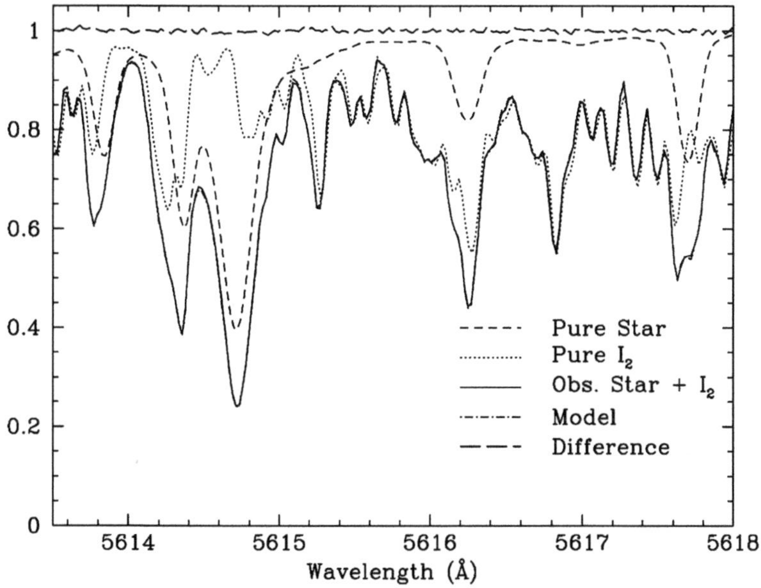

Fig. 3.9 Modeling the observed spectrum taken with an iodine cell. The pure stellar spectrum is Doppler shifted and multiplied by the pure iodine spectrum. This product is then convolved with the instrumental profile to give the model spectrum. The difference between the observed and model spectra is shown at the top

The key to obtaining the best RV precision is a proper modeling of the IP. Valenti et al. (1995) proposed parameterizing the IP as a sum of several Gaussian components which is the most common practice, but other functions may be used as well, including a slit function. Figure 3.10 illustrates the IP parameterization process for one echelle spectral order of the cs23 spectrograph on the 2.7 m telescope at McDonald Observatory (Tull et al. 1995). The spectrum has been divided into 20 different wavelength chunks of width typically a few Angstronms (only 16 are shown). The IP in each chunk has been parameterized by a superposition of five Gaussian functions and the sum convolved with a top-hat function giving the resulting curve.

Figure 3.11 demonstrates the value of the IP modeling procedure. It shows the RV measurements for a constant star, τ Ceti, taken with the CES spectrograph of the European Southern Observatory's 3.6 m telescope at La Silla (Endl et al. 2000). Without the treatment of the IP asymmetries the scatter of the RV measurements is about 36 m s^{-1}. By treating the IP asymmetries, the scatter of the measurements decreases by more than a factor of three to $\sigma = 11.6$ m s^{-1}. The improvement in the RV precision by modeling the IP depends, of course, on the stability of the spectrograph. If the instrument is intrinsically stable, then one may only achieve small improvements in the RV precision. If the spectrograph has large thermal and mechanical variations, then the IP modeling should improve the RV precision. However, if these

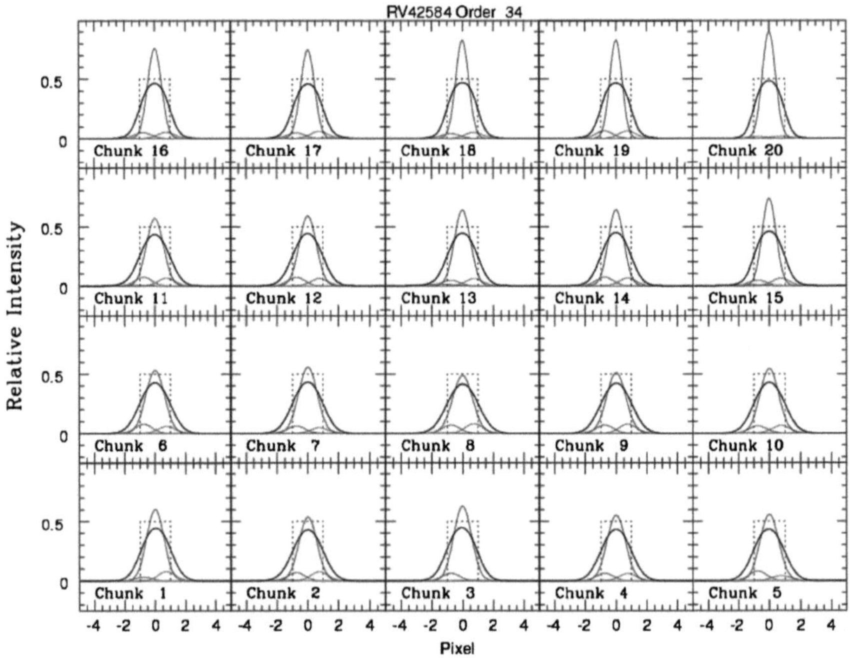

Fig. 3.10 An example of the instrumental profile determination for one echelle order of a spectrum taken with the McDonald Observatory 2.7 m cs23 spectrograph. The *grey curves* are the individual Gaussian functions and the *dotted line* is the slit function. The *heavy line* shows the final IP

variations are too large, it may be difficult to achieve a proper parameterization of the IP, making it difficult to achieve the best quoted precisions of 1–3 $\mathrm{m\,s^{-1}}$.

3.5 Simultaneous Thorium-Argon Calibration

Use of an absorption cell is not the only technique capable of achieving a precise measurement of the stellar radial velocity. An alternative method is to use a traditional hollow-cathode lamp such as Thorium-Argon, but to record this simultaneously with the stellar spectrum. This is done using one fiber optic to feed the light from the star into the spectrograph, and a second to feed the light from the Th-Ar lamp during the exposure. The Th-Ar spectrum is recorded on the CCD detector adjacent to the stellar spectrum. Since the wavelength calibration spectrum is recorded at the same time as the stellar observation, instrumental shifts like those shown in Fig. 3.2 are minimized. This technique was pioneered with the ELODIE spectrograph (Baranne et al. 1996). Subsequent improvements to the techniques were incorporated into CORALIE (Queloz et al. 2000) and finally HARPS (Pepe et al. 2000).

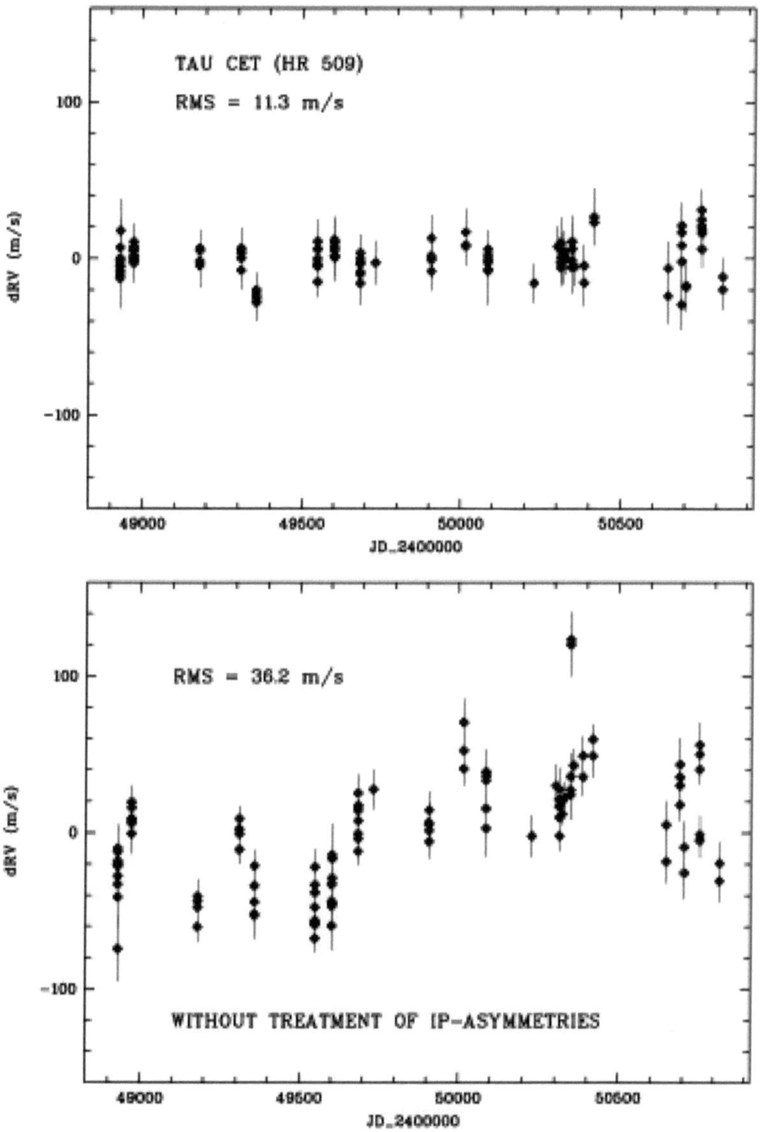

Fig. 3.11 RV measurements of the constant star τ Ceti taken with the CES spectrograph of the ESO 3.6 m telescope at La Silla. (*Top*) With modeling of the IP asymmetries. (*Bottom*) Without modeling of the IP asymmetries. Figure from Endl et al. (2000)

Figure 3.12 shows a stellar spectrum recorded using the HARPS spectrograph of ESO's 3.6 m telescope at La Silla using the simultaneous Th-Ar technique. The continuous bands represent the stellar spectrum. In between these, one can see the emission spectrum from the Th-Ar fiber. This method has also been quite successful at discovering extrasolar planets such as 51 Peg b using the CORALIE spectrograph (Mayor and Queloz 1995).

Fig. 3.12 A spectrum recorded with the HARPS spectrograph. The solid bands are from the star fiber. The emission line spectrum of Th-Ar above the stellar one comes from the calibration fiber

Although simultaneous Th-Ar technique can eliminate overall instrumental shifts, it is unable to monitor any changes in the IP. The reason is that the Th-Ar emission lines have relatively large intrinsic widths compared to iodine lines. Minute changes in the voltage applied to the hollow cathode lamp can cause variations in the intrinsic width of the Th-Ar emission lines not associated with changes in the IP. For these reasons, the HARPS spectrograph was designed with thermal and mechanical stability in mind. HARPS is a state-of-the-art spectrograph specifically designed to achieve very high precision. It housed in a vacuum chamber where the temperature is held constant to within several mK. Care was also taken to minimize mechanical vibrations. The thermal and mechanical stability that the IP remains constant. A key improvement to the HARPS spectrograph is the use of two sequential fiber optics in a double scrambler mode to ensure a stable illumination of the spectrograph that is insensitive to variations due to seeing and guiding errors. With such stability, HARPS has been able to achieve a short-term precision better than 1 $\mathrm{m\,s^{-1}}$.

The key to obtaining the very highest RV precision lies in minimizing the instrumental and IP shifts of the spectrograph, and there are two philosophical approaches to this. The HARPS approach is to stabilize the spectrograph as best as possible, thermally, mechanically, and optically. These ensure that the IP does not change over time. Furthermore, to guarantee that a stable optical image is provided to the spectrograph, double scrambling using two optical fibers is employed. However, since Th-Ar emission lines are inadequate for monitoring changes in the IP, one has to have faith that if one maintains the spectrograph at the same temperature and pressure everything in the spectrograph, including the optics, do not change with time. The iodine cell approach is to use a permanently sealed cell of molecular iodine that is temperature regulated to produce a constant reference spectrum that can be used to model any temporal changes to the IP of the spectrograph.

3.5.1 Iodine Cells Versus Simultaneous Th-Ar

As mentioned earlier, there are currently two techniques for achieving very high RV precision: iodine cells and simultaneous Th-Ar calibration. The practitioners in each camp can be rather passionate about their respective technique; each thinks their technique is superior. So which one is, indeed?

Both methods have had tremendous success at discovering extrasolar planets. Both have also demonstrated velocity precision of 1–3 $\mathrm{m\,s^{-1}}$. With all these, one could argue that the question of superiority may be moot. However, we do not think so. We have employed the iodine cell technique for over 15 years and yes, we are passionate about the technique. We are possibly not in the best position to make unbiased judgments on the methods, but here we will try our best to give a dispassionate assessment of the advantages and disadvantages of the Th-Ar method compared to gas absorption cells.

3.5.1.1 Advantages of Th-Ar

- Computational Simplicity
  Computing RVs using the Th-Ar method is relatively simple and fast. This usually entails computing a cross-correlation function after putting all the spectra on the same wavelength scale (done using the simultaneous Th-Ar). The iodine cell method is rather computationally intensive. If one wants to model adequately the IP in order to eke out the highest precision, then this requires dividing the spectrum into several hundred wavelength chunks and calculating the IP in each chunk. The IP function requires five or more Gaussian functions each with their own position, amplitude, and width. The computational time required by the iodine cell method can be more than a factor of 10 greater than the simple cross-correlation required by the Th-Ar method.
- Higher Efficiency
  In principle the Th-Ar method has a higher efficiency than the iodine cell method. Losses due the reflections off the glass cell as well as absorption by the iodine gas can be as high as 50%. Furthermore, the Th-Ar method is not restricted to just the 5,000–6,000 Å covered by iodine absorption lines. In principle, one can use the full spectral range offered by modern echelle spectrographs, typically 3,000–9,000 Å. Because more spectral lines can be used for the velocity measurement,

a higher RV precision can be obtained. However, a large part of this increased efficiency is offset by the need to use optical fibers, often two in a double scramble mode. Iodine cells can achieve very high precision using slits without the need to feed the spectrograph with a fiber. Furthermore, telluric atmospheric lines beyond about 6,500 Å largely make the red regions of the stellar spectrum less ideal for stellar RV measurements. In reality the efficiency of both methods may be comparable.

- Uncontaminated Spectral Features
  The Th-Ar technique does not contaminate the stellar spectrum (not strictly true, see below) as does the absorbing gas of a cell. This means that each stellar observation can be used for measurements of the spectral line shapes or other kinds of spectral analyses (abundance, temperature, etc.). These are important if they are required to provide confirmation of planet detections with the RV method. (RV variations due to a planet are not accompanied by other forms of spectral variability.) Since the iodine spectrum contaminates a broad wavelength range of spectral data, one either needs to remove the iodine spectrum from the stellar observation, which may introduce significant errors in the analysis, or one needs to take an additional spectrum of the star without the cell, which requires increased observing overhead.

3.5.1.2 Disadvantages of Th-Ar

In spite of these advantages there are several significant disadvantages of the Th-Ar method that may make it inappropriate for some applications.

- Finite Lifetime of Th-Ar Lamps
  If one is interested in obtaining precise stellar radial velocities over a 10–20 year time span or longer, then Th-Ar may not be the appropriate method to use. An individual iodine cell is sealed permanently for the entire life of the cell. The absorption cell is a passive device and the molecular iodine in the cell does not change with time. The wavelength reference and the radial velocity stability is provided by molecular physics in a device that is easy to stabilize. The authors have used several iodine cells for over 15 years and the resulting iodine absorption spectrum has not changed in the interim. Unless an accident destroys a cell, it never has to be replaced. By comparison Th-Ar cathode lamps are active devices where one has to apply high voltages. These lamps are known to change with time and ultimately will fail through regular use. Figure 3.13 shows a spectrum of a Th-Ar lamp taken 7 years apart. The intensity ratio of many of the lines change. Some weak emission features disappear and others reappear. It almost seems as though one has used a different lamp. Changes in the wavelength reference will almost certainly translate into RV instrumental variations. (The lamp used for the data of Fig. 3.13 failed in 2005 and had to be replaced.)
- Inability to Monitor Changes in the Instrumental Profile
  Although simultaneous Th-Ar can monitor mechanical instrumental shifts and changes in the wavelength scale, it is unable to monitor changes in the asymmetry

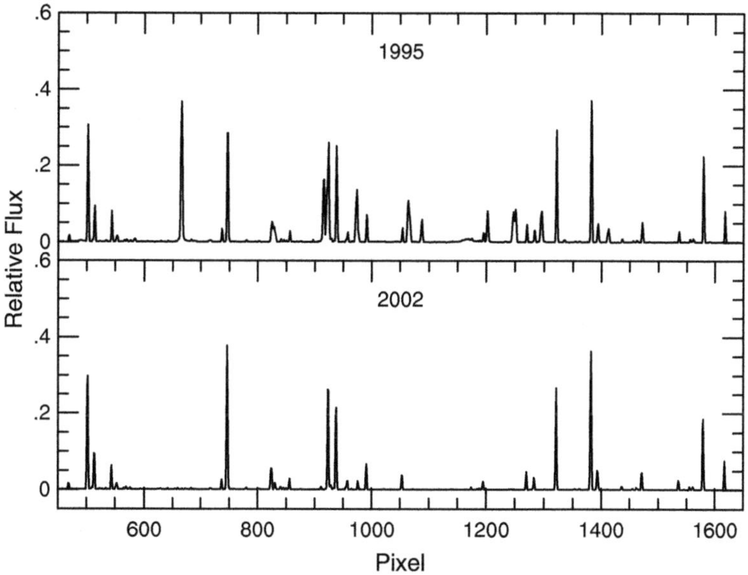

Fig. 3.13 Changes in Th-Ar spectrum due to aging of the hollow cathode lamp. *Top*: Spectrum from 1995. *Bottom*: Spectrum with same lamp from 2002

in the IP of the spectrograph. The Th-Ar emission lines are inadequate for this because they almost always have an intrinsic width comparable to the resolution of the spectrograph. Plus, changes in their shape may be intrinsic to the lamp and not the spectrograph. HARPS minimizes changes in the IP by stabilizing the spectrograph thermally and mechanically. There is no guarantee, however, that the IP remains constant over many years and it is difficult, if not impossible, to monitor these changes with Th-Ar emission lines alone. Furthermore, even if one could monitor changes in the IP of a HARPS-like instrument, it would be difficult to incorporate these into the modeling process. On the other hand, the iodine cell provides a natural way to monitor changes in the IP *in situ* and these changes are included as part of the RV calculation.

- Contamination of the Stellar Spectrum by Th-Ar
  There can be "cross talk" between the stellar and Th-Ar fibers, and for strong emission lines, light from the Th-Ar channel can spill over into the stellar spectrum. This can be seen in Fig. 3.12 where light from the strong Th-Ar lines cross into the stellar spectrum. This contamination is difficult to treat in the analysis. For long exposures, it may be difficult to adjust the intensity level of the Th-Ar lamps to minimize such contamination. The contamination of the stellar spectrum by iodine lines by comparison is treated in the modeling process that is used to compute the RV. In fact for HARPS, a common practice is to *not* use the simultaneous Th-Ar calibration, only the non-simultaneous calibration taken at the beginning or end of the night (e.g., Galland et al. 2006).

- Costs
  The simultaneous Th-Ar method is considerably more expensive than an iodine cell. It requires a fiber optic feed, often in a double scramble mode (two separate fiber optics for each channel). To stabilize the IP, which cannot be measured in-situ, requires that the spectrograph to be sealed in a vacuum chamber that is temperature stabilized. The resulting increase in the cost of the design of the spectrograph then becomes substantial. By comparison, an iodine cell, including the heater foil, insulation, and temperature controller can be constructed for a few hundred U.S. Dollars. This inexpensive device can be used to convert any existing high resolution spectrograph into an RV machine.

In summary, both methods produce comparable results, but if costs, and long-term stability (>10 years) is a prime consideration, then an iodine absorption cell is the only viable option.

3.5.2 Spectrograph Requirements for Precise RV Measurements

Suppose you have a very stable spectrograph, or at least a measurement technique that has eliminated all sources of instrumental errors. How does the RV precision depend on such parameters as signal-to-noise ratio, wavelength coverage, and spectral resolution?

- Signal-to-Noise Ratio
  The RV precision should be inversely proportional to the amount of noise in the data. The more noise, the less precise the measurement.
- Wavelength Coverage
  Each spectral line represents an individual measurement of the Doppler shift of the star. If a total of N lines are used for the Doppler measurement, then the error will be decreased by a factor of $\sqrt{N}$ over a single line measurement.
- Spectral Resolution
  An RV measurement can be thought of as basically measuring the centroid of the line. If there are more points sampling the spectral line profile, one can determine this centroid more accurately. If we can measure the position of a spectral line to some fraction of pixel (say 0.001), then higher velocity resolution per pixel will result in more accurate determination of RV. In other words, higher the spectral resolution will produce more accurate RV measurement. It is, however, important to note that this goes counter to the first two points. First, higher resolution spectrographs usually have more limited wavelength coverage than lower resolution ones, which decreases the number of spectral lines that can be used for the RV measurement. Second, high resolution spectrographs spread the light over more detector pixels thus decreasing the signal-to-noise ratio per pixel for a fixed exposure time.

The RV precision also depends on the type of star. For a precise RV measurement, a star with a plethora of stellar absorption lines that are narrow and not broadened by stellar rotation are more desired. This implies late-type stars with spectral type later than about F6. Early-type stars are hot and thus have few spectral lines for good RV measurements. They also rotate very rapidly which results in spectral lines that are very shallow and broad – difficult for determining a line position that is needed to measure the Doppler shift. Hatzes and Cochran (1992) performed numerical simulations to determine how the RV precision depended on these various parameters. An updated version of their expression is

$$\sigma(m/s) = C(S/N)^{-1}R^{-3/2}B^{-1/2}[f(Sp.T)]^{-1/2}(v\sin i/2)^{-1} \tag{3.2}$$

where (S/N) is the signal-to-noise ratio of the data, R is the resolving power $(=\lambda/\delta\lambda)$ of the spectrograph, B is the wavelength coverage in Å of the stellar spectrum used for the RV measurement, and C is a constant of proportionality.

The quantity $v \sin i$ in Eq. (3.2) is the projected rotational velocity of the star in kilometers per second. The RV precision is roughly proportional to the inverse of the rotational velocity of the star scaled to a nominal slowly rotating star that has a equatorial velocity of 2 km s^{-1}. (For stars rotating less than this nominal value the $v \sin i$ term should just be discarded.) The function $f(Sp.T)$ represents the relative line density for the star as a function of stellar type. If we take $f = 1$ for a G-type star, then $f \approx 0.1$ for an A-type star, and $f \approx 10$ for an M-type star.

The value of the constant C can be estimated using the performance of the HARPS spectrograph. For $S/N \approx 150$, $R = 110{,}000$, $B = 2000$, Å, the quantity σ will be equal to 1 m s^{-1} for a G-type star. This results in $C \approx 2.4 \times 10^{11}$. With this expression one should be able to estimate the expected RV precision of a spectrograph to within a factor of a few. Note that for a fixed detector size, σ should be proportional to R^{-1}, as noted by Hatzes and Cochran (1992).

3.6 Extrasolar Planets in Binary Systems

As of this writing there are over 400 known extrasolar planets that have been detected via the RV method. Most of these planets orbit stars that are single. This is certainly a bias effect as early RV surveys chose targets that were believed to be single stars. Several factors influence one's decision to search for planets in binary systems. If the binary is a wide pair in the sky and thus in a long period orbit, there would be no problems. The orbital period will be long so that RV binary motion will be small. If the binary pair has a small separation of say less than about 2 arcsec, then there is a danger that light from both components would enter the spectrograph slit. This will seriously affect the RV measurement. Such close "visual" pairs are avoided by most, if not all surveys. Short period ($P < 10$ years) single-line spectroscopic binary stars have RV variations that show complex variations resulted from the combination of the large binary motion and much smaller motion of a planetary companion, if present. It is still possible to detect planetary companions in such systems, but one needs enough RV measurements to completely parameterize the binary orbit. Improper removal of the binary motion could result in a spurious detection. For relative long binary orbits (∼years) this would, of course, delay the detection of planetary companions. Sometimes when stars in such planet surveys are found to be new binary systems, the targets are often removed from the program (e.g., Vogt et al. 2002). Finally, if the binary system is a double-lined spectroscopic binary (mass-ratios near unity), the RV analysis would be greatly complicated as one has to measure accurately the RV variations of two spectral components, plus any additional variation due to planets around either or both (i.e., circumbinary) stars. To-date there has been no RV survey that has included a large sample of double-lined spectroscopic binary stars.

In spite of these challenges, there have been extrasolar planets discovered in binary systems, and these can give us important clues as to the process of planet formation. Here we give a few representative examples of extrasolar planets in binary stars, and highlight their importance for planet formation theory.

3.6.1 16 Cyg B

The 16 Cyg A and B system consists of two main sequence stars with spectral types of G1.5 and G2.5. Friel et al. (1993) derived the stellar parameters for this system. To within the measurement errors, these stars have similar effective temperatures [$T_{eff}(A) = 5785 \pm 25$ K, $T_{eff}(B) = 5760 \pm 20$ K], surface gravities [Log $g(A) = 4.28 \pm 0.07$, Log $g(B) = 4.35 \pm 0.07$], and abundances ([Fe/H]$_A = +0.06 \pm 0.04$, [Fe/H]$_B = +0.02 \pm 0.04$). Essentially these stars are "identical twins". The only apparent difference is that 16 Cyg B shows a slight depletion in Li abundance [Log $N(Li)_A = 1.27 \pm 0.04$, Log $N(Li)_B = 0.48$ (King et al. 1997). The two stars have an apparent separation of 39 arcsec which corresponds to a separation of about 835 AU. The orbit for this system is very poorly known.

The planetary companion to 16 Cyg B was discovered by Cochran et al. (1997) and was one of the earliest discovered extrasolar planets. Figure 3.14 shows more recent RV measurements for both 16 Cyg A and B taken at McDonald Observatory. The top panel shows the RV variations for 16 Cyg B. At the time of its discovery, this planet was the only extrasolar planet with the highest orbital eccentricity ($e = 0.63$) and gave the first indication that extrasolar planets can have very eccentric orbits. The highly eccentric orbit is readily apparent as displayed by the sawtooth pattern of the RV variations. The lower panel of Fig. 3.14 shows the RV measurements for 16 Cyg A. These are constant to a level of $\sigma = 5.34\,\mathrm{m\,s^{-1}}$. Note that there is clear downward trend in the 16 Cyg A velocities which may be due to the binary motion. It has been proposed that the high eccentricity of this object might be caused by the presence of the binary companion (Holman et al. 1997). Table 3.1 lists the orbital properties of this extrasolar planet.

The planetary system of 16 Cyg is a remarkable system. Here are two nearly identical stars in a binary system with the same effective temperature, mass, age, and chemical composition. Presumably they were born in essentially the same

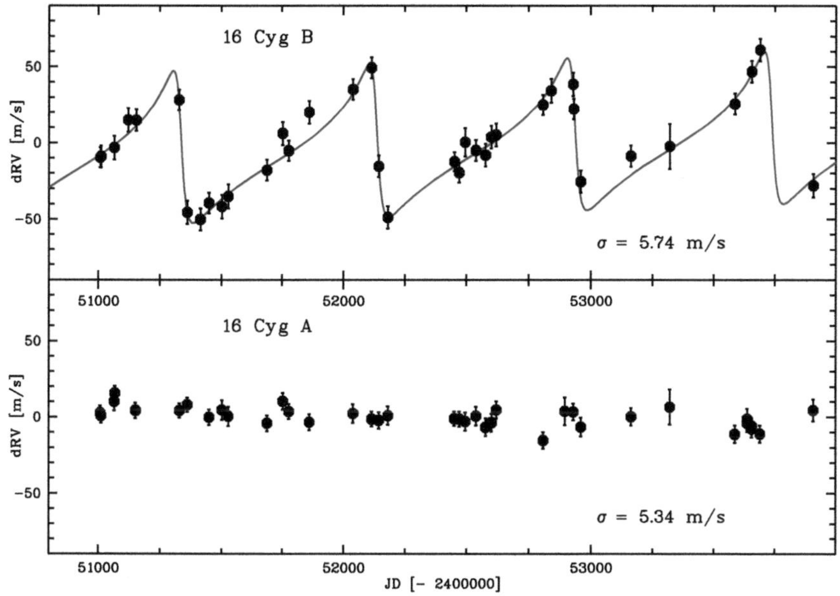

Fig. 3.14 RV measurements for 16 Cyg A (*bottom*) and B (*top*). 16 Cyg B shows RV variations due to the presence of a planet in an eccentric orbit. The orbital solution is shown as a line. Over the same time 16 Cyg A has shown no RV variations to a level less than $6\,\mathrm{m\,s^{-1}}$

Table 3.1 Orbital elements for the planet around 16 Cyg B

Element	Value
Period (days)	802.22 ± 2.29
T (Julian Day)	$2{,}450{,}533.10 \pm 7.18$
Eccentricity	0.665 ± 0.031
ω (deg)	80.82 ± 7.18
K1 ($\mathrm{m\,s^{-1}}$)	46.5 ± 3.1
$f(m)$ (solar masses)	$(3.48 \pm 0.92) \times 10^{-9}$
Semi-major axis (AU)	20.18 ± 0.66
m sin i (M_{Jupiter})	1.51 ± 0.32

environment. One has a giant extrasolar planet and the other clearly does not. Our full RV measurements for 16 Cyg A span over 15 years and these can exclude Jupiter mass companions of 2.45 M_{Jupiter} masses out to 5.2 AU (Wittenmyer et al. 2006). What is that extra ingredient that caused 16 Cyg B to form a planet and 16 Cyg A not? One hint may be the different lithium abundance which may point to a different angular momentum history for the stars (Cochran et al. 1997). Clearly, searching for planets around similar binary "twins" may help shed light on the precise conditions required for extrasolar planet formation.

3.6.2 *γ Cep*

The pioneering radial velocity survey of Campbell and Walker (1979) first detected the Doppler wobble of a possible planetary companion to γ Cep A. These authors noted that their detected wobble was superimposed on a long-term trend that was clearly due to a stellar companion and suggested that it might have been caused by a planetary object (Campbell et al. 1988). In a subsequent paper these authors retracted their planetary idea concluded that their observed Doppler wobble was most likely due to stellar rotation (Walker et al. 1992). This conclusion was made based on possible variations in the Ca II 8662 equivalent width with the RV period. Later Hatzes et al. (2003) combined over 20 years of radial velocity data for this star and by performing a careful re-analysis of the Ca II data, concluded that the RV variations in the γ Cep A were in fact due to a planetary companion.

Figure 3.15 shows the RV measurements of γ Cep from four different data sets: the original data from HF survey (Walker et al. 1995), the survey on the Canada France Hawaii Telescope (CFHT), and three phases of the McDonald Observatory Planet Search Program. The large variations (top panel) are due to the binary companion. One can see from this figure that superimposed on the binary orbit is a subtle "wobble" due to the planetary companion. The bottom panel shows the RV variations after removing the orbital contribution of the stellar companion and phased to the orbital period for the four data sets. The RV variations of γ Cep have been coherent and in phase for over 20 years.

Recently the binary companion of γ Cep A was imaged using adaptive optics by Neuhäuser et al. (2007). By combining the imaging, radial velocity, and astrometric data, these authors were able to derive a refined orbit for both the binary system and the planetary companion. Table 3.2 list the orbital elements and masses of the binary, and Table 3.3 lists the orbital elements for the planetary companion from Neuhäuser et al. (2007). The orbital inclination for the A/B system is $i = 119.3 \pm 1.0$ degrees (Neuhäuser et al. 2007). Assuming that the planet and binary orbits are co-planar, the true mass for the stellar companion would be $m_b = 1.83\ M_{\mathrm{Jupiter}}$.

Figure 3.16 shows the γ Cep system compared to our solar system. The entire binary+planet system would fit within the orbit of Neptune. With that, what are the implications of γ Cep b for planet formation theories? The standard scenario for planet formation suggests that giant planets like Jupiter must form in a region beyond the so-called snowline, the minimum distance from the star where solid material can condense. This snow line is typically beyond about 3 AU (Sasselov and Lecar 2000). Once a Jupiter-mass planet is formed, it may open a gap in the circumstellar disk and tidal forces may cause it to migrate from its birth location (e.g., Ward 1997). In the γ Cep binary system, given the mass-ratio and the binary separation (approximately 20 AU), any truncation of the proto-planetary disk would have occurred outside the snowline. This implies that the planet around γ Cep A could thus have formed at the snowline and only migrated a short distance.

Although the existence of a planet around γ Cep A is consistent with the standard model for planet formation (Jang-Condell et al. 2008), the suggestion of a planet in the HD 188,753 binary system (Konacki 2005) is not (Jang-Condell 2007). In 2005,

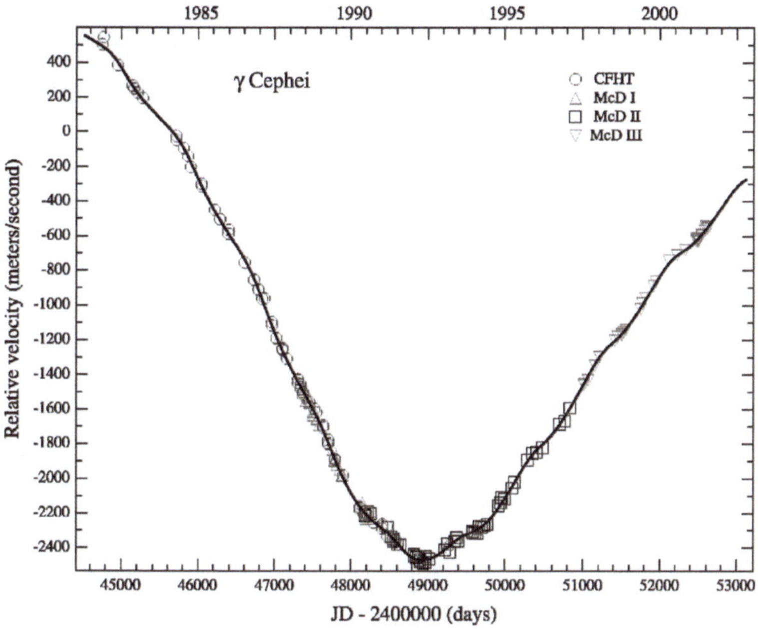

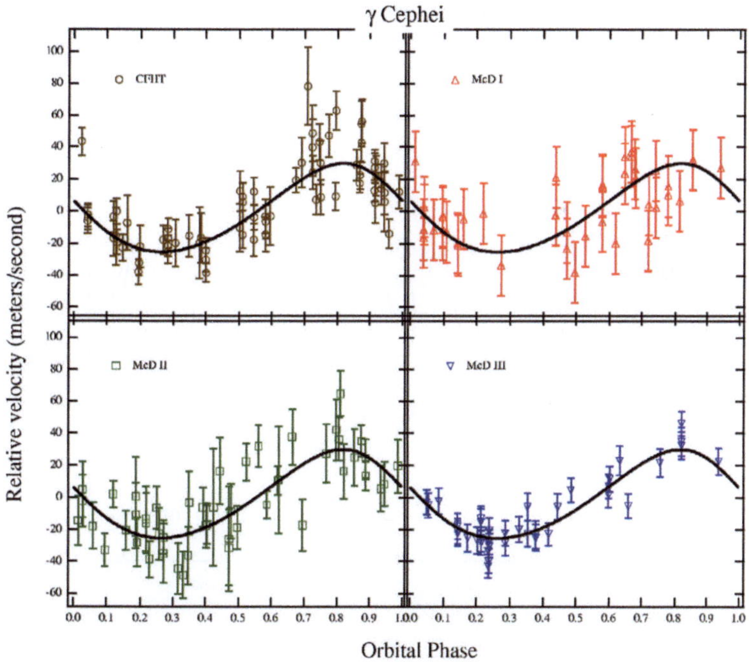

Fig. 3.15 (*Top*) The RV measurements for γ Cep using four different data sets. The large variations are due to the binary motion. (*Bottom*) RV variations after removal of the binary motion and phased to the orbital period of the planetary companion. The orbital solution is for the binary+ planet (*top*) and planet only (*bottom*)

Table 3.2 Binary orbital elements for γ Cep

Element	Value
Period (years)	67 ± 1.4
T (year)	$253{,}156.8 \pm 52.4$
Eccentricity	0.4085 ± 0.0065
ω (deg)	160.96 ± 0.4
i (deg)	119.3 ± 1.0
K1 ($\mathrm{m\,s^{-1}}$)	$1{,}925 \pm 14$
Semimajor axis (AU)	20.18 ± 0.66
Mass_A	1.40 ± 0.12
Mass_B	0.409 ± 0.018

Table 3.3 Orbital elements for the planet around γ Cep

Element	Value
Period (days)	902.96 ± 5.99
T (Julian Day)	$1{,}991.606 \pm 0.032$
Eccentricity	0.200 ± 0.069
ω (deg)	75.6 ± 18.8
K1 ($\mathrm{m\,s^{-1}}$)	26.3 ± 5
$f(m)$ (solar masses)	$(1.60 \pm 0.92) \times 10^{-9}$
Semimajor axis (AU)	2.13 ± 0.05
m sin i (M_{Jupiter})	1.46 ± 0.32

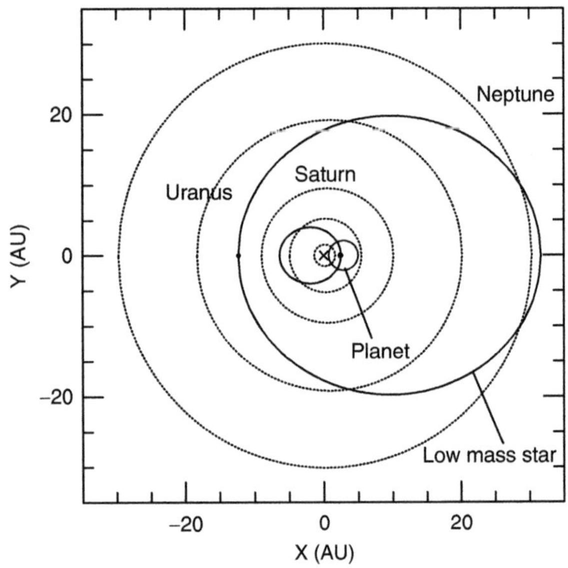

Fig. 3.16 The γ Cep binary plus planet system in comparison to the solar system

Konacki (2005) reported the discovery of a giant planet with an orbital period of 3.35 days (i.e., a 'hot Jupiter') around the primary star in the triple system HD 188,753. The primary of this system has a mass of 1.06 $M_{\odot}$ and its secondary is itself a binary with a total mass of 1.63 $M_{\odot}$. The separation between the primary and secondary is only 12.3 AU. Given this separation and mass-ratio the proto-planetary disk of the primary star would have been truncated to a radius of only 1.3 AU, well within the snowline. Based on the standard theory of giant planet formation, the planet around HD 188,753 A, if indeed is present, should not exist (Jang-Condell 2007). More recent RV measurements for this system agree with these theoretical model and refuted the presence the hot Jupiter (Eggenberger 2007). So for now, migration theory is safe, but the initial excitement generated by the purported discovery of HD 188,753 Ab demonstrates that future discoveries of exoplanets in binary systems may provide important clues to understanding the process of planet formation.

References

A. Baranne, D. Queloz, M. Mayor, G. Adrianzyk, G. Knispel, D. Kohler, D. Lacroix, J.P. Meunier, G. Rimbaud, A. Vin, A&AS, 119, 373 (1996)

J.M. Beckers, Nature, 260, 227 (1976)

R.P. Butler, G.W. Marcy, E. Williams, C. McCarthy, P. Dosanjh, S.S. Vogt, PASP, 108 (1996)

B. Campbell, G.A.H. Walker, PASP, 91, 540 (1979)

B. Campbell, G.A.H. Walker, S. Yang, ApJ, 331, 902 (1988)

W.D. Cochran, A.P. Hatzes, T.J. Hancock, ApJL, 380, 35L (1991)

W.D. Cochran, A.P. Hatzes, R.P. Butler, G.W. Marcy, ApJ, 483, 457 (1997)

W.D. Cochran and 15 other authors, ApJ, 611, L133 (2004)

A. Eggenberger, S. Udry, T. Mazeh, Y. Segal, M. Mayor, A&A, 466, 1179 (2007)

M. Endl, M. Kürster, S. Els, A&A, 362, 585 (2000)

E. Friel, G.C. Cayrel de Strobel, Y. Chmielewski, M. Spite, A. Lebre, C. Bentolila, A&A, 274, 825 (1993)

F. Galland, A.-M. Lagrange, S. Udry, J.-L. Beuzit, F. Pepe, M. Mayor, A&A, 452, 709 (2006)

R. Griffin and R. Griffin, MNRAS, 162 (1973)

R. F. Griffin, J.-M. Carquillat, N. Ginestet, The Observatory, 122, 90 (2002)

E.W. Guenther, G. Wuchterl, A&A, 401, 677 (2003)

A.P Hatzes, W. Cochran, in ESO Workshop on High Resolution Spectroscopy with the VLT. Editor, M.-H. Ulrich, Publisher, European Southern Observatory, Garching bei Munchen, Germany, p 275 (1992)

A.P. Hatzes, W.D. Cochran, M. Endl, B. McArthur, D.B. Paulson, G.A.H. Walker, B. Campbell, S. Yang, ApJ, 599. 1383 (2003)

A.P. Hatzes, E.W. Guenther, M. Endl, W.D. Cochran, M.P. Döllinger, A. Bedalov, A&A, 437, 743 (2005)

M. Holman, J. Touma, S. Tremain, Nature, 386, 254 (1997)

W. Huggins, Philosophical Transactions of the Royal Society of London, 158, 529 (1868)

H. Jang-Condell, ApJ, 654, 641 (2007)

H. Jang-Condell, M. Mugrauer, T. Schmidt, ApJ, 683, L191 (2008)

K.M. Kim, D.E. Mkrtichian, B.-C. Lee, I. Han, A.P. Hatzes, A&A, 454, 839 (2006)

J.R. King, C.P. Deliyannis, D. Hiltgen, A. Stephens, K. Cunha, A.M. Boesgaard, AJ, 113, 1871 (1997)

A. Koch, H. Wöhl, A&A, 134 (1984)
M. Konacki, Nature, 436, 230 (2005)
M. Kürster, M. Endl, S. Els, A.P. Hatzes, W.D. Cochran, S. Döbereiner, K. Dennerl, A&A, 353, L33 (2000)
D. Latham, R.P. Stefanik, T. Mazeh, M. Mayor, G. Burki, Nature, 339. 38L (1989)
G. Marcy, R.P. Butler, PASP, 104 (1992)
M. Mayor, D. Queloz, Nature, 378, 355 (1995)
R. Neuhäuser, M. Mugrauer, M. Fukagawa, G. Torres, T. Schmidt, A&A, 462, 777 (2007)
F. Pepe, M. Mayor, B. Delabre, D. Kohler, D. Lacroix, D. Queloz, S. Udry, W. Benz, J.-L. Betraux, J.-P. Sivan, SPIE, 4008, 582 (2000)
D. Queloz, M. Mayor, L. Weber, A. Blécha, M. Burnet, B. Confino, D. Naef, F. Pepe, N. Santos, S. Udry, A&A, 354, 99 (2000)
D. Sasselov, M. Lecar, ApJ, 528, 995 (2000)
C.G. Tinney, R.P. Butler, G.W. Marcy, H.R.A. Jones, A.J. Penny, S.S. Vogt, K. Apps, G.W.Henry, ApJ, 551, 507 (2001)
R.G. Tull, P.J. MacQueen, C. Sneden, D.L. Lambert, PASP, 107, 251 (1995)
J.A. Valenti, R.P. Butler, G.W. Marcy, PASP, 107, 966 (1995)
H. Vogel, Astronomische Nachtrichten, 82, 291 (1872)
S.S. Vogt, G.W. Marcy, R.P. Butler, K. Apps, ApJ, 536, 902 (2000)
S.S. Vogt, R.P. Butler, G.W. Marcy, D. Fischer, D. Pourbaix, K. Apps, G. Laughlin, AJ, 568, 352 (2002)
G.A.H. Walker, D.A. Bohlender, A.R. Walker, A. W. Irwin, S.L. Yang, A. Larson, ApJ, 396, 91 (1992)
G.A.H. Walker, A.R. Walker, A.W. Irwin, A.M. Larson, S.L.S. Yang, D.C. Richardson, Icarus, 116, 359 (1995)
W. R. Ward, ApJ, 482, 211 (1997)
R.A. Wittenmyer, M. Endl, W.D. Cochran, A.P. Hatzes, G.A.H. Walker, S.L.S. Yang, D.B. Paulson, AJ, 132, 177 (2006)
A. Wolszczan, D.A. Frail, Nature, 355, 145 (1992)

Chapter 4
Observational Techniques for Detecting Planets in Binary Systems

Matthew W. Muterspaugh, Maciej Konacki, Benjamin F. Lane, and Eric Pfahl

4.1 Why Focus Planet Searches on Binary Stars?

Searches for planets in close binary systems explore the degree to which stellar multiplicity inhibits or promotes planet formation. There is a degeneracy between planet formation models when only systems with single stars are studied – several mechanisms appear to be able to produce such a final result. This degeneracy is lifted by searching for planets in binary systems; the resulting detections (or evidence of non-existence) of planets in binaries isolates which models may contribute to how planets form in nature. Some models in which giant planet formation occurs over large amounts of time (e.g., the core-accretion scenario) predict that an extra-turbulent environment, such as those around binary stars, will disrupt planet formation. If the timescale is short (as in the gravitational instability theory), the process may continue, or even be enhanced due to additional instabilities in the planet-forming disks. It may be that multiple mechanisms contribute to giant planet formation in nature. Establishing the rate at which giant planets exist in binaries will distinguish the relative frequencies at which different processes contribute.

M.W. Muterspaugh (✉)
Department of Mathematics and Physics, College of Arts and Sciences, Tennessee State University, Boswell Science Hall, Nashville, TN 37209
and
Tennessee State University, Center of Excellence in Information Systems, 3500 John A. Merritt Blvd., Box No. 9501, Nashville, TN 37203-3401
e-mail: matthew1@coe.tsuniv.edu

M. Konacki (✉)
Nicolaus Copernicus Astronomical Center, Polish Academy of Sciences, Rabianska 8, 87-100 Torun, Poland
and
Astronomical Observatory, A. Mickiewicz University, Sloneczna 36, 60-286 Poznan, Poland
e-mail: maciej@ncac.torun.edu

B.F. Lane
Kavli Institute, MIT Physics Department, 70 Vassar Street, Cambridge, MA 02139
e-mail: blane@mit.edu

E. Pfahl
Institute for Defense Analyses, 4850 Mark Center Dr., Alexandria, VA 22311

N. Haghighipour (ed.), *Planets in Binary Star Systems*, Astrophysics and Space Science Library 366, DOI 10.1007/978-90-481-8687-7_4,

Studying relatively close pairs of stars, where dynamic perturbations are the strongest, provides the most restrictive constraints of this type (see, for example, Thébault et al. 2004). Exoplanet searches targeting these systems will determine whether the planet formation mechanisms found in nature are sensitive to such dynamics or not, a property which must be matched by theoretical models (Hatzes and Wuchterl 2005).

In this chapter, we consider observational efforts to detect planetary companions to binary stars in two types of hierarchical planet-binary configurations: first "S-type" planets which orbit just one of the stars, with the binary period being much longer than that of the planet's; second, "P-type" or circumbinary planets, where the planet simultaneously orbits both stars, and the planetary orbital period is much longer than that of the binary (Dvorak 1982). Simulations show that each of these two types has a large range of stable configurations (see, e.g., also, this book, Chapter by Haghighipour et al.; Rabl and Dvorak 1988; Benest 1988, 1989, 1993, 1996, 2003; Holman and Wiegert 1999; Broucke 2001; Pilat-Lohinger and Dvorak 2002, 2003). We review the use of dual-star astrometry and iodine-cell referenced radial velocimetry for detecting S-type planets in wide binaries, PHASES astrometry, modified iodine velocimetry, and eclipse-timing for finding S-type planets in closer binaries, and velocimetry and eclipse-timing for finding P-type planets.

4.2 S-Type Planets

S-Type planets orbit just one of the stars in a binary, and the binary separation is much larger than that between the star and planet. Some of the binaries are so widely separated (projected semimajor axis $a_b \gtrsim 1$ arcsec, where a_b is the semimajor axis of the binary) that they can be spatially resolved by ground-based telescopes without active image correction. For these binaries, traditional planet-finding techniques can be used. In fact, astrometric methods often perform best in this regime, as the secondary star serves as a convenient reference for the primary, and vice versa. Here, astrometric and radial velocity (RV) programs are considered as the most versatile search methods. (While transit searches might also be possible, these typically have very limited spatial resolutions, and the second star can act as a photometric "contaminant.") When the binaries are not spatially resolved with simple imaging, modifications must be made to meet the measurement precisions required for detecting extrasolar planets. These modifications are detailed in Section 4.2.2.

4.2.1 *Wide Binaries*

Traditional planet-finding techniques can be used to study the environments of widely separated binaries. We review the dual-star astrometry and iodine-cell referenced velocimetry techniques here. From an observational standpoint, "wide" binaries are considered to be those that can be resolved by traditional (uncorrected) imaging techniques. Due to atmospheric seeing, this sets the projected sky separation at larger than roughly 1 arcsec.

4.2.1.1 Dual-Star Astrometry

Interferometric narrow-angle astrometry (Shao and Colavita 1992; Colavita 1994) promises astrometric performance at the 10–100 μas level for pairs of stars separated by 1–60 arcsec. The lower limit of the allowable binary separation for this technique is that the binary is resolved by the individual telescopes in the interferometer; the upper limit is set by the scale over which the effects of atmospheric turbulence are correlated. This technique was first demonstrated with the Mark III interferometer for short integrations (Colavita 1994). It was later on extended to longer integrations, and shown to work at the 100 μas level at the Palomar Testbed Interferometer (PTI, Lane et al. 2000).

Achieving such performance requires simultaneous measurement of the interferometric fringe positions of both stars, greatly complicating the instrument (two beam combiners and metrology throughout the entire array are required). In addition, the instrumental baseline vector $\vec{B}$ connecting the unit telescopes must be known to high precision ($\approx$100 μm).

In an optical interferometer, light is collected at two or more apertures and brought to a central location where beams are combined and a fringe pattern is produced on a detector. For a broadband source of central wavelength λ and optical bandwidth $\Delta\lambda$, the fringe pattern is limited in extent, and appears only when the optical paths through the arms of the interferometer are equalized to within a coherence length $\Lambda = \lambda^2/\Delta\lambda$. For a two-aperture interferometer, neglecting chromatic dispersion by unequal air paths, the intensity measured at one of the combined beams is given by

$$I(x) = I_0 \left[1 + V \frac{\sin(\pi x/\Lambda)}{(\pi x/\Lambda)} \sin(2\pi x/\lambda) \right], \tag{4.1}$$

where V is the fringe contrast or "visibility", which can be related to the morphology of the source, and x is the optical path difference between arms of the interferometer (Fig. 4.1). More detailed analysis of the operation of optical interferometers can be found in Lawson (2000).

The location of the resulting interference fringes are related to the position of the target star and the observing geometry via

$$d = \vec{B} \cdot \vec{S} + \delta_a \left(\vec{S}, t \right) + c. \tag{4.2}$$

In this equation d is the optical path-length that one must introduce between the two arms of the interferometer to find fringes (often called the "delay"), $\vec{S}$ is the unit

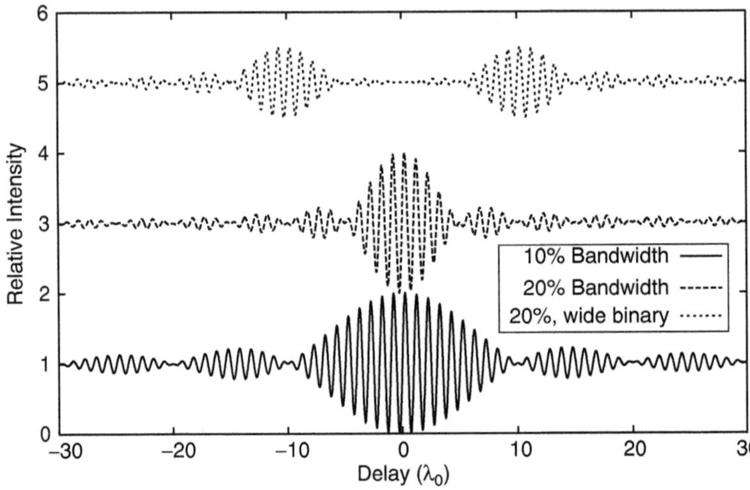

Fig. 4.1 The response of an interferometer. The top two curves have been offset by 2 and 4 for clarity. The widths of the fringe packets are determined by the bandpass of the instrument, and the wavelength of fringes are determined by an averaged wavelength of starlight. The top curve shows the intensity pattern obtained by observing two stars separated by a small angle on the sky – the observable is the distance between the fringe packets

vector in the source direction, and c is a constant that represents additional scalar delay introduced by the instrument. The term $\delta_a\left(\overrightarrow{S}, t\right)$ in Eq. (4.2) is related to the differential amount of path introduced by the atmosphere over each telescope due to variations in refractive index.

If the other quantities are known or small, measurement of the instrumental path length d (required to observe fringes) determines the position of the star $\overrightarrow{S}$. For a 100-m baseline interferometer, an astrometric precision of 10 μas corresponds to knowing d to 5 nm, a difficult but not impossible proposition for all terms except that related to the atmospheric delay. Atmospheric turbulence, which changes over distances of tens of centimeters and on millisecond timescales, forces one to use very short exposures to maintain fringe contrast, and hence limits the sensitivity of the instrument. It also severely limits the astrometric accuracy of a simple interferometer, at least over large sky-angles.

However, in narrow-angle astrometry, one is concerned with a close pair of stars, and the observable is a differential astrometric measurement. In other words, one is interested in knowing the angle between the two stars ($\overrightarrow{\Delta_s} = \overrightarrow{s_2} - \overrightarrow{s_1}$). The atmospheric turbulence is correlated over small angles. If the measurements of the two stars are simultaneous, or nearly so, the atmospheric term subtracts out making high precision "narrow-angle" astrometry possible.

The requirement that the target and reference stars be observed simultaneously, results in a significant instrumental complexity. That means, essentially two

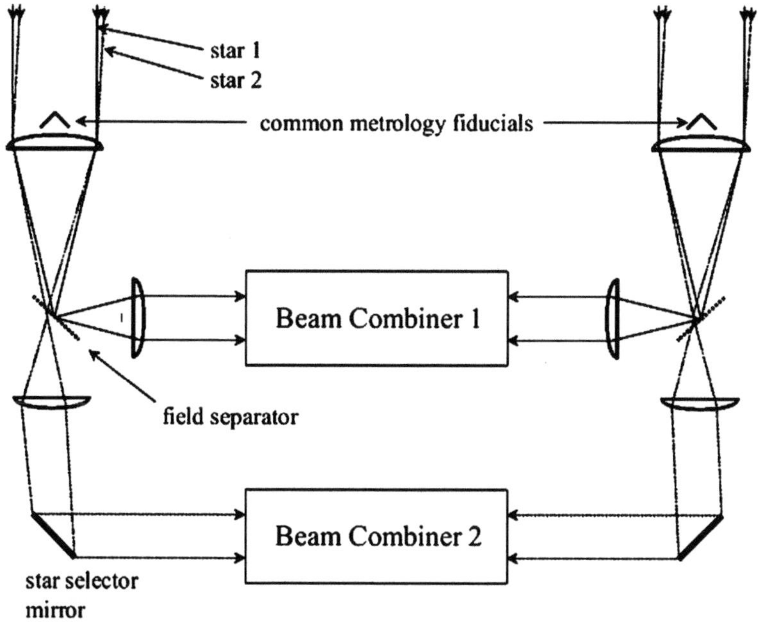

Fig. 4.2 Schematic of splitting the light in a dual-star interferometer. Figure from Colavita (1999)

complete interferometers are required to share the same set of apertures (Fig. 4.2). The splitting of light from the stars into two separate sets of delay lines, beam transport systems, and beam combiners, is done in a "dual-star module" located just after the apertures, with the split generally being accomplished using a beam-splitter. Considerable care must be taken in designing the system in order to avoid small pathlength measurement errors.

The exact level of astrometric precision that can be achieved depends on many factors, including the separation of the target/reference pair, the size of the interferometric baseline, and the levels and distribution of atmospheric turbulence. For a typical Mauna Kea seeing profile, the astrometric precision is given by

$$\sigma_a \simeq 300 \frac{\theta}{\sqrt{t} B^{2/3}} \text{ (arcsec)}, \tag{4.3}$$

where B is the baseline length in meters, θ is the target/reference separation in radians, and t is the integration time in seconds. For typical baselines of $\sim$100 m, and an angular separation of $\sim$30 arcsec, Eq. (4.3) implies an astrometric precision of 30 microarcsecond in an hour (Fig. 4.3). Better performance is possible from space; the Space Interferometry Mission (SIM-Planetquest) (Shao et al. 1995) will be capable of astrometry at the single microarcsecond level for a variety of science topics, including the characterization of rocky planets around nearby stars.

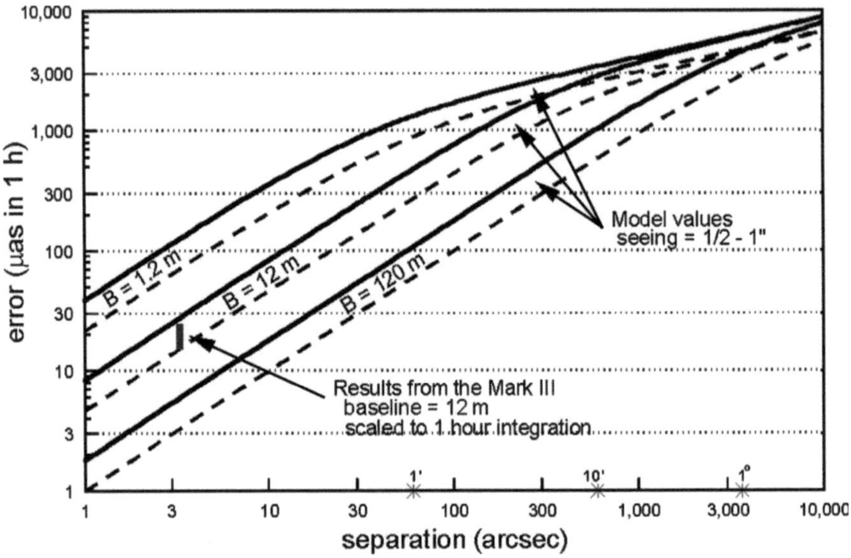

Fig. 4.3 Astrometric accuracy vs. star separation in a 1-h integration for different baseline lengths. Model atmospheres providing 1/2 and 1.0 arcsec seeing are shown. These results assume an infinite outer scale. Better results are achieved when the baseline exceeds the outer scale, as would be expected with a 100 m baseline at most sites. Measurements with the Mark III interferometer of a 3.3 arcsec binary star are consistent with the model. Figure from Colavita (1999)

The magnitude of the astrometric signal of the star's motion about its center of mass (CM) with its planet is given by:

$$\Delta a_{CM} = 2\frac{M_p}{M_s}a_p = \frac{M_p/M_J}{M_s/\mathrm{M}_\odot}\frac{a_p}{524}, \tag{4.4}$$

where M_p, M_s, M_J, $\mathrm{M}_\odot$ are masses of the planet, star, Jupiter, and the Sun, and a_p is the semimajor axis of the planet's orbit.

The minimum mass that can be detected is thus roughly given by

$$M_p/M_J \gtrsim 524\frac{\sigma_a}{a_p}\frac{M_s}{\mathrm{M}_\odot} \tag{4.5}$$

$$\gtrsim 0.1\frac{\sigma_a/20\,\mu\text{ arcsec}}{a_p/1(\text{AU})}\frac{d'}{10\text{pc}}\frac{M_s}{\mathrm{M}_\odot} \tag{4.6}$$

$$M_p/M_\oplus \gtrsim 1.6\frac{\sigma_a/1\,\mu\text{ arcsec}}{a_p/1(\text{AU})}\frac{d'}{10\text{pc}}\frac{M_s}{\mathrm{M}_\odot} \tag{4.7}$$

where $M_\oplus$ is the mass of the Earth, and d' is the distance to the target star.

4.2.1.2 Radial Velocities

When the stars in a binary can be spatially resolved without active image correction on ground based telescopes, the spectrum of each star can be recorded separately without contamination from the other, and the standard precision RV method described below can be used (see, for example, Campbell et al. 1988; Butler et al. 1996). Similarly, if the secondary is much fainter than the primary, precision RV might be performed on the brighter star as though it were single, although there is concern about the influence of the fainter lines. Several (~30) exoplanet candidates in binaries have been discovered in this manner. In some cases, the stars were not previously known to be binaries, and their natures were only discovered by long-term RV trends, or follow-up adaptive optics imaging. Some of these efforts to detect planets in binary stellar systems include that of Toyota et al. (2005) for single-lined and wide binaries (Desidera et al. 2006) targeting wide binaries, and the program targeting single-lined and wide binaries by Udry et al. (2004).

The highest precision RV observations are obtained either from the I_2 (molecular iodine) absorption cell, or the use of carefully designed spectrographs with fiber scrambling. In order to achieve an RV precision of ~1 m s^{-1}, an iodine absorption cell is used to superimpose a reference spectrum on the stellar spectrum by sending a starlight through the cell. The spectrum provides a fiducial wavelength-scale against which radial velocity shifts are measured.

Thanks to its conceptual simplicity, the iodine technique is the most commonly adopted way to obtain precision radial velocities. Iodine absorption cells are available on many spectrographs for the purpose of planet detection, including: HIRES at the 10 m Keck I (Keck Observatory), Hamilton at the 3 m Shane (Lick Observatory), SARG at the 3.6 m TNG (Canary Islands), UCLES at the 3.9 m Anglo-Australian Telescope (Anglo-Australian Observatory), HRS at the 9 m HET (McDonald Observatory), MIKE at the 6.5 m Magellan (Las Campanas Observatory), UVES at the 8 m Kueyen (Cerro Paranal), and HDS at the 8.2 m Subaru (National Astronomical Observatory of Japan).

In the iodine absorption cell technique, the Doppler shift of a star spectrum is determined by solving the following equation (Marcy and Butler 1992)

$$I_{obs}(\lambda) = [I_s(\lambda + \Delta\lambda_s)\, T_{I_2}(\lambda + \Delta\lambda_{I_2})] \otimes PSF, \tag{4.8}$$

where $\Delta\lambda_s$ is the shift of the star spectrum, $\Delta\lambda_{I_2}$ is the shift of the iodine transmission function T_{I_2}, $\otimes$ represents a convolution, and *PSF* is the spectrograph's point-spread function. The parameters $\Delta\lambda_s$ and $\Delta\lambda_{I_2}$, as well as the parameters describing the PSF, are determined by performing a least-squares fit to the observed spectrum, I_{obs}, as seen through the iodine cell. To this end, one also needs a high SNR (signal-to-noise) star spectrum taken without the cell, I_s, which serves as a template for all the spectra observed through the cell, as well as the I_2 transmission function, T_{I_2}. The latter can be obtained with a Fourier Transform Spectrometer such as the one at the Kitt Peak National Observatory. The Doppler shift of a star spectrum is then given by $\Delta\lambda = \Delta\lambda_s - \Delta\lambda_{I_2}$.

The velocity reflex amplitude of a star due to an unseen companion is given by

$$\Delta v_b = 2\frac{2\pi a_p \sin i_p}{P_p}\frac{M_p}{M_s+M_p} = \frac{2\sqrt{G}M_p \sin i_p}{\sqrt{(M_s+M_p)\,a_p}}$$
$$= 56.9\,\mathrm{m\,s^{-1}} \times \frac{M_p \sin i_p/M_J}{\sqrt{\left[(M_s+M_p)/\mathrm{M}_\odot\right](a_p/1\mathrm{AU})}}. \quad (4.9)$$

Here P_p is the period of the planet's orbit, G is the gravitational constant, and i_p is the inclination of the planet's orbit with respect to the sky.

For an RV precision of σ_{rv}, the minimum mass that can be detected is roughly equal to

$$M_p \sin i_p/M_J \gtrsim 0.018\frac{\sigma_{rv}}{1\,\mathrm{m\,s^{-1}}}\sqrt{\left[(M_s+M_p)/\mathrm{M}_\odot\right](a_p/1\mathrm{AU})}, \quad (4.10)$$

$$M_p \sin i_p/M_\oplus \gtrsim 5.6\frac{\sigma_{rv}}{1\mathrm{m\,s^{-1}}}\sqrt{\left[(M_s+M_p)/\mathrm{M}_\odot\right](a_p/1\mathrm{AU})}. \quad (4.11)$$

4.2.1.3 Observational Precisions

Astrometry is most sensitive to long period planets, RV to short period ones. Figure 4.4 shows the companion masses one can detect for each method, assuming 20 μas ground-based astrometry, 1 μas space-based astrometry, and 1 m s^{-1} RV precisions.

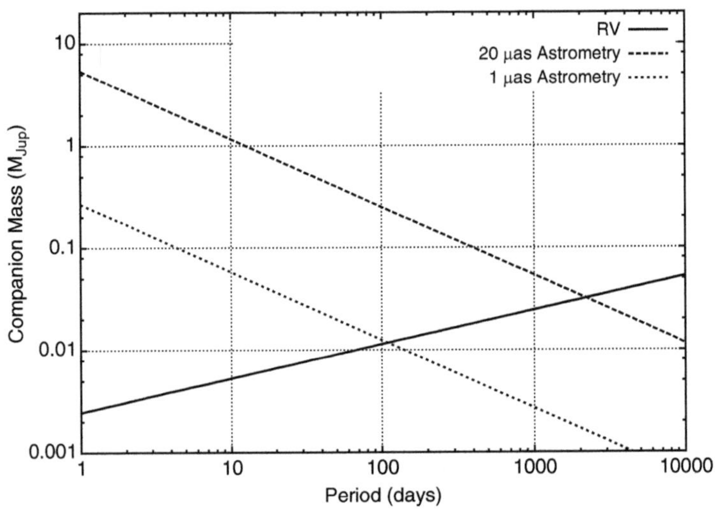

Fig. 4.4 Sensitivity to S-type planets in wide binaries, comparing astrometric and radial velocity techniques. All calculations assume solar-mass stars. Astrometric sensitivity assumes a distance of 10 pc to the target system

4.2.2 Close Binaries

Radial velocity surveys for extrasolar planets have been restricted largely to stars within $\simeq$100 pc due to sensitivity limits. The majority of detected exoplanets are at distances of 10–50 pc. Therefore, our observational definition of a "wide" binary corresponds to projected orbital separations of $\gtrsim$50 AU. More compact systems form the complementary class of "close" binaries. From a theoretical standpoint, binaries with semimajor axes of $\lesssim$50 AU pose important challenges to standard ideas about the formation of giant planets (Norwood and Haghighipour 2002).

Imagine a protoplanetary disk around one star in a newly formed binary. Suppose that the binary orbit has semimajor axis a_b and eccentricity e_b, and assume, for simplicity, that the stars have equal masses. The tidal gravitational field of the companion star truncates the circumprimary disk at a radius of $R_t \simeq 0.26a_b(1-e_b{}^2)^{1.2}$ (Pichardo et al. 2005). If $R_t \lesssim 3$ AU (i.e., inside the so-called "ice line"), it seems unlikely that icy grains could form and grow into planetesimals, thus precluding the embryonic stage of giant planet formation in the core-accretion scenario (Lissauer 1993). In more extreme cases, when $R_t \lesssim 1$ AU, there may be insufficient material in the disk to yield a Jovian-mass planet (Jang-Condell 2006). Even when R_t is as large as $\simeq$10 AU, stirring of the disk by the tidal field and the thermal dissipation of spiral waves may inhibit planetesimal formation, as well as stabilize the disk against fragmentation (Nelson 2000; Thébault et al. 2004; Thébault et al. 2006). In this case, it may be that neither the core-accretion picture nor gravitational instability (Boss 2000) are accessible modes of giant planet formation. We adopt $R_t = 10$ AU ($a_b \lesssim 40$ AU for modest e_b) as a fiducial upper limit for which Jovian planet formation is significantly perturbed and perhaps strongly inhibited. This provides a simple theoretical definition of a close binary that roughly matches our observational measure.

A handful of planets in binaries with $a_b \lesssim 20$ AU have already been discovered (see Table 4.1). The tightest of these candidate systems (HD 188753 Konacki 2005b)

Table 4.1 Close binaries with planets. When no eccentricity is given, only the projected binary separation is known. The quantity M_1/M_2 is the planetary host mass divided by companion mass. In HD 188753, the secondary is a binary with semimajor axis 0.67 AU. In GJ 86, the secondary is a white dwarf; to estimate the tidal truncation radius R_t, an original companion mass of 1M$_\odot$ is assumed. (1) Konacki (2005a), (2) Campbell et al. (1988), (3) Hatzes et al. (2003), (4) Queloz et al. (2000), (5) Mugrauer and Neuhäuser (2005), (6) Lagrange et al. (2006), (7) Zucker et al. (2004), and (8) Chauvin et al. (2006)

Object	a_b (AU)	e	M_1/M_2	R_t (AU)	Refs.
HD 188753	12.3	0.50	1.06/1.63	1.3	1
γ Cephei	18.5	0.36	1.59/0.34	3.6	2, 3
GJ 86	~20		0.7/1.0	~5	4, 5, 6
HD 41004	~20		0.7/0.4	~6	7
HD 196885	~25		1.3/0.6	~7	8

has a periastron separation of only $\simeq$6 AU, which seems severely at odds with the conventional lore on Jovian planet formation. While further study of this system is needed due to the non-detection of the giant planet as reported by Eggenberger et al. 2007 (also see Jang-Condell 2006 for a supporting theoretical argument), (it should be noted that the lower spectral resolution and technique used by Eggenberger et al. (2007), which does not deconvolve the instrumental PSF, may be a factor in their lower RV precision and thus lack of detection), this and other systems have inspired considerations into how giant planets might come to be in close binaries. Pfahl (2005) and Portegies Zwart and McMillan (2005) suggested that the planetary host star may have been acquired in a dynamical exchange interaction in a star cluster *after* the planet formed, thus circumventing the complicating factors listed above. As most stars are born in clustered environments, one wonders how often dynamics can account for planets in close binaries. This idea was explored in Pfahl and Muterspaugh (2006), where it was found that exchange interactions can account for only $\sim$0.1% of close binaries hosting planets. However, the (admittedly small) sample of systems in Table 4.1 seems to indicate that a larger fraction of $\sim$1% of close binaries harbor giant planets (see Pfahl and Muterspaugh (2006) for details), and perhaps planets do somehow form frequently in these hostile environments. It is crucial that we begin to develop a census of planets in close binaries in order to test the different theories about planet formation and dynamics.

4.2.2.1 PHASES Astrometry

The dual-star astrometry method can be modified to be applicable to binaries that are so close that the individual telescopes of an interferometer cannot resolve the pair (Lane and Muterspaugh 2004). The interferometer itself over-resolves the binary (Fig. 4.1) and its high spatial resolution will then allow for precision astrometric measurements. In this mode, the small separation of the binary results in both components being in the field of view of a single interferometric beam-combiner. The fringe positions are measured by modulating the instrumental delay with an amplitude large enough to record both fringe packets.

Since the fringe position measurement of the two stars is no longer truly simultaneous, it is, however, possible for the atmosphere to introduce path-length changes (and hence positional error) in the time between measurements of the separate fringes. To reduce this effect, a fraction of the incoming starlight is redirected to a separate beam-combiner. This beam-combiner is used in a "fringe-tracking" mode (Shao and Staelin 1980; Colavita et al. 1999), where it rapidly (10 ms) measures the phase of one of the starlight fringes, and adjusts the internal delay to keep that phase constant. The fringe tracking data is used both in real-time as a feed-back servo, after which a small residual phase error remains, and in post-processing, where the measured residual error is applied to the data as a feed-forward servo. This technique – known as phase referencing – has the effect of stabilizing the fringe measured by the astrometric beam-combiner. For this observing mode, laser metrology is only required between the two beam-combiners through the location of the

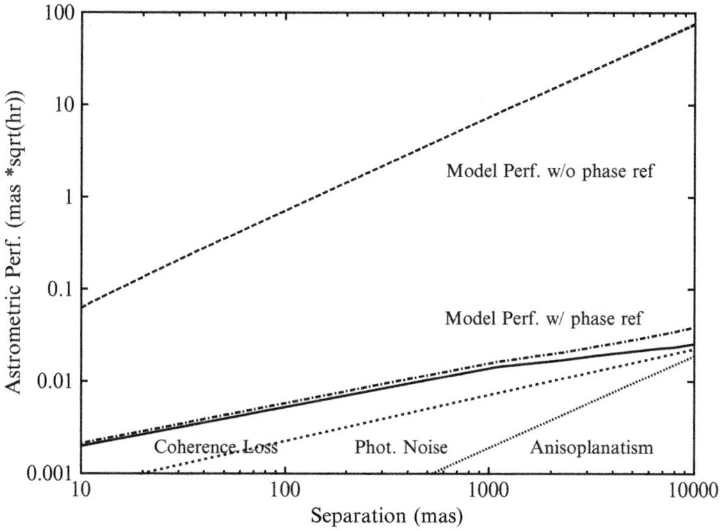

Fig. 4.5 The expected narrow-angle astrometric performance in mas for the phase-referenced fringe-scanning approach, for a fixed delay sweep rate, and an interferometric baseline of 110 m. Also shown is the magnitude of the temporal loss of coherence effect in the absence of phase referencing, illustrating why stabilizing the fringe via phase referencing is necessary

light split (which occurs after the optical delay has been introduced), rather than throughout the entire array. Without phase referencing, the obtainable astrometric precision is a factor of a hundred worse (see Fig. 4.5).

The first beam-combiner is now stabilized against atmospheric motions, and modulates the interferometric delay to observe the fringe packets formed by each star in the binary. To analyze this data, a double fringe packet based on Eq. (4.1) is then fit to the data, and the differential optical path between fringe packets is measured. A grid is constructed in differential right ascension and declination over which the search is made. For each point in the search grid, the expected differential delay is calculated based on the interferometer location, baseline geometry, and time of observation for each scan. A model of a double-fringe packet is calculated and compared to the observed scan to derive a χ^2 value. This is repeated for each scan, co-adding all of the χ^2 values associated with that point in the search grid, and making a χ^2 surface as a function of differential right ascension (R.A.) and declination. The best-fit astrometric position is found at the minimum χ^2 position. The uncertainties associated with this position are defined by the appropriate χ^2 contour, which depends on the number of the degrees of freedom in the problem and the value of the minimum χ^2. The final product is a measurement of the apparent vector between the stars and associated uncertainty ellipse. Because the data were obtained with a single-baseline instrument, the resulting error contours are very elliptical, with aspect ratios that sometimes exceed 10:1.

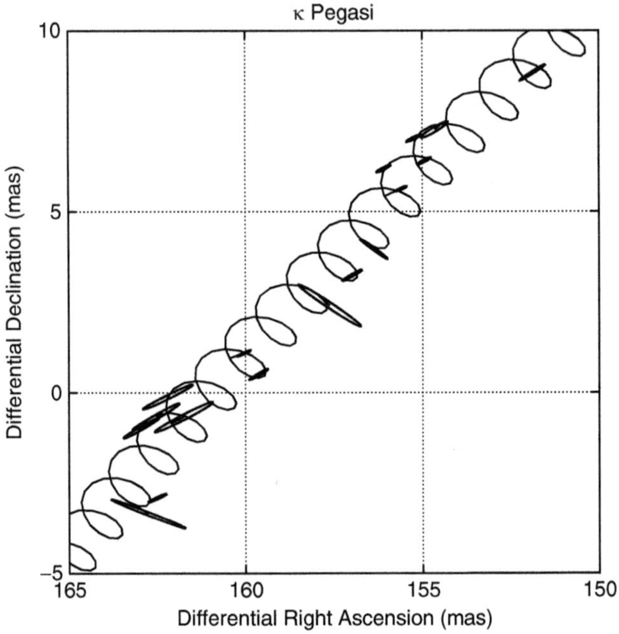

Fig. 4.6 The visual orbits of κ Pegasi showing perturbations by the third component. The *spiral line* represents the apparent motion of the short-period pair's center of light. The *ellipses* represent the 1σ uncertainties for PHASES measurements

The Palomar High-precision Astrometric Search for Exoplanet Systems (PHASES) program uses this technique to monitor approximately 50 binaries to search for substellar companions. An example of such systems is κ Pegasi which is a well-known, nearby triple star system. This system consists of a "wide" pair with a semimajor axis of 235 mas (8.14 AU). One component of this system is a single-line spectroscopic binary and has a semimajor axis of 2.5 mas and a physical separation of 0.087 AU (Fig. 4.6). The perturbation due to the unseen (faint) short-period component of this system is evident from Fig. 4.6. Similar sized perturbations with longer orbital periods would indicate the presence of planetary companions. Figure 4.7 shows the mass-period phase space in which PHASES observations show companions do not exist in face-on, circular orbits in the 13 Pegasi system.

4.2.2.2 Radial Velocities

In the case when a composite spectrum of a binary star is observed, the classical approach with the iodine cell (described in Section 4.2.1.2) cannot be used since it is not possible to observationally obtain two separate template spectra of the binary components. To resolve this problem, one can proceed as follows. First, one always takes two sequential exposures of each (binary) target – one with and the

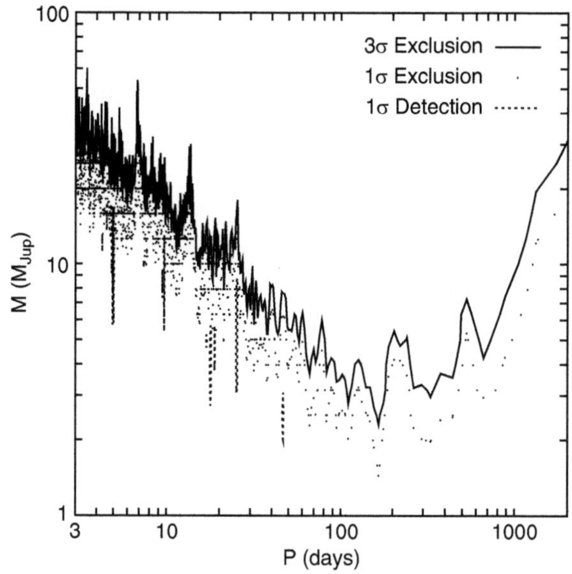

Fig. 4.7 The 13 Pegasi mass-period companion phase-space shows PHASES observations can rule out tertiary objects as small as two Jupiter-masses. A few mass-period combinations introduce slight improvements over the single-Keplerian model, but none of these are more significant than 1.7σ, and are probably not astrophysical in origin. There is a long-period cutoff in sensitivity due to the finite span of the observations. Similar detection limits have recently been published on other PHASES targets (Muterspaugh et al. 2006)

other without the cell. This is contrary to the standard approach for single stars where an exposure without the cell (a template) is taken only once. This way one obtains an instantaneous template that is used to model only the adjacent exposure taken with the cell. Next, one performs the usual least-squares fit and obtains the parameters described in Eq. (4.8). Obviously, the derived Doppler shift, $\Delta\lambda_i$ (where i denotes the epoch of the observation), carries no meaning since each time a different template is used. Moreover, it describes a Doppler "shift" of a composed spectrum that is typically different at each epoch. However, the parameters – in particular the wavelength solution and the parameters describing PSF – are accurately determined and can be used to extract the star spectrum, $I^{\star,i}_{obs}(\lambda)$, for each epoch i, by inverting Eq. (4.8). That is,

$$I^{\star,i}_{obs}(\lambda) = [I^{i}_{obs}(\lambda) \otimes^{-1} PSF^{i}]/T_{I_2}(\lambda), \tag{4.12}$$

where $\otimes^{-1}$ denotes deconvolution, and PSF^i represents the set of parameters describing PSF at the epoch i. Such a star spectrum has an accurate wavelength solution, and is free of the I_2 lines and the influence of a varying PSF. In the final step, the velocities of both components of a binary target can be measured with the well-known two-dimensional cross-correlation technique TODCOR (Zucker and Mazeh 1994) using as templates the synthetic spectra derived with the ATLAS 9 and

ATLAS 12 programs (Kurucz 1995) and matched to the observed spectrum, $I_s(\lambda)$. The formal errors of the velocities can be derived from the distribution between the velocities from different echelle orders or using the formalism of TODCOR (Zucker and Mazeh 1994). The technique currently produces RVs of binary stars with an average precision of 20 m s^{-1} (Konacki 2005b). Improvements to the technique are being introduced to reach the level below 10 m s^{-1} (Konacki 2009).

With this modified iodine technique, Konacki (2005b) initiated the first RV survey for circumprimary or circumsecondary planets of binary or multiple stars (mainly hierarchical triples). The survey's sample of $\sim$450 binaries (northern and southern hemisphere) was selected based on the following criteria. (1) The apparent separation of the components had to be smaller than the width of the slit (e.g., 0.6 arcseconds for Keck-I/HIRES) to avoid possible systematic effects. Such systems will remain unresolved under most seeing conditions. (2) The brightness ratio between the components should not be too large (at the order of 10 or less) to be able to clearly identify spectra of both components. (3) The orbits of the binaries should be well known to constitute a firm ground on which one can discuss possible detection (or lack) of planets and substellar companions in the context of the binary characteristics. All these requirements could be satisfied by targeting a subset of speckle binaries that have determined orbits from the Catalog of Orbits of Visual Binary Stars (Hartkopf et al. 2001). This catalog contains 1,700 binaries, of which 1,300 have projected semimajor axes smaller than 1 arcsec. The sample is sufficiently large to produce meaningful statistics. Also, such binaries have been ignored by previous RV studies.

The survey was initiated in 2003 at the Keck-I/HIRES and continued there until mid-2007. It was also carried out for a year (2006–2007) at the TNG/SARG. Approximately 150 speckle binary and triple stellar systems were observed. The first candidate planet in a triple star system was announced by Konacki (2005b). This and a few other interesting targets, including several new triple star systems, were followed up with the TNG/SARG and HET/HRS. All collected spectra are now being re-reduced with a new data pipeline and a publication summarizing this effort will be prepared in the near future.

4.2.2.3 Eclipse Timing

Should a binary happen to be oriented with its orbital plane in the line of sight, it will exhibit eclipses as one star passes in front of another. It has long been recognized that periodic shifts in the observed times of photometric minima of eclipsing binaries can indicate the presence of an additional component to the system (see, for example, Woltjer 1922; Irwin 1952; Frieboes-Conde and Herczeg 1973; Doyle et al. 1998). For a binary with separation wide enough to allow for stable planetary systems to exist around just one component, the probability of such an alignment is extremely small, and very few targets are accessible. However, should one find such a fortunate happenstance, one can detect the planetary companions by precision timing of the eclipses. Clearly, this method has no direct analog for single systems.

This method can detect S-type planets or similarly moons of transiting planets. For this evaluation, it is assumed that the depth of the planet (or moon) eclipse is sufficiently small as to be ignored (in such a detection, one could then re-evaluate light curves to look for such transit signals) and the binary orbit is circular. The velocity of the stars orbiting each other and the offset of the star-planet center of mass from the center of the star itself determines the timing variation observed as

$$\begin{aligned}\Delta t &= x_{CM}/v_b \\ &= \left(\frac{a_p M_p \sin\phi}{M_2 + M_p}\right)\left(\frac{P_b}{2\pi a_b}\right) \\ &\approx 57\,\text{sec} \times (P_b/\text{month})\,\frac{a_p}{(a_b/7)}\,\frac{(M_p/M_J)}{(M_2/\mathrm{M}_\odot)}\sin\phi \end{aligned} \tag{4.13}$$

where M_2 is the mass of the secondary star (or the transiting planet, assumed to host the S-type companion), M_p is the mass of the S-type object orbiting M_2, ϕ is angle between the planet's orbital angular momentum vector and the direction of motion of the host star during eclipse, and P_b is the period of the binary orbit. The eclipse timing delays due to orbit of M_2 about the center of mass of M_2-M_p system. The factor of 7 in Eq. (4.13) is an approximate criteria for stability, implying that the planet's semimajor axis is seven times smaller than that of the binary (this factor varies by the system, and can be determined through detailed simulations). Equation (4.13) is thus an upper limit for the timing effect. Converting the semimajor axis to orbital periods, this equation becomes

$$\begin{aligned}\Delta t &\approx 41\,\text{sec} \times (P_b/\text{month})^{1/3}\,(P_p/\text{day})^{2/3}\,\frac{M_p/M_J}{\left(M_b^{1/3} M_2^{2/3}\right)/\mathrm{M}_\odot}\sin\phi \\ &\approx 65\,\text{sec} \times (P_b/\text{month})^{1/3}\,(P_p/\text{day})^{2/3}\,\frac{M_p/M_J}{M_b/\mathrm{M}_\odot}\sin\phi. \end{aligned} \tag{4.14}$$

In this equation, $M_b = M_1 + M_2 + M_p$. here it has been assumed that $M_1 \approx M_2$, in which case the maximum stable planet period is 1/13 times that of the binary period, implying days and months are the natural units for each respectively (S-type planets cannot exist in much shorter period systems, and longer period systems are even less likely to show eclipses). The equivalent relationship for a moon orbiting an eclipsing Jupiter is

$$\Delta t \approx 13.3\,\text{sec} \times (P_b/\text{month})^{1/3}\,(P_p/\text{day})^{2/3}\,\frac{M_p/M_\oplus}{(M_b/\mathrm{M}_\odot)^{1/3}\,(M_2/M_J)^{2/3}}\sin\phi \tag{4.15}$$

where now the b subscript refers to the star-Jupiter analog system and p to the Jupiter analog's moon.

The precision with which eclipse minima can be timed is derived using standard χ^2 fitting techniques. Assume a photometric data set $\{y_i\}$ occurring at times

$\{t_i\}$ with measurement precisions $\{\sigma_i\}$, and a model photometric light curve of flux $F(t-t_0)$. The corresponding intensity is $I(t-t_0) = fF(t-t_0)\pi D^2 \Delta t/4$, where f $(0 \le f \le 1)$ is the fractional efficiency and throughput of the telescope, D is the telescope diameter, and Δt is the sample integration time. The quantity $F(t-t_0)$ might be determined to high precision by observing multiple eclipse events. The fit parameter t_0 is uncertain by an amount equal to the difference between the value for which χ^2 is minimized and the value for which it is increased by one: $1 + \chi^2(t_0) = \chi^2(t_0 + \sigma_{t_0})$,

$$1+\chi^2 = 1 + \sum_{i=1}^{N}\left[\frac{y_i - I\,(t_i - t_0)}{\sigma_i}\right]^2 = \sum_{i=1}^{N}\left[\frac{y_i - I\,\left(t_i - t_0 - \sigma_{t_0}\right)}{\sigma_i}\right]^2$$

$$\approx \sum_{i=1}^{N}\left[\frac{y_i - I\,(t_i - t_0) - \left(\frac{\partial I(t)}{\partial t}\right)_{t_i - t_0}\sigma_{t_0}}{\sigma_i}\right]^2$$

$$= \sum_{i=1}^{N}\left(\left[\frac{y_i - I\,(t_i - t_0)}{\sigma_i}\right]^2 + \left[\frac{\left(\frac{\partial I(t)}{\partial t}\right)_{t_i - t_0}\sigma_{t_0}}{\sigma_i}\right]^2 \right.$$

$$\left. -2\frac{\left(\frac{\partial I(t)}{\partial t}\right)_{t_i - t_0}\left[y_i - I\,(t_i - t_0)\right]\sigma_{t_0}}{\sigma_i^2}\right).$$

Because t_0 is the minimizing point, the first derivative of χ^2 at t_0 is zero. That is,

$$\left[\frac{\partial \chi^2\,(t)}{\partial t}\right]_{t_0} = 2\sum_{i=1}^{N}\frac{\left(\frac{\partial I(t)}{\partial t}\right)_{t_i - t_0}\left[y_i - I\,(t_i - t_0)\right]\sigma_{t_0}}{\sigma_i^2} = 0. \tag{4.16}$$

Rearrangement of terms in Eq. (4.16) leads to

$$\sigma_{t_0} = \left[\sum_{i=1}^{N}\left(\frac{\left(\frac{\partial I(t)}{\partial t}\right)_{t_i - t_0}}{\sigma_i}\right)^2\right]^{-1/2} \approx \sigma_I\left[\sum_{i=1}^{N}\left(\frac{\partial I\,(t)}{\partial t}\right)^2_{t_i - t_0}\right]^{-1/2}. \tag{4.17}$$

An eclipse of length τ is approximated as a trapezoid-shape light curve (Fig. 4.8) with maximum and minimum photon fluxes F_0 and $F_0(1-h/2)$. Here h is a dimensionless positive number producing an eclipse depth of $hF_0/2$. In the case of a faint secondary, h is roughly twice the ratio of the squares of the stellar radii $(2R_2^2/R_1^2)$. The ingress and egress are each assumed to be of length $k\tau/2$. The quantity $k \approx 2R_2/(R_1+R_2)$ is unity in the case of an eclipsing binary with components

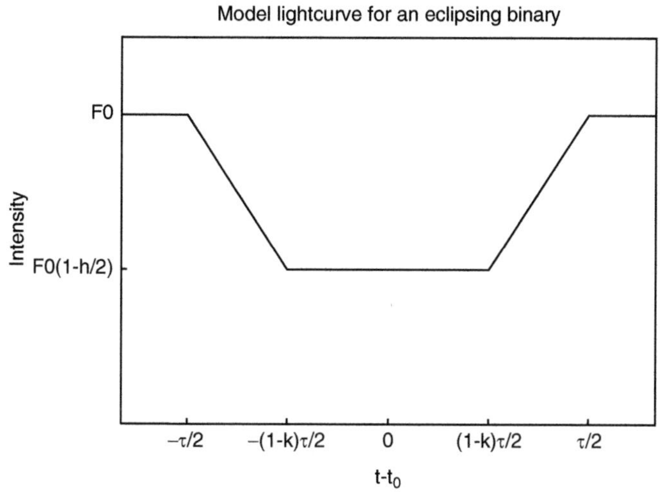

Fig. 4.8 Eclipsing binary model light curve

of equal size, when the trapezoid becomes a "V"-shape. In a functional form, the above-mentioned model is given by (also, see Fig. 4.8)

$$F(t-t_0)=\begin{cases} F_0 & t-t_0 \le -\tau/2 \\ F_0\left(1-ht/(k\tau)-h/(2k)\right) & -\tau/2 \le t-t_0 \le -\tau/2+k\tau/2 \\ F_0\left(1-h/2\right) & -\tau/2+k\tau/2 \le t-t_0 \le \tau/2-k\tau/2 \\ F_0\left(1+ht/(k\tau)-h/(2k)\right) & \tau/2-k\tau/2 \le t-t_0 \le \tau/2 \\ F_0 & \tau/2 \le t-t_0 \end{cases}$$

Only portions of the light curve during ingress and egress have nonzero slope $\left|\frac{\partial F(t)}{\partial t}\right| = hF_0/(k\tau)$. In more accurate models, the slope of the light curve will be non-zero but small in other regions, and will not contribute much to the sum in Eq. (4.16).

The number of data points contributing to the sum is thus $N = gk\tau/\Delta t$, where $0 \le g \le 1$ is the fraction of the observed eclipses and also accounts for the fraction of time that was lost (e.g., to camera readout). The quantity Δt is the integration time for each measurement. The measurement noise σ_I is given by

$$\sigma_I = \left(I + \sigma_{sc}^2 + I_{bg} + n_{dark}\Delta t + \sigma_{rn}^2\right)^{1/2} \tag{4.18}$$

where $I_{bg} = fF_{bg}\pi D^2\Delta t/4$ is the sky background, n_{dark} is detector dark current, σ_{rn} is detector read noise, and σ_{sc} is scintillation noise given by Young (1967) as

$$\sigma_{sc} = 0.09I\,(D/1\,\mathrm{cm})^{-2/3}\,Xe^{-h/(8000\,\mathrm{m})}/\left(2\Delta t/1\,\mathrm{sec}\right)^{1/2} \tag{4.19}$$

$$\approx 0.003I\,(D/1\,\mathrm{m})^{-2/3}/\left(\Delta t/1\,\mathrm{sec}\right)^{1/2} \tag{4.20}$$

where X is the airmass and h is the altitude of the observatory. The drop in noise during eclipse is ignored (a factor less than $\approx$1.4) and Eq. (4.18) is combined with Eq. (4.17) to obtain an overall timing precision (in seconds) of

$$\sigma_{t_0} = \sqrt{\frac{k\,(\tau/1\,\mathrm{sec})}{gh^2}} \left[\frac{4}{fF_0\pi\,(D/1\,\mathrm{m})^2} + \frac{9\times 10^{-6}}{(D/1\,\mathrm{m})^{4/3}} \right.$$
$$\left. + \frac{\pi D^2 fF_{bg}/4 + n_{dark} + \sigma_{rn}^2/\,(\Delta t/1\,\mathrm{sec})}{f^2F_0^2\pi^2\,(D/1\,\mathrm{m})^4/16} \right]^{1/2} \quad (4.21)$$

$$\approx 0.18\,\mathrm{sec} \times \sqrt{\frac{k\,(\tau/1\,\mathrm{hr})}{fgh^2}} \left[\frac{10^{(V-12)/2.5}}{(D/1\,\mathrm{m})^2} + \frac{f}{(D/1\,\mathrm{m})^{4/3}} \right]^{1/2}. \quad (4.22)$$

The last term in the first line (associated with dark current, read noise, and background) is generally smallest and will be ignored. In most cases, the second term – associated with scintillation – is dominant (though zero in the case of space-based observatories). The exponent of $(V-12)/2.5$ in Eq. (4.21) shows that for meter-sized telescopes, photon noise is only dominant for stars fainter than twelfth magnitude.

Systematic and astrophysical noise sources may have effects that limit the actual precisions achieved. Mass transfer between stars can cause drifts in orbital periods. Variations of this type are non-periodic, distinguishing themselves from companion signals. Applegate (1992) has shown that gravitational coupling to the shapes of magnetically active stars can cause periodic modulations over decade timescales. This gravitational/magnetic coupling requires the star to be inherently variable (false positives can be removed using the overall calibrations of photometric data). It is possible that star spots will have large effects on timing residuals that are particularly difficult to calibrate (Watson and Dhillon 2004). Due to orbit-rotation tidal locking, the effect of a starspot on the light curve can be detected from the light curves of several orbits, and fitting the starspot can potentially remove the introduced timing biases.

The timescale for eclipses of such long period binaries is of the order of 12 h. Comparing Eqs. (4.14) and (4.21), one finds that S-type planets with periods of a few days can be detected around either star in eclipsing binaries with month-long periods if their masses are more than

$$M_p/M_\oplus \gtrsim 3\frac{(M_b/\mathrm{M}_\odot)}{(P_b/\mathrm{month})^{1/3}\,(P_p/\mathrm{day})^{2/3}} \sqrt{\frac{k\,(\tau/12\,\mathrm{hr})}{fgh^2}} \left[\frac{10^{(V-12)/2.5}}{(D/1\,\mathrm{m})^2} + \frac{f}{(D/1\,\mathrm{m})^{4/3}} \right]^{1/2}. \quad (4.23)$$

Equation 4.21 indicates that a meter-class ground-based telescope can time a giant planet transit ($h \approx 0.02$, $k \approx 0.18$) to approximately 9.4 s in the regime where the photometric precision is dominated by scintillation noise, assuming a Jupiter-sized planet orbiting a star of solar-size and mass with period of a month (implying 6-h

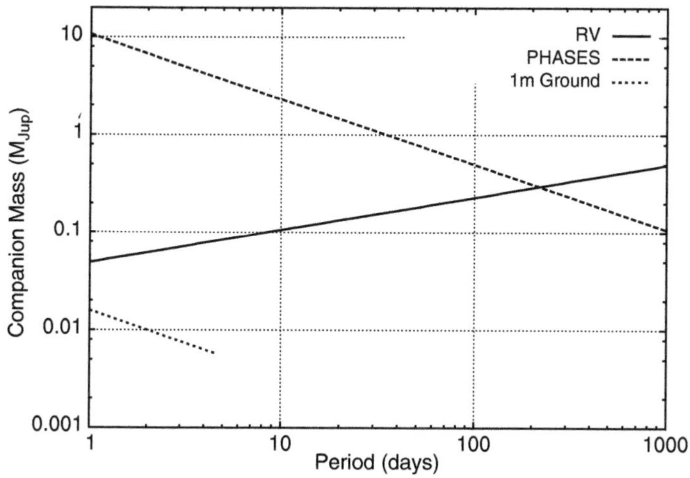

Fig. 4.9 Sensitivity to S-Type planets in narrow binaries, comparing astrometric, radial velocity, and eclipse timing techniques. All calculations assume solar-mass stars. The PHASES sensitivity assumes 20 μas precision and a distance to the system of 20 pc. Eclipse timing assumes a 1 m photometric telescope observing a $V = 10$ magnitude system with binary orbital period of 2 months (longer period systems are even less likely to show eclipses). The eclipse timing sensitivity curve only extends to the region where planets are likely to have stable orbits

duration eclipses). Figure 4.9 shows the companion masses one can detect for each method, assuming 20 μas ground-based astrometry, 20 m s^{-1} RV, and 1 m ground-based photometric telescope for eclipse timing. This precision is sufficient to find Earth-mass moons. For bright stars, space-based observatories offer even better precisions. The planetary system of HD 80606 is one of such long-period transiting systems (Gillon 2009, arXiv:0906.4904v1). However, the eccentricity of its giant planet is so large (0.934) that moons might not be realistic to expect.

Space-based photometric missions such as Kepler have, as their primary goal, the detection of Earth-like planets via transits of the planet across the star. However, such photometric events can be explained by other astrophysical phenomena, such as a transiting Jupiter blended with a background star, otherwise these results may be unreliable. However, Earth-like moons of transiting Jupiters might be identified through timing, and it is possible to confirm the nature of such a system. In such a scenario, a transiting Jupiter can be positively confirmed by ground-based radial velocity observations. Once this has been established, variations in the transit times would be used to detect Earth-sized moons. Because these photometric missions have limited lifetimes ($\approx$3 years), detections of moons are only possible for short period (few months or less) Jupiters, for which many transit events can be observed (unless a follow-up ground-based campaign is pursued with large telescopes). If the planet/moon are to be in the habitable zone, one must look for such systems around late-type (cool) stars. It is possible that such systems have the greatest likelihood of being habitable; tidal-locking of the Earth-sized moon to the Jupiter-like planet would ensure that the moon has day/night cycles and stabilize its rotational axis similar to the way in which the Earth's is stabilized by its own moon. Both of these conditions have been argued as favorable for life (see, for example, Laskar et al. 1993).

4.3 P-Type (Circumbinary) Planets

All the confirmed planets found in binary systems thus far are in S-type orbits. Discovery of circumbinary planets would constitute a new class of planetary system and would inspire new considerations to the interplay between system dynamics and planet formation.

4.3.1 Radial Velocities

A circumbinary planet will exhibit two indirect effects on the velocities of the stellar components of the system. First, the apparent system velocity will vary in a periodic manner due to the motion of the binary about the system barycenter. Second, the finite speed of light will cause apparent changes in the phase of the binary orbit. These effects may be detectable using modern observational techniques.

The first effect, that is, periodic changes in the apparent system velocity, is the same effect as seen in a single star. However, it may be harder to detect for three reasons: (1) the binary system is usually more massive than a single star of the same magnitude, (2) extremely short-period planetary orbits (to which system velocity measurements are most sensitive) are unstable around binaries, and (3) the presence of two sets of spectral lines may complicate the measurement, as in Section 4.2.2.2. Equation (4.9) shows that a Jupiter-mass planet with the shortest period stable orbit around a 10-day period binary causes a $2M_\odot$ binary to move about its barycenter by $\approx 40\,\mathrm{m\,s^{-1}}$, with the amplitude decreasing as the square root of planet's semimajor axis. Radial velocity observations with $20\,\mathrm{m\,s^{-1}}$ precision as demonstrated by Konacki's method (Konacki 2005a) can detect Jupiter-like planets in orbits of the size $\approx$4 AU or less, down to the critically stable orbit (Holman and Wiegert 1999).

The second observable effect is the additional light travel time as the binary system undergoes reflex motion caused by the planet. The magnitude of this effect is given by

$$\begin{aligned}\Delta t &= 2\frac{a_p M_p \sin i_p}{c M_b}\\ &= 0.95\,\mathrm{sec}\,\frac{\left(a_p/1\,\mathrm{AU}\right)\left(M_p/M_J\right)\sin i_p}{M_b/\mathrm{M}_\odot}.\end{aligned}$$

Following a similar derivation as that for finding the expected precision of eclipse-timing, the precision with which one can estimate the orbital phase of a binary, based on radial velocity measurements, is

$$\sigma_\phi = \frac{\sigma_{rv}}{\sqrt{\sum_i \left(\frac{\partial v_i}{\partial \phi}\right)^2}}. \tag{4.24}$$

The quantity σ_{rv} in this equation presents the precision of the radial velocity measurement and $\partial v_i / \partial \phi$ is the derivative of the model radial velocity curve with respect to orbital phase, evaluated at times t_i. The timing precision corresponding to the phase precision above is given by $\sigma_\phi / 2\pi = \sigma_t / P_b$, where P_b is the binary period.

Assuming a circular orbit for the binary, $v(t) \approx K \cos(2\pi t / P_b + \phi)$ where K is the semiamplitude of the RV signal. If N measurements (each with two measured velocities, one for each star) are approximately evenly distributed in phase,

$$\sigma_\phi = \frac{\sqrt{2}\sigma_{rv}}{\sqrt{(2N-12)}K}, \tag{4.25}$$

$$\sigma_t = \frac{P_b \sigma_{rv}}{\sqrt{2(2N-12)}\pi K}, \tag{4.26}$$

where 12 is the number of degrees of freedom for the model. If the lines from both stars are observed, the effective K is $K1 + K2$ and the resulting (1σ) minimum detectable mass is given by

$$M_p = 41.4\, M_J \frac{(\sigma_{rv}/20\,\mathrm{m\,s^{-1}})\,(P_b/10\,\mathrm{days})^{4/3}\,(M_b/\mathrm{M}_\odot)^{2/3}}{\sqrt{2N-12}\sin i_b\, \sin i_p\, (a_p/1\,\mathrm{AU})}, \tag{4.27}$$

where i_b and i_p are the inclinations of the binary and planet orbits with respect to the sky, respectively. Twenty-five 20 m s^{-1} radial velocity measurements of a 10-day binary of Sun-like stars detect moderate-mass brown dwarfs ($\approx$30 M_J) at innermost stable orbit. Objects at the planet/brown dwarf threshold of 13 M_J are only detectable in orbits larger than 0.82 AU. Alternatively, if only one set of lines are observed, the resulting expression becomes

$$M_p = 41.4\, M_J \left(1 + \frac{M_1}{M_2}\right) \frac{(\sigma_{rv}/20\,\mathrm{m\,s^{-1}})\,(P_b/10\,\mathrm{days})^{4/3}\,(M_b/\mathrm{M}_\odot)^{2/3}}{\sqrt{N-11}\sin i_b\, \sin i_p\, (a_p/1\,\mathrm{AU})}. \tag{4.28}$$

Here M_1 is the mass of the star whose lines are observed, and M_2 is that of the faint star.

High precision radial velocity observations are only possible on slowly rotating ($v \sin i < 10\,\mathrm{m\,s^{-1}}$) stars. Here v is the stellar rotational velocity, and i is the inclination of its spin vector. Measurements of more rapidly rotating stars are limited

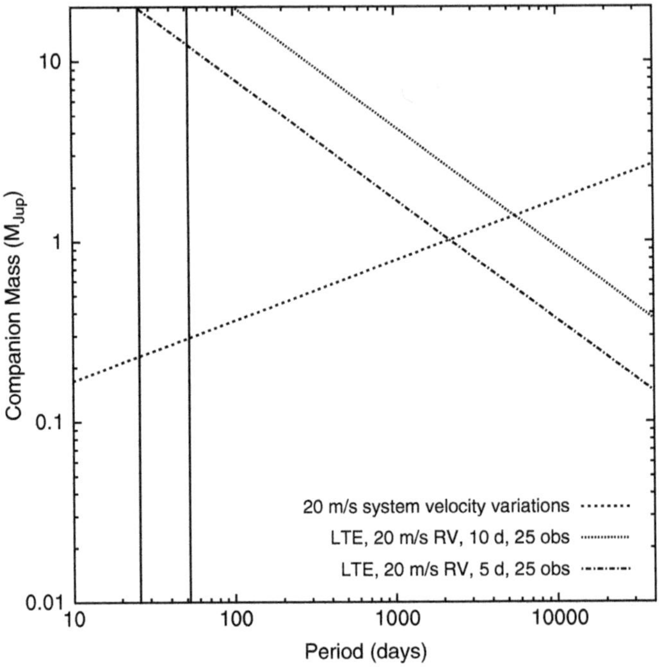

Fig. 4.10 Sensitivity of radial velocity measurements to circumbinary planets. The two *vertical lines* at the left represent the approximate critical orbits around 5-day (*to the left*) and 10-day period binaries. Shorter period companions have unstable orbits. Stars whose rotation rates are tidally locked to orbital periods less than about 5 days show sufficient rotational line broadening to prevent 20 m s^{-1} radial velocity precisions. The calculations assume the binary consists of two stars each with a mass of 1 M$_{\odot}$

by line broadening to levels worse than the nominal 20 m s^{-1} that has been used in the present chapter. This effect is particularly important for finding planets around short-period binaries, in which the stars' rotation rates are often tidally locked to the binary orbital period (Fig. 4.10). These rotation rates limit the observed precisions for systems with periods approximately 5 days or less.

4.3.2 Eclipse Timing

Similar to the case of S-type binaries, Eclipse Timing can be used to detect planetary objects in P-type binary systems.

The amplitude of the eclipse timing variation effect is given by Eq. (4.13). As with RV measurements, there is a mass/inclination ambiguity; the following derivation assumes no correlation between binary and planet inclinations (Fig. 4.11).

Dividing the precision of an individual measurement by $N_{obs} - 6$, where N_{obs} is the number of observed eclipses and there are six parameters to a timing perturbation

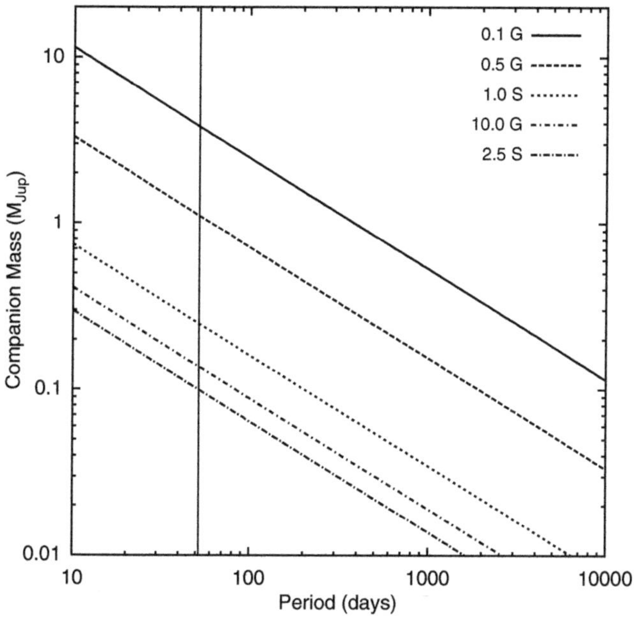

Fig. 4.11 Sensitivity of eclipse timing measurements to circumbinary planets. The *vertical line* represents the approximate critical orbit around a 10-day period binary. The calculations assume the binary consists of two stars each massing 1 $M_\odot$, 6 h eclipses, $N_{obs} = 25$ observations (150 total hours of data), $V = 10$ magnitudes, and 1σ detections. From top to bottom, lines show sensitivity for $D = 0.1$ m on the ground, $D = 0.5$ m on the ground, $D = 1.0$ m in space (i.e., Kepler), $D = 10$ m on the ground, and $D = 2.5$ m in space (HST, SOFIA)

fit (two periods, the orbital eccentricity, angle of periastron, epoch of periastron, and the ratio of the mass of the planet to the total mass of the binary), converting F_0 to V magnitude, and combining Eq. (4.21) with that for the timing effect of reflex motion (Eq. 4.24) gives a minimum detectable companion mass of

$$M_p = 0.19 M_J \sqrt{\frac{k\,(\tau/1\,\mathrm{hr})}{fgh^2\,(N_{obs}-6)}\,\frac{M_b/\mathrm{M}_\odot}{(a_p/1\,\mathrm{AU})\sin i_p}\left[\frac{10^{(V-12)/2.5}}{(D/1\,\mathrm{m})^2}+\frac{f}{(D/1\,\mathrm{m})^{4/3}}\right]^{1/2}}. \tag{4.29}$$

Figure 4.12 compares the sensitivity of RV and eclipse timing for circumbinary planets. Eclipse timing is more sensitive to long period planets, while RV studies can be more generally applied, as they do not require the special circumstances of an edge-on binary orbit.

One might also inquire about the sensitivity of this technique to outer planets in systems comprised of a single star and a transiting "hot" Jupiter. In this case, $h \approx 2R_p^2/R_{star}^2 \approx 0.02$ and $k \approx 2R_p/(R_{star} + R_p) \approx 0.18$, the "binary" is half as massive, and the eclipse duration is half as long. The companion sensitivity drops by a factor of 8, and the technique is (barely) in the range of detecting additional

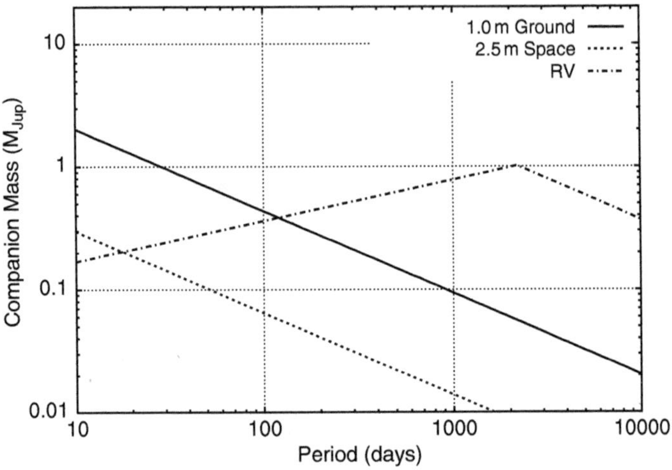

Fig. 4.12 Sensitivity to circumbinary planets, comparing radial velocity and eclipse timing techniques. All calculations assume solar-mass stars. Radial velocity assumes 20 m s^{-1} precision. Both the system velocity and apparent period variation observables are included for the sensitivity curve, the latter assumes a 5 day binary orbital period. Eclipse timing assumes either a 1 m ground-based photometric telescope or 2.5 m space-based telescope (such as HST or SOFIA), observing a $V = 10$ magnitude system, and 25 observations

companions of planet mass. However, for the typically $V = 10$ magnitude transiting planet systems being discovered, 3 m s^{-1} radial velocity observations are more sensitive than half-meter telescope transit timing for companions with periods up to 60 years; even for observatories such as HST and SOFIA (for which scintillation noise is small or zero), this transition occurs at 15 year period companions.

It should be noted that the above description does not account for the possibility of resonant orbits, for which timing perturbations can be greatly enhanced by many-body dynamics – we have assumed independent Keplerian orbits for the subsystems. Resonant effects on timing perturbations have been considered by Holman and Murray (2005) and Agol et al. (2005).

References

E. Agol, J. Steffen, R. Sari, and W. Clarkson. On detecting terrestrial planets with timing of giant planet transits. *MNRAS*, 359:567–579, May 2005. doi: 10.1111/j.1365-2966.2005.08922.x.

J. H. Applegate. A mechanism for orbital period modulation in close binaries. *ApJ*, 385:621–629, February 1992.

D. Benest. Planetary orbits in the elliptic restricted problem. I – The Alpha Centauri system. *A&A*, 206:143–146, November 1988.

D. Benest. Planetary orbits in the elliptic restricted problem. II – The Sirius system. *A&A*, 223: 361–364, October 1989.

D. Benest. Stable planetary orbits around one component in nearby binary stars. II. *Celestial Mechanics and Dynamical Astronomy*, 56:45–50, June 1993.

D. Benest. Planetary orbits in the elliptic restricted problem. III. The η Coronae Borealis system. *A&A*, 314:983–988, October 1996.

D. Benest. Planetary orbits in the elliptic restricted problem. V. The ADS 11060 system. *A&A*, 400:1103–1111, March 2003.

A. P. Boss. Possible rapid gas giant planet formation in the solar nebula and other protoplanetary disks. *ApJ*, 536:L101–L104, June 2000. doi: 10.1086/312737.

R. A. Broucke. Stable orbits of planets of a binary star system in the three-dimensional restricted problem. *Celestial Mechanics and Dynamical Astronomy*, 81:321–341, December 2001.

R. P. Butler, G. W. Marcy, E. Williams, C. McCarthy, P. Dosanjh, and S. S. Vogt. Attaining Doppler precision of 3 M s-1. *PASP*, 108:500–509, June 1996.

B. Campbell, G. A. H. Walker, and S. Yang. A search for substellar companions to solar-type stars. *ApJ*, 331:902–921, August 1988. doi: 10.1086/166608.

G. Chauvin, A. Lagrange, S. Udry, T. Fusco, F. Galland, D. Naef, J. Beuzit, and M. Mayor. Probing long-period companions to planetary hosts. VLT and CFHT near infrared coronographic imaging surveys. *A&A*, 456:1165–1172, September 2006. doi: 10.1051/0004-6361:20054709.

M. M. Colavita. Measurement of the atmospheric limit to narrow angle interferometric astrometry using the Mark-III Stellar interferometer. *A&A*, 283:1027–1036, March 1994.

M. M. Colavita. Fringe visibility estimators for the Palomar testbed interferometer. *PASP*, 111: 111–117, January 1999. doi: 10.1086/316302.

M. M. Colavita, J. K. Wallace, B. E. Hines, Y. Gursel, F. Malbet, D. L. Palmer, X. P. Pan, M. Shao, J. W. Yu, A. F. Boden, P. J. Dumont, J. Gubler, C. D. Koresko, S. R. Kulkarni, B. F. Lane, D. W. Mobley, and G. T. van Belle. The Palomar testbed interferometer. *ApJ*, 510:505–521, January 1999.

S. Desidera, R. Gratton, R. Claudi, M. Barbieri, G. Bonanno, M. Bonavita, R. Cosentino, M. Endl, S. Lucatello, A. F. Martinez Fiorenzano, F. Marzari, and S. Scuderi. Searching for planets around stars in wide binaries, pages 119–126, February 2006.

L. R. Doyle, H. Deeg, J. M. Jenkins, J. Schneider, Z. Ninkov, R. Stone, H. Gotzger, B. Friedman, J. E. Blue, and M. F. Doyle. Detectability of jupiter-to-brown-dwarf-mass companions around small eclipsing binary systems. In *ASP Conf. Ser. 134: Brown Dwarfs and Extrasolar Planets*, pages 224–231, 1998.

R. Dvorak. Planetary orbits in double star systems. *Oesterreichische Akademie Wissenschaften Mathematisch naturwissenschaftliche Klasse Sitzungsberichte Abteilung*, 191:423–437, 1982.

A. Eggenberger, S. Udry, T. Mazeh, Y. Segal, and M. Mayor. No evidence of a hot Jupiter around HD 188753 A. *A&A*, 466:1179–1183, May 2007. doi: 10.1051/0004-6361:20066835.

H. Frieboes-Conde and T. Herczeg. Period variations of fourteen eclipsing binaries with possible light-time effect. *A&AS*, 12:1–78, October 1973.

W. I. Hartkopf, B. D. Mason, and C. E. Worley. Sixth catalog of orbits of visual binary stars. http://www.ad.usno.navy.mil/wds/orb6/orb6.html, 2001.

A. P. Hatzes and G. Wuchterl. Astronomy: Giant planet seeks nursery place. *Nature*, 436:182–183, July 2005. doi: 10.1038/436182a.

A. P. Hatzes, W. D. Cochran, M. Endl, B. McArthur, D. B. Paulson, G. A. H. Walker, B. Campbell, and S. Yang. A Planetary Companion to γ Cephei A. *ApJ*, 599:1383–1394, December 2003. doi: 10.1086/379281.

M. J. Holman and N. W. Murray. The use of transit timing to detect terrestrial-mass extrasolar planets. *Science*, 307:1288–1291, February 2005. doi: 10.1126/science.1107822.

M. J. Holman and P. A. Wiegert. Long-term stability of planets in binary systems. *AJ*, 117: 621–628, January 1999.

J. B. Irwin. The determination of a light-time orbit. *ApJ*, 116:211–217, July 1952.

H. Jang-Condell. Constraints on the formation of the planet around HD188753A. *ApJ*, 654: 641–649, January 2007. doi: 10.1086/509494.

M. Konacki. Precision radial velocities of double-lined spectroscopic binaries with an iodine absorption cell. *ApJ*, 626:431–438, June 2005a. doi: 10.1086/429880.

M. Konacki. An extrasolar giant planet in a close triple-star system. *Nature*, 436:230–233, July 2005a. doi: 10.1038/nature03856.
M. Konacki. Precision radial velocities of double-lined spectroscopic binaries with an iodine absorption cell. *ApJ*, 626:431–438, June 2005b.
M. Konacki. Precision radial velocities of double-lined binary stars and the spectroscopic follow-up of circumbinary transiting planet candidates. 253:141–147, February 2009. doi: 10.1017/S1743921308026331.
R. L. Kurucz. The Kurucz Smithsonian atomic and molecular database. pages 205–+, 1995.
A. Lagrange, H. Beust, S. Udry, G. Chauvin, and M. Mayor. New constrains on Gliese 86 B. *A&A*, 459:955–963, December 2006. doi: 10.1051/0004-6361:20054710.
B. F. Lane and M. W. Muterspaugh. Differential astrometry of subarcsecond scale binaries at the Palomar testbed interferometer. *ApJ*, 601:1129–1135, February 2004.
B. F. Lane, M. M. Colavita, A. F. Boden, and P. R. Lawson. Palomar testbed interferometer: update. In *Proc. SPIE Vol. 4006, p. 452–458, Interferometry in Optical Astronomy, Pierre J. Lena; Andreas Quirrenbach; Eds.*, pages 452–458, July 2000.
J. Laskar, F. Joutel, and P. Robutel. Stabilization of the earth's obliquity by the moon. *Nature*, 361: 615–617, February 1993.
P. R. Lawson, editor. *Principles of Long Baseline Stellar Interferometry*, 2000.
J. J. Lissauer. Planet formation. *ARA&A*, 31:129–174, 1993. doi: 10.1146/annurev.aa.31.090193.001021.
G. W. Marcy and R. P. Butler. Precision radial velocities with an iodine absorption cell. *PASP*, 104:270–277, April 1992.
M. Mugrauer and R. Neuhäuser. Gl86B: a white dwarf orbits an exoplanet host star. *MNRAS*, 361: L15–L19, July 2005. doi: 10.1111/j.1745-3933.2005.00055.x.
M. W. Muterspaugh, B. F. Lane, S. R. Kulkarni, B. F. Burke, M. M. Colavita, and M. Shao. Limits to tertiary astrometric companions in binary systems. *ApJ*, 653, 1469–1479, November 2006. doi: 10.1086/508743.
A. F. Nelson. Planet formation is unlikely in equal-mass binary systems with A ˜ 50 AU. *ApJ*, 537: L65–L68, July 2000.
J. Norwood and N. Haghighipour. On the stability of υ Andromedae extrasolar planetary system; an S-Type binary-planetary system with more than one planet. *BAAS*, 34:892, 2002.
E. Pfahl. Cluster origin of the triple star HD 188753 and its planet. *ApJ*, 635:L89–L92, December 2005. doi: 10.1086/499162.
E. Pfahl and M. Muterspaugh. Impact of stellar dynamics on the frequency of giant planets in close binaries. *ApJ*, 652:1694–1697, December 2006. doi: 10.1086/508446.
B. Pichardo, L. S. Sparke, and L. A. Aguilar. Circumstellar and circumbinary discs in eccentric stellar binaries. *MNRAS*, 359:521–530, May 2005. doi: 10.1111/j.1365-2966.2005.08905.x.
E. Pilat-Lohinger and R. Dvorak. Stability of S-type orbits in binaries. *Celestial Mechanics and Dynamical Astronomy*, 82:143–153, 2002.
E. Pilat-Lohinger, B. Funk,, and R. Dvorak. Stability limits in double stars. A study of inclined planetary orbits. *A&A*, 400:1085–1094, March 2003.
S. F. Portegies Zwart and S. L. W. McMillan. Planets in triple star systems: The case of HD 188753. *ApJ*, 633:L141–L144, November 2005. doi: 10.1086/498302.
D. Queloz, M. Mayor, L. Weber, A. Blécha, M. Burnet, B. Confino, D. Naef, F. Pepe, N. Santos, and S. Udry. The CORALIE survey for southern extra-solar planets. I. A planet orbiting the star Gliese 86. *A&A*, 354:99–102, February 2000.
G. Rabl and R. Dvorak. Satellite-type planetary orbits in double stars – A numerical approach. *A&A*, 191:385–391, February 1988.
M. Shao and M. M. Colavita. Potential of long-baseline infrared interferometry for narrow-angle astrometry. *A&A*, 262:353–358, August 1992.
M. Shao and D. H. Staelin. First fringe measurements with a phase-tracking stellar interferometer. *Applied Optics*, 19:1519–1522, May 1980.

M. Shao, T. R. Livermore, D. M. Wolff, J. W. Yu, and M. M. Colavita. An overview of the space interferometry mission, SIM. *Bulletin of the American Astronomical Society*, 27:1384, December 1995.

P. Thébault, F. Marzari, H. Scholl, D. Turrini, and M. Barbieri. Planetary formation in the γ Cephei system. *A&A*, 427:1097–1104, December 2004. doi: 10.1051/0004-6361:20040514.

P. Thébault, F. Marzari, and H. Scholl. Relative velocities among accreting planetesimals in binary systems: The circumprimary case. *Icarus*, 183:193–206, July 2006. doi: 10.1016/j.icarus.2006.01.022.

E. Toyota, Y. Itoh, H. Matsuyama, S. Urakawa, S. Kimura, Y. Oasa, T. Mukai, and B. Sato. Search for extrasolar planets in binary systems. In *Protostars and Planets V*, eds. B. Reipurth, D. Jewitt & K. Keil, p. 8247, 2005.

S. Udry, A. Eggenberger, M. Mayor, T. Mazeh, and S. Zucker. Planets in multiple-star systems:properties and detections. pages 207–214, August 2004.

C. A. Watson and V. S. Dhillon. The effect of star-spots on eclipse timings of binary stars. *MNRAS*, 351:110–116, June 2004.

J. Woltjer. On a special case of orbit determination in the theory of eclipsing variables. *Bulletin of the Astronomical Institutes of the Netherlands*, 1:93–94, June 1922.

A. T. Young. Photometric error analysis. VI. Confirmation of Reiger's theory of scintillation. *AJ*, 72:747–753, August 1967. doi: 10.1086/110303.

S. Zucker and T. Mazeh. Study of spectroscopic binaries with TODCOR. 1: A new two-dimensional correlation algorithm to derive the radial velocities of the two components. *ApJ*, 420:806–810, January 1994. doi: 10.1086/173605.

S. Zucker, T. Mazeh, N. C. Santos, S. Udry, and M. Mayor. Multi-order TODCOR: Application to observations taken with the CORALIE echelle spectrograph. II. A planet in the system HD 41004. *A&A*, 426:695–698, November 2004. doi: 10.1051/0004-6361:20040384.

Chapter 5
The SARG Planet Search

S. Desidera, R. Gratton, M. Endl, A.F. Martinez Fiorenzano, M. Barbieri, R. Claudi, R. Cosentino, S. Scuderi, and M. Bonavita

5.1 Introduction

The search for planets in multiple systems allows to improve our knowledge of planet formation and evolution. On one hand, the frequency of planets in binary systems has a strong effect on the global frequency of planets, as more than half of solar-type stars are in binary or multiple systems (Duquennoy and Mayor 1991). On the other hand, the properties of planets in binaries, and their differences with the properties of the planets orbiting single stars, would shed light on the effects caused by the presence of the companion stars. Indeed, the first analysis of the properties of planets in binaries showed the occurrence of some differences with respect to those orbiting single stars (Zucker and Mazeh 2002; Eggenberger et al. 2004).

The search for planets in binaries can follow two complementary approaches. The first approach is to perform dedicated surveys looking for planets in binary systems. Several programs currently in progress, focusing on different types of binaries

S. Desidera (✉), R. Gratton, R. Claudi, and M. Bonavita
INAF, Osservatorio Astronomico di Padova, Padova 35122, Italy
e-mail: silvano.desidera@oapd.inaf.it; raffael.gratton@oapd.inaf.it; riccardo.claudi@oapd.inaf.it; mariangela.bonavita@oapd.inaf.it

M. Endl
McDonald Observatory, The University of Texas at Austin, Austin, TX 78734
e-mail: mike@astro.as.utexas.edu

A.F.M. Fiorenzano and R. Cosentino
INAF, Fundacion Galileo Galilei, Santa Cruz de La Palma, Tenerife 38700, Spain
e-mail: fiorenzano@tng.iac.es; rco@ct.astro.it

M. Barbieri
LAM, Observatoire de Marseille, Marseille 13388, cedex 13, France
e-mail: marco.barbieri@institutoptique.fr

R. Cosentino and S. Scuderi
INAF, Osservatorio Astrofisico di Catania, Catania 78 95123, Italy
e-mail: Salvo.Scuderi@oact.inaf.it

M. Bonavita
Dipartimento di Astronomia, Universitá di Padova, Padova 35122, Italy

N. Haghighipour (ed.), *Planets in Binary Star Systems*, Astrophysics and Space Science Library 366, DOI 10.1007/978-90-481-8687-7_5,

are described in this book. The second approach is to study the binarity of the hosts of planets discovered in general surveys, which include many binary stars in their lists in spite of some selection biases against them (Patience et al. 2002; Chauvin et al. 2006; Mugrauer et al. 2006).

In this chapter, we describe the first planet search entirely dedicated to binary systems, the survey on-going at TNG (Telescopio Nazionale Galileo) using the high resolution spectrograph SARG (Spettrografo Alta Risoluzione Galileo). The main goal of the project is to search for planets in binaries with similar stellar components and typical separations between 50 and 500 AU. The second major goal of the SARG survey is to search for abundance anomalies caused by the ingestion of planetary material by the central star.

5.2 Properties of Planets in Binary Systems

More than 40 planets have been found in binary or multiple systems (see Desidera and Barbieri (2007) for a recent compilation). We performed a statistical analysis of the properties of known planets in binaries and made a comparison with those orbiting single stars, based on the planet and stellar parameters listed in the Catalog of Nearby Exoplanets by Butler et al. (2006). Figure 5.1 shows the mass-ratio[1]

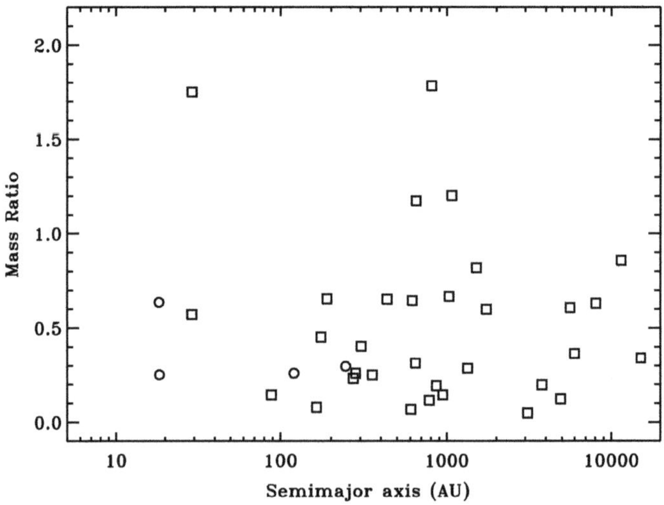

Fig. 5.1 Mass-ratio vs semimajor axis of binaries with planets. *Open circles* represent binaries for which orbits are available. *Open squares* correspond to those for which only the separation is available. Figure from Desidera and Barbieri (2007)

[1] The mass-ratio is defined here as $M_{\text{companion}}/M_{\text{planethost}}$. In nearly all cases, the companion is less massive than the planet host.

vs semimajor axis for stars with planets in binary (or multiple star) systems. For hierarchical triple systems in which the planet orbits the isolated companion,[2] the masses of the farther companions are summed. It results from Fig. 5.1 that planets might exist in binaries with very different properties. In the case of low-mass companions at a very wide separation (e.g., larger than 1,000 AU), the dynamical effects of the companion on the formation and evolution of the planetary system might be very limited, while in the cases of very tight binaries with separation smaller than about 20 AU (e.g., γ Cep, GL 86, HD 41004) the presence of the planet represents a challenge for the current models of planet formation (Hatzes and Wuchterl 2005). Figure 5.1 also shows a broad range of companion mass, from masses similar or even larger than the planet-hosting star down to substellar masses in a few cases. In most cases, the companions of planet hosts are low-mass stars.

Figure 5.2 shows the projected mass vs orbital period of extrasolar planets orbiting single stars as well as the components of binary stars. As shown in this figure, broad ranges of masses and periods exist for these objects. It appears that massive short period planets are mostly found in binary systems.

The large variety of the binary properties, in terms of separation and mass-ratio, makes necessary the use of some quantitative estimate of the dynamical effects of the binary companion(s). We use the critical semimajor axis a_{crit} (Holman and Wiegert 2001) as a reference value to determine the dynamical stability of a planet

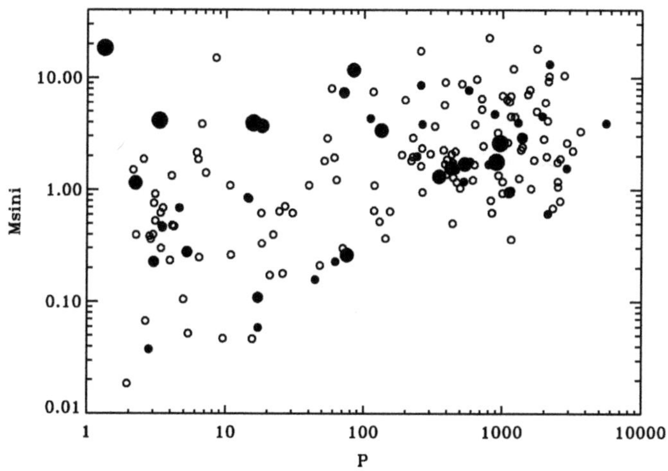

Fig. 5.2 Projected mass vs orbital period of extrasolar planets. *Open circles*: single stars; *filled circles*: binary stars. The size of the symbol is proportional to the critical semimajor axis for dynamical stability (larger symbols refer to the tighter binaries). Figure from Desidera and Barbieri (2007)

[2] 16 Cyg (Patience et al. 2002), HD 178911 (Tokovinin et al. 2000) plus HD 40979 (Mugrauer et al. 2007a), HD65216 (Mugrauer et al. 2007b) discovered after the publication of Desidera and Barbieri (2007).

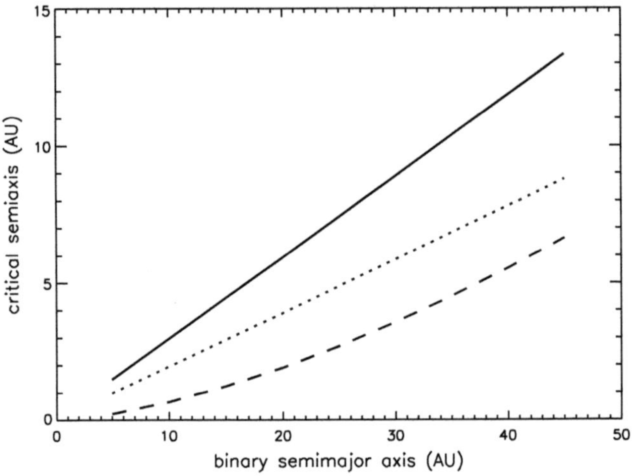

Fig. 5.3 Values of the semimajor axis for dynamical stability $a_{\rm crit}$ (*solid line*) and limit of the region in which the encounter velocities of planetesimal is small enough to allow the accretion of kilometer-sized planetesimals $a_{\rm cross}$ (*dashed line*) and the radius of tidal truncation of the circumstellar disk $a_{\rm tid}$ (*dotted line*), versus binary semimajor axis, for e = 0.31, $M_{\rm com}$ = 0.5 $M_\odot$, and $M_{\rm obj} = 1 M_\odot$. Figure from Desidera and Barbieri (2007)

in a binary system. This quantity is a physical quantity that represents the dynamical effects due to a companion on planet formation and stability, and includes the orbital parameters and mass-ratio.[3] For this reason, $a_{\rm crit}$ is more useful than the projected binary separation. Other useful quantities are the radius of tidal truncation of the circumstellar disk $a_{\rm tid}$ (Pfhal and Mutherspaugh 2006) and the limit of the region in which the encounter velocities of planetesimal is small enough to allow the accretion of kilometer-sized bodies $a_{\rm cross}$ (Thebault et al. 2006). Figure 5.3 shows a comparison between these quantities.

The critical semimajor axis $a_{\rm crit}$ was used by us (Desidera and Barbieri 2007) to divide the sample of binary stars with planets, according to the relevance of the dynamical effects. We define a 'tight' (or 'close') binary as one with $a_{\rm crit} < 75$ AU and a 'wide' binary as one with $a_{\rm crit} > 75$ AU. The boundary between tight and wide binaries (i.e., $a_{\rm crit} = 75$ AU) corresponds to a projected separation of about 200–300 AU depending on the mass-ratio.

[3] The critical semimajor axis for dynamical stability of a planet in a binary as defined by Holman and Wiegert (2001) is given by

$$a_{\rm crit} = (0.464 - 0.380\mu - 0.631e_{\rm bin} + 0.586\mu e_{\rm bin})\, a_{\rm bin} + \left(0.150e_{\rm bin}^2 - 0.198\mu e_{\rm bin}^2\right) a_{\rm bin}. \quad (5.1)$$

In this equation, $a_{\rm bin}$ and $e_{\rm bin}$ are the binary semimajor axis and eccentricity, and μ is the mass-ratio defined as $\mu = M_{\rm comp}/(M_{\rm obj} + M_{\rm comp})$. The quantity $M_{\rm obj}$ is the mass of the object for which the critical semimajor axis for dynamical stability is computed and $M_{\rm comp}$ is the mass of its companion. The critical semimajor axis given by Eq. (5.1) refers to a planet in a circular orbit on the same plane as that of the binary.

As discussed in Section 5.1, the study of the properties of planets in binaries and the search for differences between the characteristics of these objects and those of planets orbiting single stars, is crucial to the understanding of the role of binarity in planet formation and evolution. Such a study was first carried out by Zucker and Mazeh (2002) and Eggenberger et al. (2004), and more recently on an expanded sample of planet by us (Desidera and Barbieri 2007). We have tested the hypothesis that the parameters of planets (mass, period, eccentricity) in tight and wide binaries, as well as in single stars, can be drawn from the same parent distribution, using the Kolmogorov–Smirnov and Mann–Whitney U tests.

The following results were found (see Fig. 5.4):

- The mass distribution of short period ($P < 40$ days) planets in tight binaries is significantly (>99%) different from that of planets orbiting single stars as well as the components of wide binaries. Massive, short period planets are mostly found in tight binaries (Figs. 5.2–5.4a). This somewhat resembles the evidence that short-period spectroscopic binaries have, in most cases, a farther companion (Tokovinin et al. 2006).
- The mass distribution of planets with periods longer than 40 days in tight and wide binaries and around single stars are not significantly different (Fig. 5.4b).
- The differences in period distribution are also not highly significant (Fig. 5.4c). However, there is a marginal indication of the lack of long period planets in tight binaries (no planet with period longer than 1,000 days).
- The eccentricity distribution of planets in tight binaries with periods longer than 40 days is not significantly different from those orbiting single stars. On the other hand, there is a marginal indication of larger eccentricity of planets in wide binaries (Figs. 5.4d–5.5).[4]
- The frequency of systems with more than one planet around the components of wide binaries is similar to that of planets orbiting single stars. No multiple planets have been yet discovered around the components of tight binaries. The small number of planets in tight binaries (and their 15% probability of occurrence) makes the lack of multiple planets in tight binaries insignificant.

We conclude that planets in close binaries have different characteristics compared to those orbiting single stars and components of wide binaries.

The understanding of the formation mechanism of the planets in close binaries is a key problem. One possibility is that these planets formed before the binary configuration was modified by stellar encounters in the native star cluster (Pfhal and Mutherspaugh 2006). The alternative is that planets do form in close binaries in spite of the seemingly unfavorable conditions.

The mass and period distributions of planets in wide binaries do not have statistically significant difference with to those of planets orbiting single stars. The only

[4] The very recent discovery of planets in highly eccentric orbits, orbiting stars in binary systems (Tamuz et al. 2007), adds further support to the link between binarity and extreme planet eccentricities.

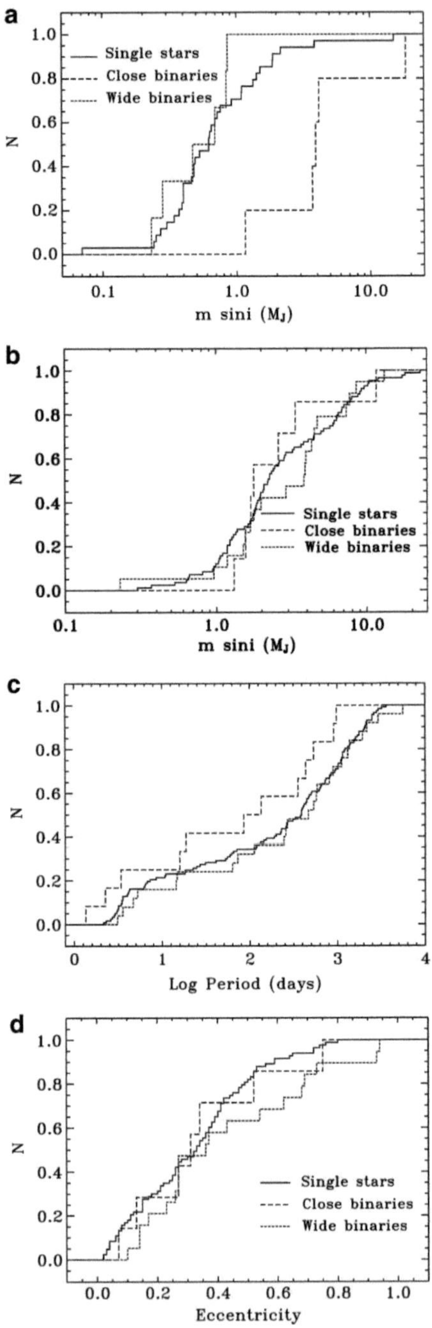

Fig. 5.4 Cumulative distributions of planets parameters for planets orbiting single stars (*continuous lines*), components of wide binaries (*dotted lines*), and components of tight binaries (*dashed lines*). (**a**): mass distribution of planets with periods shorter than 40 days. (**b**): mass distribution of planets with periods longer than 40 days. (**c**): period distribution. (**d**): eccentricity distribution of planets with periods longer than 40 days. Figure from Desidera and Barbieri (2007)

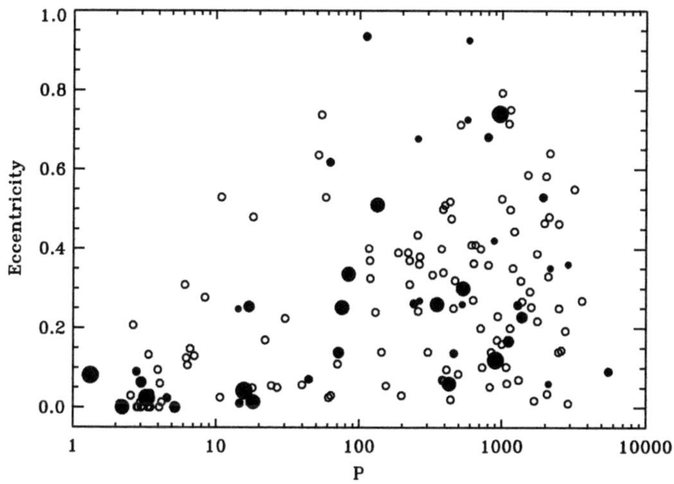

Fig. 5.5 Eccentricity vs orbital period for planets in binaries (*filled circles*) and those orbiting single stars (*empty circles*). Different sizes of filled circles refer to different periastron distances of the binary orbit (larger sizes: closer orbits). Figure from Desidera and Barbieri (2007)

marginally significant difference between planets orbiting single stars and those orbiting the components of wide binaries is in the planet's orbital eccentricity. In any case, large eccentricities are not unique to planets in binaries, and the possible differences in eccentricity appears to be limited to the range $e \geq 0.5$–0.6. This indicates that there are mechanism(s) generating planet eccentricity up to 0.4–0.5 that are independent of the binarity of the planet-hosting star, and are characteristic of formation and evolution of a planetary system (e.g., disk–planet interactions (Tremaine and Zakamska 2004), planet–planet scattering (Marzari and Weidenschilling 2002)). These probably act during or shortly after planet formation. Further eccentricity enhancements, possibly linked to the presence of a companion, might take place at later epochs. In fact, Takeda et al. (2007) noted that most high-eccentricity planets orbit old stars (ages > 5 Gyr). Mechanisms that require long time scales to modify planetary orbits, such as Kozai oscillations (Wu and Murray 2003) and chaotic evolution of planetary orbits induced by dynamical perturbations (Benest and Gonczi 2003; Marzari et al. 2005) then seem favored.

In summary, these results indicate that a companion at large separation (≥300–500 AU, exact limit not well constrained) probably does not affect planet formation around the other component too much, whereas the effect of the companion is much more relevant at small separations, causing significant differences in the physical properties of the planets. The exploration of the frequency and properties of planets at intermediate binary separations (100–300 AU), the range of a large fraction of the binaries of the SARG planet search, is important to establish the separation required to show the peculiar features of planet frequency and characteristics. This is a separation range where some planet formation models predicts little or no effect (e.g., Thebault et al. 2006, a prediction to be tested observationally. This is the main goal of the SARG planet search project.

5.3 Binary Systems as a Tool to Study the Ingestion of Planetary Material by the Central Star

The evidence for a high metal content in stars harboring planets is becoming stronger as planet discoveries cumulate and suitable planet hosts and control samples are studied using homogeneous procedures (Santos et al. 2004; Fischer and Valenti 2005). Two alternative hypotheses have been proposed to explain these observations: either the high metallicity is responsible for the presence of planets, making their formation easier (see, e.g., Ida and Lin 2004), or the planets are the cause of the high metallicity, because of pollution of metal-rich planetary material onto the (outer region of the) central star (Gonzalez 1997).

Infall of planetesimals on the star during planet formation phase is generally expected on the basis of current models. The orbital migration proposed to explain the occurrence of the close-in giant planets found by radial velocity surveys also points to the infall of portions of the proto-planetary disk on the star.

Most of the accretion of planetary metal-rich material is expected to take place during the early phases of the planetary system lifetime. However, when a star is still in the phase of gravitational contraction, its convective zone is much thicker than that of main sequence stars (see, e.g., Murray et al. 2001). In this case, the metal-rich material should be uniformly distributed by convective mixing over a large portion of the star, resulting in a negligible photospheric chemical alteration even for rather large amounts of accreted material. Accretion at later times, when the star is approaching or has already reached the main sequence (stellar age larger than about 10 Myr for a solar-type star) is likely required to produce observable differences. The ingestion of planets scattered toward the star by dynamical interactions (Marzari and Weidenschilling 2002) might also produce metallicity enhancements at late phases.

Murray et al. (2001) found that the Sun should have ingested some $2M_{\oplus}$ of meteoritic material (about $0.4M_{\oplus}$ of iron) during its main-sequence lifetime, considering the drop of iron density in the asteroid region and the time distribution of the impact craters. This corresponds to a metallicity enhancement of 0.017 dex. Such a small difference in abundance is not detectable when considering a field star, for which no proper reference for the original unpolluted abundance is available. In binary systems and star clusters, instead such a reference is provided by the other companion/members of the system. Therefore, the comparison of the chemical composition of wide binaries is a very powerful approach to study the occurrence of planetary pollution, provided that differential abundance analysis with a precision of about 0.02 dex can be obtained. The occurrence of large amounts of accretion of metal-rich material is required if the high metallicity is the result of planets or planetesimal ingestion (Gonzalez 1997). In this case, some systematic difference is expected between members of a binary system with and without planetary companions. The SARG planet search, that includes both the radial velocity monitoring to search for planets and the differential abundance analysis, is ideal to test this hypothesis.

5.4 The SARG Sample

With the two science goals identified in Sections 5.1–5.3 (i.e., the search for planets in a sample of binaries at a separation at which dynamical effects may start to be relevant, and the search for abundance anomalies caused by the ingestion of metal-rich planetary material), we started a radial velocity (RV) survey of the components of wide binaries, a few years ago. We are using SARG, the high resolution spectrograph of the TNG (Gratton et al. 2001), equipped with an iodine cell to derive high precision RVs.

Our sample binaries were selected from the Hipparcos Multiple Star Catalog. We considered binaries in the magnitude range of $7.0 < V < 10.0$ and with a magnitude-difference-between-the-components of $\Delta V < 1.0$. The projected separations of our sample binaries were larger than 2 arcsec (to avoid contamination of the spectra), their parallaxes were larger than 10 mas, with errors smaller than 5 mas, $B - V > 0.45$, and spectral types later than F7. About 50 pairs of such wide binaries with mass-ratios close to 1 were selected.

Considering systems with similar components is crucial for the accuracy of the differential chemical abundance analysis. Figure 5.6 shows the distribution of the projected separations in AU. For most of the pairs, the separation is between 50 and 600 AU. Figure 5.7 shows the distribution of the V band magnitude differences between the components.

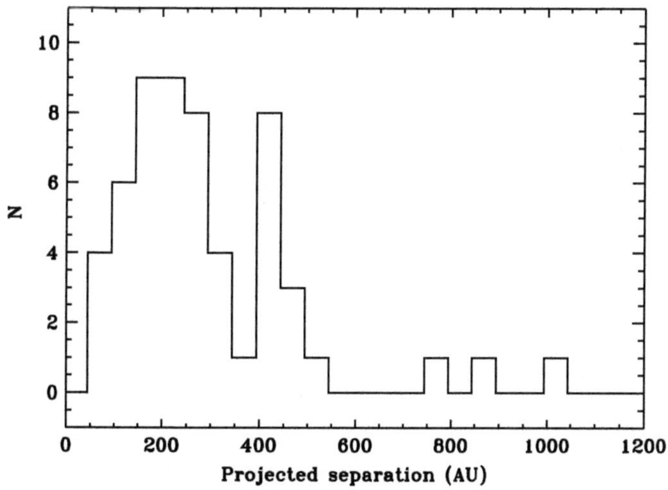

Fig. 5.6 Distribution of the projected separations of the binaries in the sample of the SARG survey

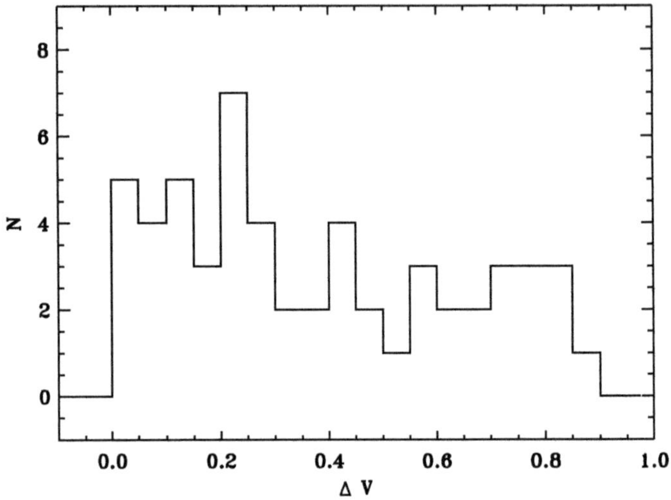

Fig. 5.7 Distribution of the visual magnitude differences of the binaries in the sample of the SARG survey

5.5 Observations

The observations used for the radial velocity determinations are acquired with the SARG spectrograph (Gratton et al. 2001) using the Yellow Grism, that covers the spectral range 4,600–7,900 Å without gaps, and the 0.25 arcsec slit. The resulting resolution is R = 150,000 (two pixels sampling). The iodine cell is inserted into the optical path, superimposing a dense forest of absorption lines used as reference spectrum for the radial velocity determination (Butler et al. 1996). Exposure times are fixed in most cases at 15 min, to reduce the errors in barycentric correction caused by the lack of knowledge of the exact flux mid time of the exposure. A high signal to noise spectrum without the iodine cell was also acquired for all the targets, to be used for the abundance analysis (see Section 5.6) and as template for the radial velocity determination (see Section 5.7).

During the observations, the slit is usually oriented perpendicularly to the separation of the components to minimize the contamination of the spectra by the companion. The closest pairs (separation 2–3 arcsec) are observed only in good seeing conditions. In spite of these efforts, some residual contamination of the spectra is present in a few cases. This issue is discussed in Section 5.8. The survey was recently completed (about 20 spectra per object acquired on average) and further observations are limited to the follow-up of the planet candidates.

5.6 Abundance Analysis

The abundance analysis of about half of the pairs of the SARG survey has been published in Desidera et al. (2004a). We also studied 33 pairs of Southern declination observed with the FEROS spectrograph at ESO-La Silla, selected with similar criteria (Desidera et al. 2006). Taking into account the small overlap between the two samples, we have in hand the results for 50 pairs.

Performing a line-by-line differential analysis (Fig. 5.8) and exploiting the physical link between the components (same distance from the Sun), we found that errors in estimating the difference of iron content between the two components of about 0.02 dex can be achieved for pairs with temperature differences smaller than 300–400 K and slow-rotating components with effective temperatures in the range 5,500–6,300 K. This is adequate for detailed study of chemical alterations in the external convective layer.

Most of the pairs have abundance difference smaller than 0.03 dex (Fig. 5.9). We found one case (HIP 64030 = HD 113984) with a large (0.25 dex) abundance difference, with the primary showing lower metal abundances. The primary of this binary appears to be a blue straggler, and the abundance difference might be due to the peculiar evolution of the star (see Section 5.6.1). A few other pairs show small abundance differences ($\leq$0.09 dex). In a few cases these differences suggest the ingestion of a small amount of metal rich material, but in others they are likely spurious, because of the large temperature difference between the components, the high level of magnetic activity, that might cause alterations in the stellar atmosphere

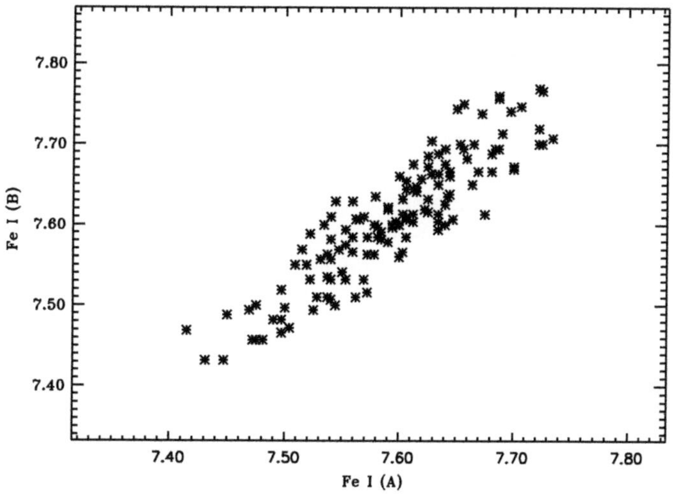

Fig. 5.8 Iron abundance derived for each line of the components of HIP 114914 A and B. A clear correlation can be seen, indicating that the use of a line-by-line differential analysis significantly reduces the errors on abundance difference between the components. Figure from Desidera et al. (2006)

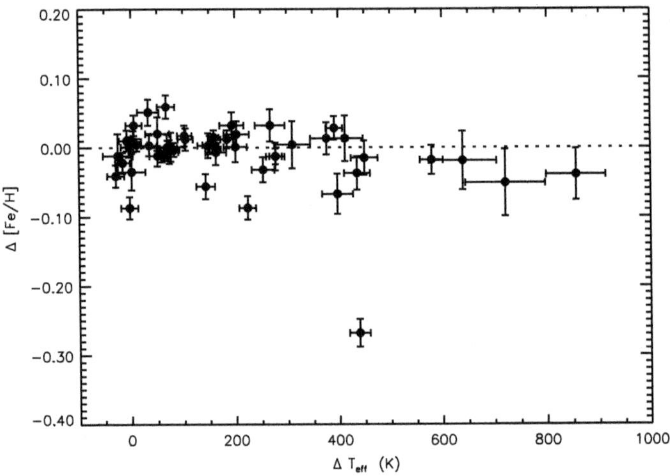

Fig. 5.9 Iron abundance difference (primary–secondary) between the components of pairs as a function of temperature difference for the pairs studied in Desidera et al. (2004, 2006)

or additional errors in our analysis because of intrinsic variability, or possible contamination of the spectra by an additional star in close orbit around one of the components. Some cases of abundance differences involving pairs with warm ($T_{\mathrm{eff}} \geq 6{,}000$ K) primaries might be due to the diffusion of heavy elements.

Figure 5.10 shows the amount of iron accreted by the nominally metal richer component of the binary to explain the observed abundance difference. For most of the slow-rotating stars warmer than 5,500 K, characterized by a thinner convective envelope and for which our analysis appears to be of higher accuracy, this is similar to the estimates of rocky material accreted by the Sun during its main sequence lifetime (about 0.4 Earth–masses of iron, Murray et al. 2001). We then conclude that the occurrence of large alterations in stellar abundances caused by the ingestion of metal rich, rocky material is not a common event. For at least 65% of the pairs with components warmer than 5,500 K, the limits on the amount of rocky material accreted by the program stars are comparable to the estimates of rocky material accreted by the Sun during its main-sequence lifetime.

5.6.1 *The Special Case of the Blue Straggler HD 113984*

The wide binary HIP64030 = HD 113984 is the only pair in our sample that shows a large (about 0.25 dex) iron content difference, with the primary showing lower metal abundances. The positions of the components on the color magnitude diagram suggest that the primary is a blue straggler (Note that this blue straggler itself is a spectroscopic binary). Therefore, the abundance difference may be somewhat linked to the peculiar evolutionary history of the system.

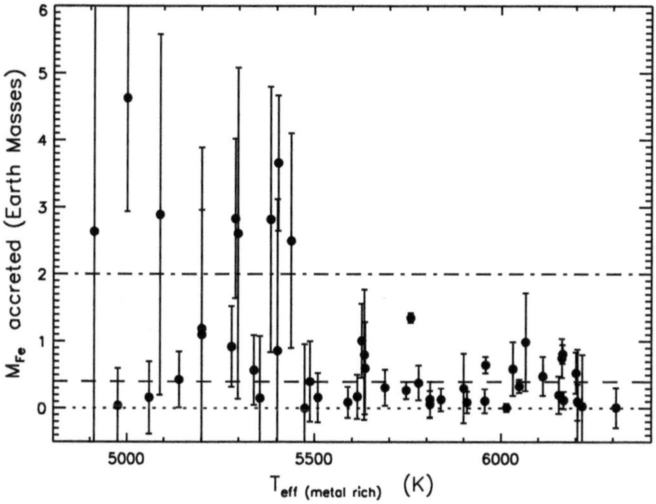

Fig. 5.10 Estimate of iron accreted by the metal-rich component of each pair as a function of its effective temperature, taking into account the mass of the mixing zone as in Murray et al. (2001). The less severe limits at lower effective temperatures are mostly due to the more massive convective zone of cool stars. The horizontal lines show the amount of iron expected to have been accreted by the Sun during the main sequence lifetime (0.4 $M_{\oplus}$, Murray et al. 2001), and the amount of iron corresponding to the upper limit on abundance difference between the inner and outer regions of the Sun according to helioseismology (2 $M_{\oplus}$, Winnick et al. 2002). The mass of meteoritic material is assumed to be about 5.5 times the mass of iron. Figure from Desidera et al. (2006)

The analysis of additional elements beside iron (Desidera et al. 2007) shows that the abundance difference for the elements studied increases with increasing condensation temperature, suggesting that accretion of chemically fractionated material might have occurred in the system (Fig. 5.11). Alteration of C and N likely due to CNO processing is also observed, as expected for the mass transfer process occurring during the formation of the blue straggler. We also showed that the blue straggler component is a spectroscopic binary with a period of 445 days and moderate eccentricity, as typical for field blue stragglers (Preston and Sneden 2000).

Two scenarios were explored to explain the observed abundance pattern. In the first, both the light-element and heavy-element abundance anomalies occur on the blue straggler. If this is the case, the dust-gas separation may have occurred in a circumbinary disk around the blue straggler and its expected white dwarf companion, as observed in several RV Tauri and post AGB binaries (Van Winckel 2003). In the second scenario, accretion of dust-rich material occurs on the secondary. This would also explain the anomalous carbon isotopic ratio of this star. Such a scenario requires that a substantial amount of mass, lost by the central star, to have been accreted by the wider component.

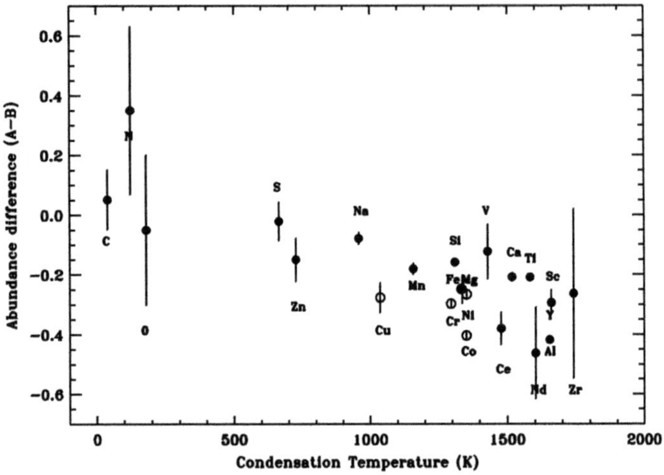

Fig. 5.11 Abundance difference for the components of HD 113984 as a function of the condensation temperature. Figure from Desidera and Barbieri (2007)

5.6.2 Abundance Difference Between Components of Binary Systems with Planetary Companions

The analysis of 50 pairs shown in Section 5.6 suggests that the frequency of pairs with large alterations in chemical composition is rather small. Therefore, it seems unlikely that the ingestion of planetary material can account for the strong correlation between the frequency of planets and metallicity. However, none of the stars studied by Desidera et al. (2004, 2006) have confirmed planet detections (most of the pairs of the FEROS sample are probably not being searched for planets). Therefore, to better test the hypothesis that the abundance of stars with planets is changed by accretion of metal-rich material (see Section 5.3), it is important to consider the abundance-difference between the components of binary systems for which one of the components has a planet. We consider only pairs with similar stars, as errors in differential chemical abundances becomes larger for large temperature difference (see discussion in Desidera et al. 2006).

Among the binary systems with planets, there are five pairs with mass-ratios between 0.8 and 1.2. Only for 16 Cyg high-precision differential abundance analysis between the components has been carried out. Laws and Gonzalez (2001) found a small abundance-difference of 0.025 dex, with the planet-hosting star (the secondary) being more metal-rich. However, Takeda (2005) did not confirm the reality of such a small abundance-difference.

For the pairs HD 80606/7, HD 99491/2 and ADS 16402, the standard abundance analysis does not reveal significant abundance-difference (see Table 5.1). For HD 20781 and its planet-hosting companion HD 20782, there are no high-resolution abundance analysis, and the abundance-difference derived from Strömgren photometry is not significant (errors about 0.1 dex).

Table 5.1 Abundance-difference between the components of binary planet hosts with similar components

System	Planet host	Δ [Fe/H]	Refs.
16 Cyg	B	-0.025 ± 0.009	Laws and Gonzalez (2001)
16 Cyg	B	0.00 ± 0.01	Takeda (2005)
HD 80606/7	A	-0.01 ± 0.11	Heiter and Luck (2003)
HD 80606/7	A	$+0.002 \pm 0.081$	Taylor (2005)
HD 99491/2	B	-0.02 ± 0.03	Valenti and Fischer (2005)
HD 99491/2	B	$+0.04 \pm 0.13$	Heiter and Luck (2003)
HD 99491/2	B	$+0.076 \pm 0.059$	Taylor (2005)
HD 20781/2	A	$+0.12 \pm 0.10$	Nordstrom et al. (2004)
ADS 16402 (HAT-P-1)	B	-0.01 ± 0.05	Bakos et al. (2007)
GSC 03413-00005 (XO-2)	B	$+0.02 \pm 0.03$	Burke et al. (2007)

In summary, there is currently no evidence for large ($\geq$0.1 dex) alterations of chemical abundances in the components of binary systems with/without planets. This supports the conclusion of our dedicated study on the abundance-difference between the components of binaries that large alteration of chemical abundance caused by the ingestion of planetary material are rare, if any.

5.7 Radial Velocities

High precision RVs for the stars in the SARG sample were determined using the AUSTRAL code (Endl et al. 2000) as described in Desidera et al. (2003). Typical errors are 1–3 m/s for bright stars observed as standards to monitor instrument performances (Fig. 5.12) and 3–10 m/s for the $V \sim$7–9 program stars.

5.7.1 Planet Candidates and Low Amplitude Variables

The RV time series were searched for periodic variations. A few interesting candidates emerged from the survey but they have not yet achieved the confidence level adequate to claim the planet detection. This can be attributed to the period longer than our current time baseline, or to the low amplitude of the radial velocity variations, which in two cases suggests multiple periods (i.e., two planets). Intensive radial velocity monitoring is still on-going on for the best candidates.

Some further stars show RV variability above internal errors. In most cases this can be explained by stellar activity jitter and residual contamination of the spectra from the companion (see Section 5.8). One case we investigated in detail is that of HD 219542B. The 2000–2002 data indicated a possible periodicity of 111 days with a significance of about 97% (Desidera et al. 2003). However, the continuation of the

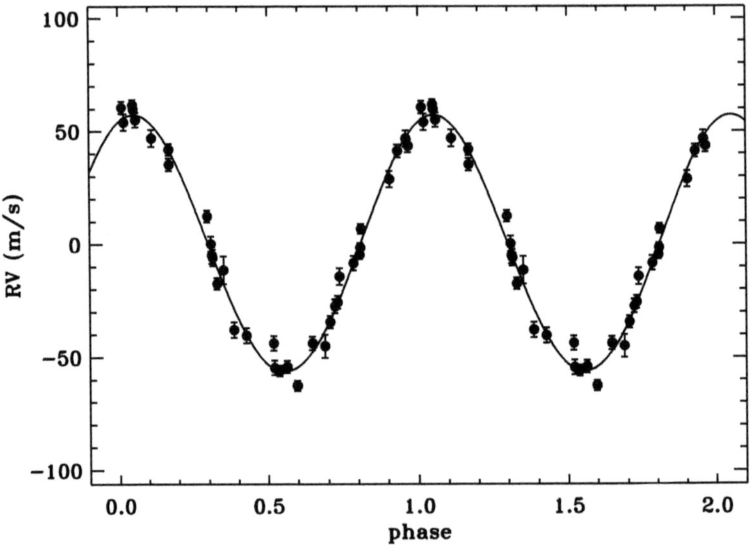

Fig. 5.12 Radial velocities of 51 Peg obtained with SARG phased to the known orbital period

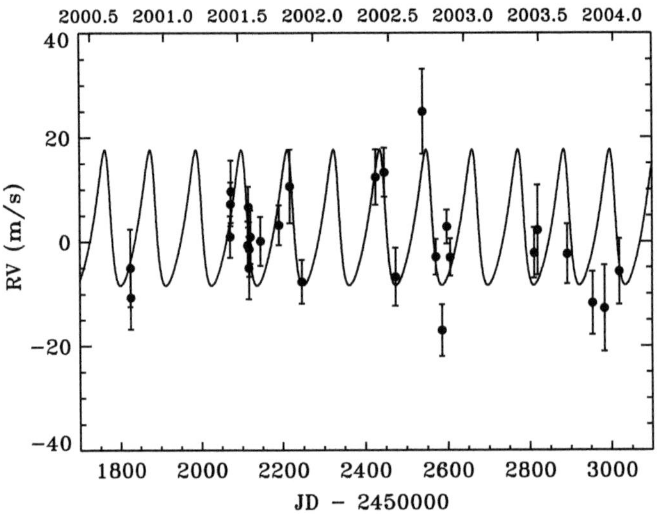

Fig. 5.13 Radial velocity curve for HD 219542 B. The data taken in the 2003 season do not follow the tentative orbital solution previously derived in Desidera et al. (2003) (overplotted as a *solid line*). Figure from Desidera et al. (2004b)

observations revealed that the RV variations are likely due to stellar activity (see Fig. 5.13, Desidera et al. 2004). In particular, the chromospheric emission measurements indicate that HD 219542 B underwent a phase of enhanced stellar activity in 2002 while the activity level has been lower in both 2001 and 2003.

5.7.2 New Triple Systems and Stars with Long Term Trends

More than 10% of the stars in the sample show long term radial velocity trends. In a few cases the trends are due to the known companion, as trends with opposite sign and nearly the same magnitude are observed for the two components. Figure 5.14 shows the case of HD 186858, for which a reliable visual + astrometric solution was presented by Soderhjelm (1999). The RV slopes of each components and their absolute RV difference follow very well the orbital solution. The full characterization of the binary orbit and individual masses of the systems that we are surveying are useful for the study of the frequency binary systems with/without planets, as described in Section 5.10.

In most cases the trends are due to new low mass stars orbiting one of the components. In one case, however, a substellar companion is compatible with the available data. One example of such RV trends with significant curvature is shown in Fig. 5.15. The continuation of the radial velocity monitoring will reveal the period and nature of these objects.

We recently started an adaptive optics program to identify the companions of stars with long-term trends using AdOpt@TNG (Cecconi et al. 2006). The direct identification of one new companion, probably responsible for the observed RV trend, is shown in Fig. 5.16.

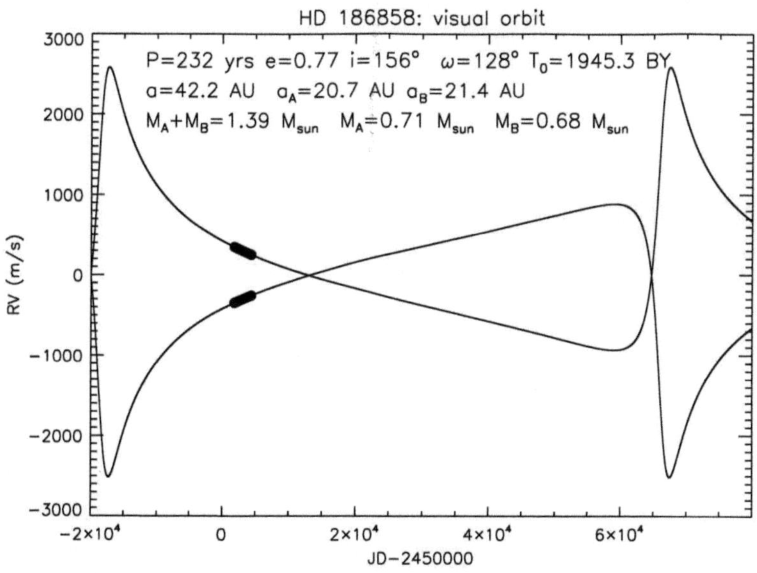

Fig. 5.14 *Solid lines*: predicted RV curve for the components of the binary system HD 186858 according to the visual + astrometric solution derived by Soderhjelm (1999). *Filled circles*: high-precision RV obtained with SARG over 6 years. The RV slopes of each components and their absolute RV difference follow the orbital solution very well

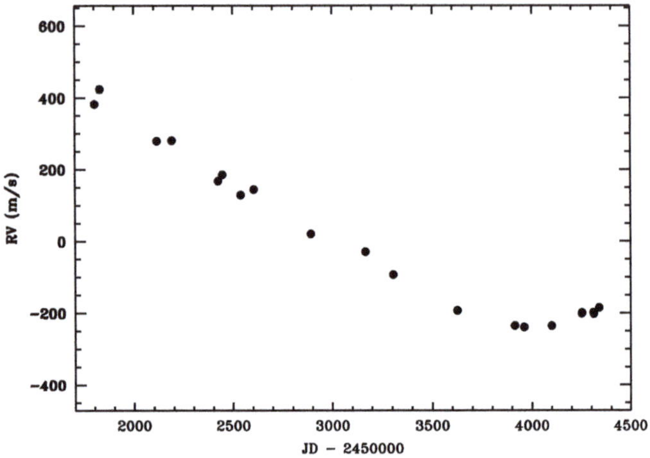

Fig. 5.15 Radial velocity curve of one of the stars showing a long-term trend, with significant curvature. The companion is likely a low mass star

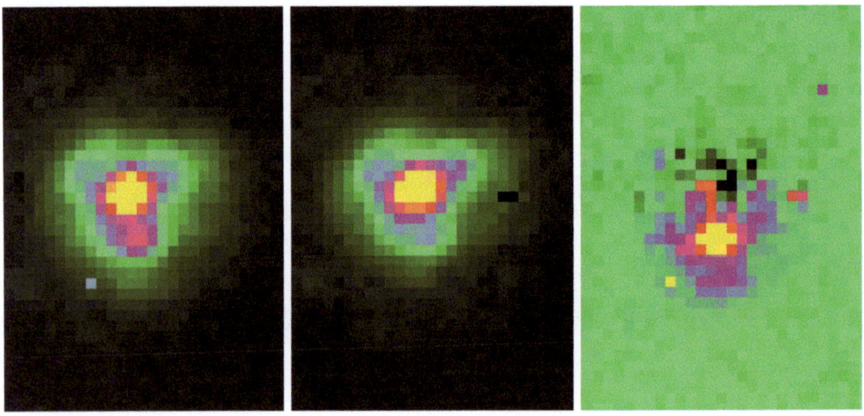

Fig. 5.16 Adaptive optics identification of a close companion around a star with a linear RV trend (Fig. 5.15). *Left panel*: image of the star with RV trend. *Middle panel*: image of the wide companion. *Right panel*: difference between the two images. PSF artifacts were removed fairly well, allowing the identification of a close companion at 0.2 arcsec from the star. This is probably responsible for the observed RV trend

The direct identification of the companions that caused the observed RV trends is of specific interest because it allows to measure the mass and separation of the companion, and then estimate its effective gravitational influence. This is relevant to the main science goal of the project, the determination of frequency of planets in binaries with different characteristics. Furthermore, direct detection of low mass stars and substellar objects orbiting stars with well determined characteristics (age and chemical composition) are crucial for the calibration of theoretical models of these objects.

Finally, we also detected a few new spectroscopic binaries among the components of the wide binaries. These systems consist of at least three components. Some of these systems are presented in Desidera et al. (2006).

5.8 Line Bisectors: A Tool to Study Stellar Activity and Contamination

The relevance of activity jitter to the interpretation of the RV data prompted us to develop a tool to measure and possibly to correct for its effect. The differential RV variations induced by stellar activity are due to changes in the profile of spectral lines caused by the presence of spots and/or the alteration of the granulation pattern in active regions. The activity jitter of a star may be predicted through of statistical relations from its chromospheric emission, rotational velocity, or amplitude of photometric variations (Saar et al. 1998; Paulson et al. 2004; Wright 2005). Simultaneous determination of RV, chromospheric emission, and/or photometry is even more powerful in disentangling the origin of the observed RV variations (Keplerian vs. stellar activity). The measurement of the line profile alterations on the same spectra (ideally on the same spectral lines) represents a direct measurement of the activity jitter. The existence of a correlation between the variations of the RV and those of the line profile is a strong indication for non-Keplerian origin for the observed RV variations.

The study of line profile as a tool to disentangle Keplerian motion and activity jitter is usually performed using a few well-isolated lines on high S/N spectra (see, e.g., Hatzes et al. 1998) or by combining the cross-correlation profiles of many spectral lines at moderate S/N ratios with a suitable template mask (see, e.g., Queloz et al. 2001). In our case, we followed the latter approach, but we had to handle the complication of having the iodine lines superimposed to the stellar spectra. On the other hand, these lines offer the opportunity to improve the wavelength calibration of the spectra, required for accurate estimates of the line bisectors. The iodine lines were removed by means of a suitable spectrum of a fast rotating early type star with the iodine cell in the optical path. The procedure is described in detail in Martinez et al. (2005).

The bisector of an absorption line is the middle point of the horizontal segment connecting points on the left and right sides of the profile with the same flux level. The line bisector is obtained by combining bisector points starting from the core toward the wings of the line. To quantify the asymmetry of the spectral lines and look for correlation with RV, it is useful to introduce the bisector velocity span (hereafter BVS, Toner and Gray 1988). This quantity is determined by considering a top zone near the wings and a bottom zone close to the core of the lines, which represent interesting regions to study the velocity given by the bisector (see Fig. 5.17). The difference of the average values of velocities in the top and bottom zones, V_T and V_B respectively, determine the bisector velocity span.

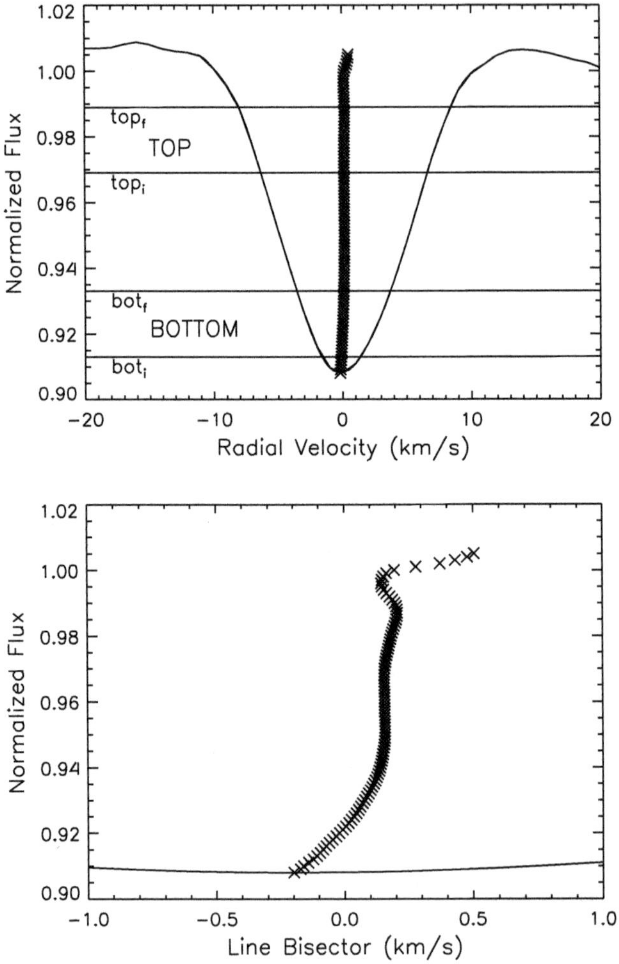

Fig. 5.17 Spectrum of HD 166435. In the *top panel* we show the normalized cross correlation profile, the line bisector, the top, and the bottom zones both with $\Delta F = \mathrm{top_f} - \mathrm{top_i} = \mathrm{bot_f} - \mathrm{bot_i} = 0.02$, where ΔF is the size (in units of relative flux) in the line profile of the top and bottom zones used for the measurement of the bisector velocity span. In the *bottom panel* we show a zoom of the profile with the RV scale increased to better display the asymmetries of the line bisector. Figure from Martinez et al. (2005)

The star HD 166435 shows evidence of RV variations, photometric variability, and magnetic activity. Furthermore, previous analysis of the variation of the line bisectors revealed a correlation between RV and line bisector orientation (Queloz et al. 2001). We used this star to test our procedure. As shown in Fig. 5.18, there is a clear anti-correlation between radial velocity and BVS variations.

The study of line shape is relevant for our program also as a diagnostic for the contamination of the spectra by the wide companion. Contaminated spectra are not easy to handle when analyzing the radial velocity curve. In fact, the internal radial

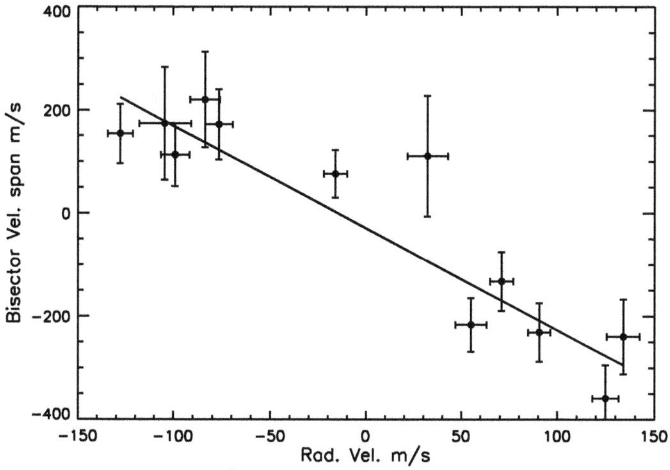

Fig. 5.18 Radial velocity – line bisector correlation for the active star HD 166435

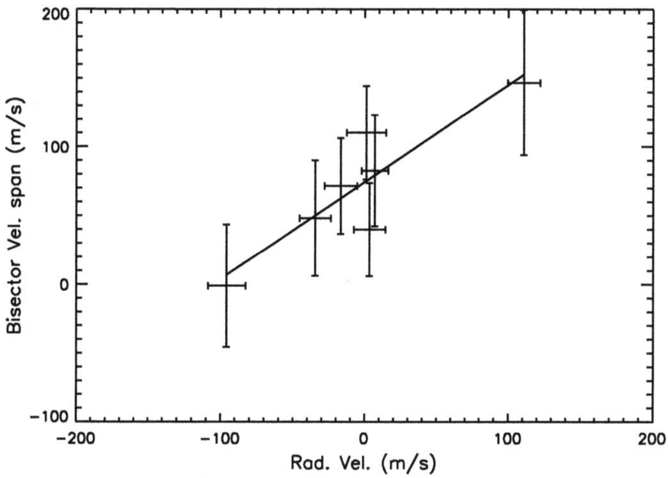

Fig. 5.19 Radial velocity – line bisector correlation for HD 8071B. This is likely due to the contamination by the companion HD 8071A

velocity errors are estimated from the scatter of individual portions of spectrum on which the spectrum is modeled separately. In the case of contamination, all these pieces deviate systematically by a similar amount (our pairs are always formed by similar stars) and the radial velocity shift might largely exceed the internal errors, causing a spurious but formally highly significant variability.

Also in the case of contamination, we observe a positive correlation between the bisector velocity span and the radial velocity. The worst case of contamination in our sample occurs for HD 8071B (see Fig. 5.19). This pair is one of the closest binaries (separation of 2.1 arcsec). The primary of this system, HD 8071 A itself is a single-lined spectroscopic binary with an RV semi-amplitude of about 7 km/s. This causes significant spectra-to-spectra variations of the contamination both in amplitude (because of the variable observing conditions) and wavelength (because of the orbital motion of HD 8071A).

5.9 Upper Limits on Planetary Companions

While no confirmed planet detection emerged up to now from our survey, a detailed analysis of the negative results would allow to constrain the frequency of planets in binary systems. Since we are focusing on a specific type of binaries, wide binaries with similar components at intermediate separations (a few hundreds AU), such a study is complementary to other studies of planets in binaries.

To this aim, we derived upper limits on the planetary companions still compatible with the observations. Our method, a Montecarlo simulation based on the evaluation of the excess of radial velocity variability caused by the presence of hypothetical planets, allows us a complete exploration of the possible orbital parameters for eccentric orbits (the real case, since most of the known planets are in eccentric orbits). Our approach is described in detail in Desidera et al. (2003).

Figure 5.20 shows the upper limits on planetary companion on short-period circular orbit for four stars representative of our sample. Figure 5.21 shows the limits for long period planets with eccentricities as large as 0.95. The average limits for the whole sample are shown in Fig. 5.22. These results indicate a lack of planets with period shorter than a few years and moderately large mass in our sample.[5] Considering the frequency of this kind of planets from other surveys (e.g., Fischer and Valenti 2005, see Section 5.10), our null result is significant at about 2σ level.

5.10 On the Frequency of Planets in Binary Systems

The lack of planets with RV semiamplitude larger than 30 m/s and period shorter than the survey duration in the SARG sample appears as an indication for a low frequency of planets in the kind of binary systems we are surveying. Since our sample includes only binaries, a reference sample is needed for a full statistical evaluation. A useful comparison sample is represented by the 'Uniform Detectability' sample identified by Fischer and Valenti (2005).

The Uniform Detectability (UD) sample has been built from the full target lists of Lick, Keck and Anglo Australian Surveys (1330 stars), satisfying the requirement of completeness for detections of planets with velocity amplitudes K > 30 m/s and orbital periods shorter than 4 years. Stars that were added after a planet was discovered

[5] The planet candidates discussed in Section 5.7 have either long periods or low velocity amplitudes and therefore are not of concern in this discussion.

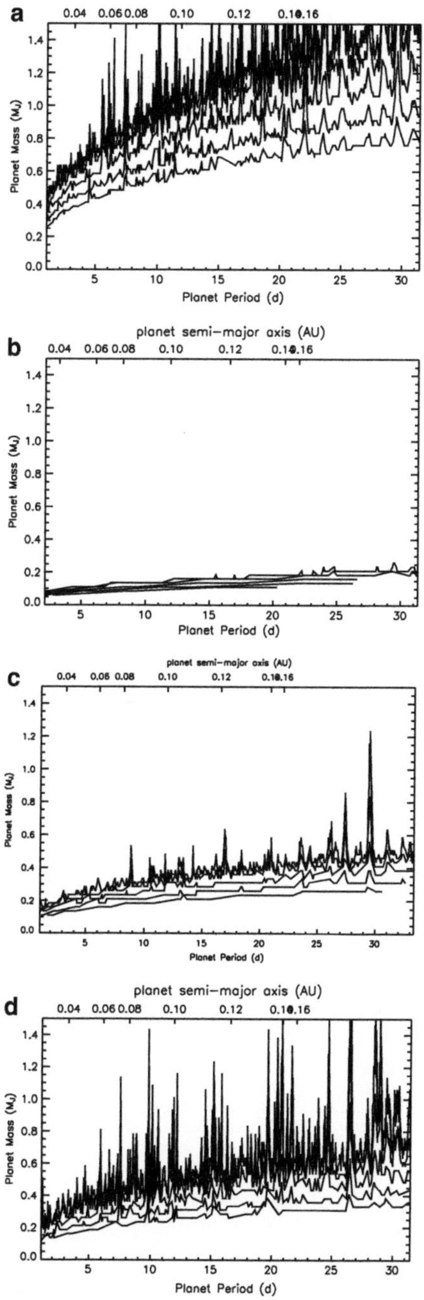

Fig. 5.20 Upper limits on planetary companions on short-period circular orbits for four stars representative of our sample. From **(a)** to **(d)**, the different lines refer to planets exclusion limits of 95%, 90%, 75%, 50%, and 25%, respectively. For the star on the **(a)** corner planet detectability is strongly limited by stellar activity. The star on the **(b)** corner is the one with the best limits, thanks to the low dispersion of RVs and the large number of measurements. The behavior of the other two stars is more typical for our survey. The 'noisy' run of exclusion limits with period for the star in the **(d)** corner is due to the small number of measurements

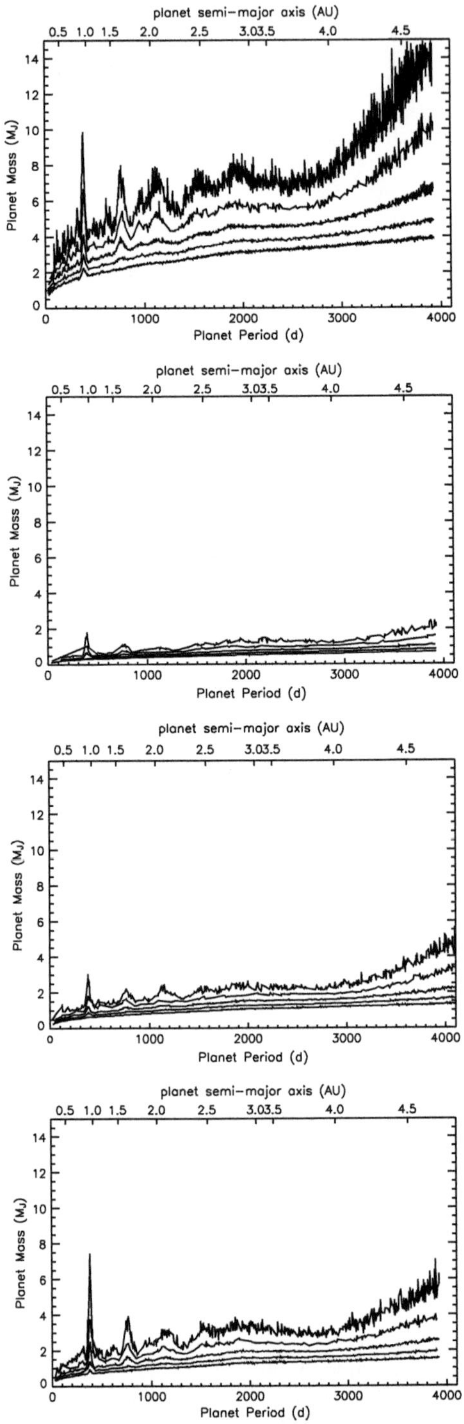

Fig. 5.21 Upper limits on planetary companions on long-period eccentric orbits for the same four stars shown in Fig. 5.20

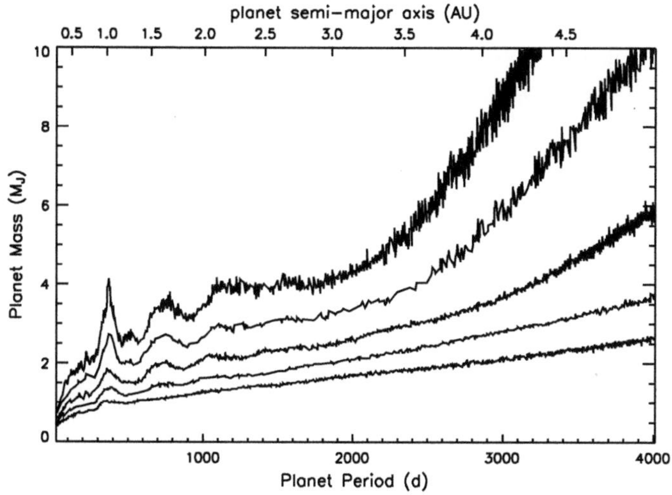

Fig. 5.22 Summary of estimates of exclusion/compatibility of planets in the SARG sample with current data for the stars with at least ten observations. For each period, the mass corresponding to the planet exclusion of (from *top* to *bottom*) 95%, 90%, 75%, 50%, and 25% (taking into account planet eccentricity) is shown. The results of individual stars were averaged to produce the plot

Table 5.2 Frequency of planets in binaries with different values of a_{crit}. Table from Bonavita and Desidera (2007)

a_{crit}	N_{stars}	$N_{planets}$	$N_{planets}/N_{stars}$
<20 AU	89	2	0.022 ± 0.018
20–50 AU	18	2	0.111 ± 0.105
50–100 AU	24	2	0.083 ± 0.076
100–250 AU	26	4	0.154 ± 0.107
>250 AU	45	5	0.111 ± 0.066
UD Singles sub-sample	647	34	0.053 ± 0.011
Entire UD binary sub-sample	202	15	0.074 ± 0.024

by other groups were not included in the sample. However, stars independently present in one of these surveys were considered even if a planet was detected first by another group. Only planets with $K > 30$ m/s and orbital periods shorter than 4 years were considered for the study of planet frequency. This corresponds to Saturn-mass planets for the shortest periods and Jupiter-mass planets for 4 year orbits.

The UD sample is biased against binaries, as the stars with companions closer than 2 arcsec known at the time of the target selection were excluded. Bonavita and Desidera (2007) performed a detailed literature search for binarity of the 850 stars in the UD sample, and identified 202 of these stars as members of binaries. For some of these stars, only long-term radial velocity and astrometric trends are available.

Among the binaries in the UD sample, 15 have planets, so the global frequency of planets in the UD binary sample is approximately 7.4%. If we consider the single-stars sub-sample, we found that 5.3% of UD single stars have planets (see Table 5.2).

The two frequencies are compatible within their errors. The slightly higher value of the global frequency in the binary sub-sample is probably due to higher completeness level of binary census in stars with planet.

Incompleteness effects are unlikely to deeply modify this picture. Even assuming that the frequency of binaries in the sample is that found by Duquennoy and Mayor (1991) (an upper limit because of the exclusion of binaries with separation less than 2 arcsec known at the time of sample selection) and that all the companions of planet hosts have been already identified, it can be seen that the global frequency of planets in binaries cannot be lower by more than a factor of three compared to that of single stars.

The rather large sample size allows us to make some sub-samples with different values of critical semimajor axis for dynamical stability of planets ($a_{\rm crit}$, see Holman and Wiegert 2001 and Section 5.2). All the stars with RV and/or astrometric trends are included in the closest bin, as it is likely that the companion responsible for the trend is at small separation.

We found that there is no significant dependence of the frequency on $a_{\rm crit}$ except for companions with $a_{\rm crit}$ less than 20 AU corresponds to a binary separation smaller than 50–100 AU, depending on the mass–ratio of the two stars. Considering also the similitude of the mass and period distribution of planets orbiting single stars and components of wide binaries (see Desidera and Barbieri 2007 and Section 5.2), we conclude that a wide companion plays a marginal role in the formation and evolution of giant planets around the primary.

For the planets in tight binaries, the results are more intriguing. On one hand, the frequency of planets in close binaries results to be lower than that of planets orbiting single stars and components of wide binaries. On the other hand, there are indications that the properties of planets in tight binaries are significantly different from those of exoplanets orbiting wide binaries or single stars (see Desidera and Barbieri 2007 and Section 5.2).

The frequency of planets in close binaries can be used to further investigate how these planets formed and the origin of their anomalous properties. Indeed, Pfhal and Mutherspaugh (2006) showed that the knowledge of the value of the frequency of planets in close binaries[6] should allow to disentangle between two alternative formation scenarios. A low frequency (less than 0.1% but with an uncertainty of about one order of magnitude, so they consider 1% as a limit-value) would be compatible with dynamical interactions that cause the formation of the tight binary after planet formation. While not fully conclusive because of the poor statistics, our results suggests that frequency of planets in close binaries probably is not as low as required to explain their presence only as the results of modifications of the binary orbit after the planet formation. Therefore, it appears that planets do form in binaries with separations smaller than 20 AU in spite of the strong gravitational interactions that might work against.

---

[6] Defined by Pfhal and Mutherspaugh (2006) as those binaries with semimajor axis less than 50 AU.

However, crucial issues still need clarification. There are some hints that the run of the frequency of planets is not characterized by a continuous decrease when moving to smaller separation: in the full list of planets in binaries by Desidera and Barbieri (2007) there is only one planet with critical semimajor axis in the range of 10–30 AU, while there are five planets with a_{crit} less than 10 AU and four planets with $30 < a_{crit} < 50$ AU. This suggests a bimodal distribution of planet frequency, with a secondary maximum at a_{crit} ~3–5 AU, but the analysis of the UD sample does not allow us to confirm it because of the small number of binaries with $10 < a_{crit} < 30$ AU and the lack of binary characterization (orbital parameters, mass–ratio) for the stars with only RV and/or astrometric trends.

The targets of the SARG planet search are crossing this range of separation (see Fig. 5.23), and therefore the completion of the survey, coupled with an estimate of planet detectability homogeneous with that of comparison samples will allow us to better address this issue. The current lack of planets in the SARG survey might suggest a relevant role of the binary mass–ratio in the occurrence of planets. A complementary very important approach is represented by a detailed characterization of the binaries in current samples of RV surveys (complete detection of binaries and, when possible, full determination of the orbital elements). The availability of a larger and more complete sample will allow us to better understand planet frequency in binaries and, at the same time, to disentangle the questions about the formation and different characteristics of the planets in these peculiar environments.

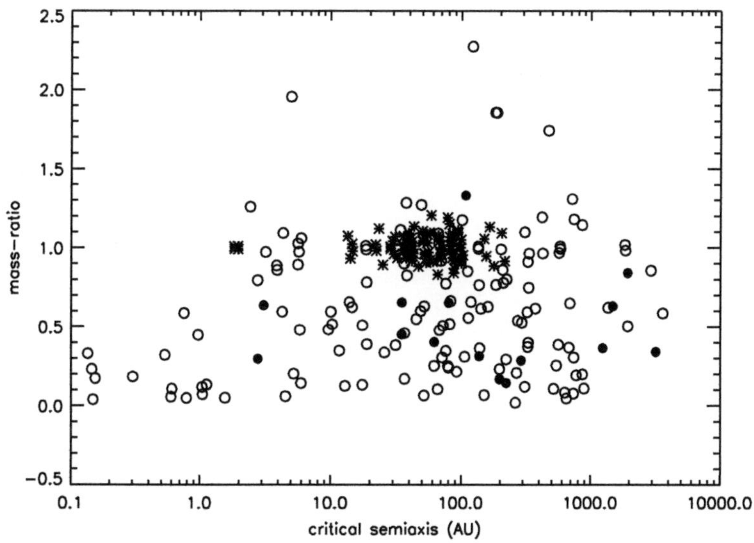

Fig. 5.23 Critical semimajor axis for dynamical stability for the binaries in the UD sample (*filled circles*: stars with planets; *empty circles*: stars without planets) and in the sample of the SARG planet search (*asterisks*)

References

G.A. Bakos, R.W. Noyes, G. Kovacs et al.: ApJ **656**, 552 (2007)
D. Benest & R. Gonczi: Earth, Moon and Planets **93**, 175 (2003)
M. Bonavita & S. Desidera: A&A **468**, 721 (2007)
C.J. Burke, P.R. McCullough, J.A. Valenti et al.: ApJ **671**, 2115
R.P. Butler, G.W. Marcy, E. Williams et al.: PASP **108**, 500 (1996)
R.P. Butler, J.T. Wright, G.W. Marcy et al.: ApJ **4646**, 505 (2006)
M. Cecconi, A. Ghedina, P. Bagnara et al.: in *proc. of the SPIE*, Volume 6272, p. 77 (2006)
G. Chauvin, A.M. Lagrange, S. Udry et al.: A&A **456**, 1165 (2006)
S. Desidera & M. Barbieri: A&A **462**, 345 (2007)
S. Desidera, R. Gratton, M. Endl et al.: A&A **405**, 207 (2003)
S. Desidera, R. Gratton, S. Scuderi et al.: A&A **420**, 683 (2004a)
S. Desidera, R. Gratton, M. Endl et al.: A&A **420**, L27 (2004b)
S. Desidera, R. Gratton, S. Lucatello, R.U. Claudi.: A & A **454**, 581 (2006)
S. Desidera, R. Gratton, S. Lucatello, M. Endl, S. Udry: A&A **462**, 1039 (2007)
A. Duquennoy & M. Mayor: A&A **248**, 485 (1991)
A. Eggenberger, M. Mayor, M. S. Udry: A&A **417**, 353 (2004)
M. Endl, M. Kürster, S. Els: A&A **362**, 585 (2000)
D. Fischer & J. Valenti: ApJ **622**, 1102 (2005)
G. Gonzalez: MNRAS **285**, 403 (1997)
R.G. Gratton, G. Bonanno, P. Bruno et al.: Exp. Astron. **12**, 107 (2001)
A.P. Hatzes, W.D. Cochran, E.J. Bakker: ApJ **508**, 380 (1998)
A.P. Hatzes & G. Wuchterl: Nat. **436**, 182 (2005)
U. Heiter & R.E. Luck: AJ **126**, 201 (2003)
M.J. Holman & P.A. Wiegert: AJ **117**, 621 (2001)
S. Ida & D.N.C. Lin: ApJ **616**, 567 (2004)
C. Laws & G. Gonzalez: ApJ **553**, 405 (2001)
A.F. Martinez Fiorenzano, R. Gratton, S. Desidera, R. Cosentino, M. Endl: A&A **442**, 775 (2005)
F. Marzari & S.J. Weidenschilling: Icarus **156**, 570 (2002)
F. Marzari, S.J. Weidenschilling, M. Barbieri, V. Granata: ApJ **618**, 502 (2005)
M. Mugrauer, R. Neuhauser, T. Mazeh et al.: AN **327**, 321 (2006)
M. Mugrauer, R. Neuhauser, T. Mazeh: A&A **469**, 755 (2007)
M. Mugrauer, A. Seifahrt, R. Neuhauser: MNRAS **378**, 1328 (2007)
N. Murray, B. Chaboyer, P. Arras, B. Hansen, R.W. Noyes: ApJ **555**, 801 (2001)
B. Nordstrom, M. Mayor, J. Andersen et al.: A&A **418**, 989 (2004)
J. Patience, R.J. White, A. ghez et al.: ApJ **581**, 684 (2002)
D.B. Paulson, S.H. Saar, W.D. Cochran, G.W. Henry: AJ **127**, 1644 (2004)
E. Pfhal & M. Mutherspaugh: ApJ **652**, 1694 (2006)
G.W. Preston & C. Sneden: AJ **120**, 1014 (2000)
D. Queloz, G.W. Henry, J.P. Sivan et al.: A&A **379**, 279 (2001)
S.H. Saar, R.P. Butler & G.W. Marcy: ApJ **498**, L153 (1998)
N.C. Santos, G. Israelian , M. Mayor: A&A **415**, 1153 (2004)
S. Soderhjelm: A&A **341**, 121 (1999)
G. Takeda, E.B. Ford, A. Sills et al.: ApJS **168**, 297 (2007)
Y. Takeda: PASJ **57**, 83 (2005)
O. Tamuz, D. Segransan, S. Udry et al.: A&A **480**, L33
B.J. Taylor: ApJS **161**, 444 (2005)
P. Thebault, F. Marzari, H. Scholl: Icarus **183**, 193 (2006)
A.A. Tokovinin, R.F. Griffin, Y.Y. Balega et al.: Astron. Lett. **26**, 116 (2000)
A.A. Tokovinin, S. Thomas, M. Sterzik, S. Udry: A&A **450**, 681 (2006)
C.G. Toner & D.F. Gray: ApJ **334**, 1008 (1988)

S. Tremaine & N. Zakamska: in *The Search for Other Worlds 14th Astrophysics Conference, AIP Conf. Proc.* **713**, 243 (2004)
J. Valenti & D. Fischer: ApJS **159**, 141 (2005)
H. Van Winckel: ARA & A **41**, 391 (2003)
R.A. Winnick, P. Demarque, S. Basu, D.B. Guenther: ApJ **576**, 1075 (2002)
J.T. Wright: PASP **117**, 657 (2005)
Y. Wu & N. Murray: ApJ **589**, 605 (2003)
S. Zucker & T. Mazeh: ApJ **568,** L113 (2002)

Chapter 6
Early Evolution of Planets in Binaries: Planet–Disk Interaction

Willy Kley and Richard P. Nelson

6.1 Introduction

6.1.1 Summary of Observations

At the time of writing, approximately 35 extrasolar planets have been discovered in binary star systems, all of which are orbiting about a single component of the binary. For a review of the global statistics see the papers by Eggenberger et al. (2004) and Mugrauer et al. (2007), as well as the relevant chapters in this book (see the chapter by Eggenberger and Udry). So far, there have been no discoveries of circumbinary planets. The binary star systems that host planets are very diverse in their properties, with binary semimajor axes ranging from $\simeq$6,400 AU down to $\simeq$20 AU. In the case where the orbits are eccentric, the binary periastron can be as small as $\simeq$12 AU, such that important dynamical effects are expected to have occurred during and after planet formation. One such example is the well studied system γ Cep (Hatzes et al. 2003) which contains a planet of mass $m_p \sin i \simeq 1.6$ Jupiter-masses with a semimajor axis of $\simeq$2.1 AU. Here the binary semimajor axis is $\simeq$20 AU and periastron is $\simeq$12 AU (see the chapter by Hatzes et al.). Another interesting case is GL86 (Mugrauer and Neuhäuser 2005), which consists of a binary system whose secondary is a $\simeq$0.55 $M_\odot$ white dwarf with a projected orbital separation of $\simeq$21 AU. GL86 is reported to host a planet with $m \sin i \simeq 4$ Jupiter-masses (Queloz et al. 2000). It is worth noting that the white dwarf progenitor was probably a Solar-mass main sequence star, such that the orbital separation was even smaller in the past. As the influence of the secondaries on the planet formation process will

W. Kley (✉)
Institut für Astronomic & Astrophysik, Universität Tübingen, Auf der Morgenstelle 10, 72076 Tübingen, Germany
e-mail: wilhelm.kley@uni-tuebingen.de

R.P. Nelson
Astronomy Unit, Queen Mary, University of London, Mile End Road, London E1 4NS, United Kingdom
e-mail: r.p.nelson@qmul.ac.uk

N. Haghighipour (ed.), *Planets in Binary Star Systems*, Astrophysics and Space Science Library 366, DOI 10.1007/978-90-481-8687-7_6,

obviously be smaller for larger binary separations, the mere existence of planets in systems with $a_{bin} \simeq 20$ AU represents a special challenge to any kind of formation process.

Clearly the close binary systems containing planets provide an excellent laboratory for testing theories of planet formation, as the presence of the companion may create conditions normally thought to be inconducive to planet formation. It is these closer systems that we mainly focus on in this article.

6.1.2 Summary of Planet Formation in Binaries

In a binary star system the early formation of planets may be strongly influenced by changes in the structure of the protoplanetary disk caused by tidal forces from the binary companion. For a circumstellar disk, significant effects will occur if the disk outer edge is tidally truncated by the binary companion, as strong spiral shock waves will be launched near the disk edge and propagate inward. For a circumstellar disk in a binary system which is not subject to strong tidal forcing, it seems likely that the effect of the companion star will be modest, unless the orbital inclinations are such that the Kozai effect becomes important (Innanen et al. 1997). In a circumbinary disk one can almost always expect strong tidal interaction between the binary and disk, and hence significant effects on planet formation. In this chapter we restrict our discussion to two basic scenarios. The first is planet formation and evolution in a circumstellar disk around the primary (most massive) star, the so called S-type configurations, although we note that of the 35 binary systems with known planets, two host planets around the secondary star (16 Cyg and HD178911). The second scenario is planet formation in circumbinary disks (P-type configuration). We restrict our discussion to those early phases of planetary formation that occur in a gas-rich environment where the young protoplanet is still embedded in the protoplanetary disk and interacts with it.

In a circumstellar disk, the tidal torques of the companion star generate strong spiral shocks, and angular momentum is transferred to the binary orbit. This in turn leads to disk truncation. Using analytical and numerical methods, Artymowicz and Lubow (1994) showed how the truncation radius r_t of the disk depends on the binary semimajor axis a_{bin}, its eccentricity e_{bin}, the mass ratio $q = M_2/M_1$ (where M_1, M_2 denote the masses of the primary and secondary stars, respectively), and the viscosity ν of the disk. For typical values of $q \approx 0.5$ and $e_{bin} = 0.3$, the disk will be truncated to a radius of $r_t \approx 1/3 a_{bin}$ for typical disk Reynold's numbers of 10^5 (Artymowicz and Lubow 1994; Larwood et al. 1996; Armitage et al. 1999). For a given mass ratio q and semimajor axis a_{bin} an increase in e_{bin} will reduce the size of the disk while a large ν will increase the disk's radius. Not only will the disk be truncated, but the overall structure and density stratification may be modified by the binary companion. In Section 6.2 we will illustrate this effect.

In a circumbinary disk, the binary creates a tidally-induced inner cavity. For typical disk and binary parameters (e.g., $e_{bin} = 0.3$, $q = 0.5$) the size of the cavity is $\simeq 2.7 \times a_{bin}$ (Artymowicz and Lubow 1994).

Whether these changes in the disk structure in circumstellar or circumbinary systems have an influence on the likelihood of planet formation in such environments has long been a matter of debate. The dynamical action of the binary has several potential consequences which may be adverse to planet formation: (i) it changes the stability properties of orbits, (ii) it increases the velocity dispersion of planetesimals, (iii) it reduces the lifetime of the disk, and (iv) it increases the disk's temperature.

In a numerical study, Nelson (2000) investigated the hydrodynamical evolution of two circumstellar disks in an equal-mass binary with a separation of 50 AU and an eccentricity of 0.3. The simulations included heating of the disk through viscous effects and radiative cooling. Nelson argued that both main scenarios of giant planet formation (i.e., core accretion and gravitational instability) are strongly handicapped in forming planets in such binaries because the eccentric companion will induce a periodic heating of the disk up to temperatures possibly above 1,200 K. Since the condensation of particles as well as the occurrence of gravitational instability require lower temperatures, planet formation will be made more difficult in both scenarios. Clearly the strength of this effect will depend on the binary separation and its mass-ratio.

In addition to the approach taken by Nelson (2000), the influence that a stellar companion may have on the evolution of a massive planet embedded in a circumstellar disk was investigated by Kley (2000). This author studied the evolution of the embedded planet through hydrodynamical simulations (see also the review article by Kley and Burkert 2000). and found that due to the compression of the disk and the increased density, the migration and accretion timescales are reduced compared to a disk that is unperturbed. It is important to note that in these preliminary simulations, only very short time spans have been covered and the initial disk configuration may have been unrealistic.

Recent numerical studies of the final stages of terrestrial planet formation in rather close binaries with separations of only 20–30 AU, that involve giant impacts between lunar-mass planetary embryos, show that it is indeed possible to form terrestrial planets in such systems (Lissauer et al. 2004; Turrini et al. 2005; Quintana et al. 2007; Haghighipour 2007), provided it is possible for the planetary embryos themselves to form.

It is already the case for planet formation around single stars that the lifetime of the disk represents a limiting factor in the formation of planets, and it has been suspected that the dynamical action of a companion will reduce the lifetime of disks substantially. However, a recent analysis of the observational data of disks in binary stars finds no or very little change in the lifetimes of the disks, at least for separations larger than about 20 AU (Monin et al. 2007).

The early phase of planetesimal formation and subsequent formation of Earth-like planets is described in more detail in other chapters of this book. Here we will concentrate on the formation and evolution of planets in a gas rich environment, where inclusion of the full dynamics of the protoplanetary disk is crucial. We consider the dynamics of planetesimals, low-mass planets, and high-mass planets in circumstellar and circumbinary disks.

6.2 Evolution of Planets in Circumstellar Disks with a Companion

The presence of a companion star influences the structure of a circumstellar disk around the primary star due to gravitational torques acting on the disk. This leads to an exchange of energy and angular momentum between the binary and the disk. For close binaries the disk becomes truncated where the truncation radius r_t depends primarily on the parameters of the binary, i.e., the mass-ratio q, the semimajor axis a_{bin}, and eccentricity e_{bin}, and the viscosity of the disk. The radius r_t has been calculated semi-analytically and numerically by Artymowicz and Lubow (1994).

The effects of the companion on planet formation are likely to be most pronounced in binaries with separations $\leq$20 AU, rather than in long period systems with $a_{bin} > 1{,}000$ AU. Among the very close binary stars containing planets is the well studied system γ Cep. Including observations taken over decades, Hatzes et al. (2003) confirmed the presence of a planet orbiting the primary star in this system. Very recently, new radial velocity measurements and additional Hipparcos data have refined the binary orbit (Torres 2007) and the direct imaging of the secondary has fixed the masses of the binary to $M_1 = 1.4M_\odot$ and $M_2 = 0.4M_\odot$ (Neuhäuser et al. 2007). This system with a binary separation of about 20 AU contains a massive planet with a minimum mass of $1.6M_{Jup}$ orbiting the primary star at a distance of approximately 2.1 AU.

Gamma Cep is the tightest binary system known to contain a Jupiter-sized planet. For this reason, it has attracted much attention in the past several years. Several studies looked at the stability and/or the possibility of (additional) habitable planets in this system (e.g., Dvorak et al. (2004); Turrini et al. (2004); Haghighipour (2006); Verrier and Evans (2006)). Assuming that the planet has not been captured at a later time, or that the binary orbit has not shrunk since planet formation, this system represents a very challenging environment for the formation of planets indeed, and we choose it to illustrate the main influence a close companion can have on the planet formation process. Note that we focus here on this particular short period system since it makes the additional challenges and problems generated by the presence of the binary to the process of planet formation most obvious. Other more distant systems will experience similar behavior, although weakened. We focus here on the planet formation process via the core-accretion scenario. The gravitational instability route is discussed the chapter by Mayer et al.

6.2.1 Disk Evolution in the Presence of a Companion

When studying the formation of planets in a protoplanetary disk in the presence of a secondary star, it is necessary to first follow the evolution of the perturbed disk without an embedded planet, and bring the system into equilibrium, before adding a planetary embryo at a later time.

We choose to model a specific system where the orbital elements of the binary have been chosen to match those of γ Cep quite closely. The data for this system have been taken from (Hatzes et al. 2003) which do not include the most recent improvements mentioned above (Neuhäuser et al. 2007). These newest refinements primarily concern the mass of the primary and do not alter our conclusions at all. We are interested here in demonstrating the principle physical effects rather than trying to achieve a perfect match with all the observations of this particular system.

For this study we choose a binary with $M_1 = 1.59 M_\odot$, $M_2 = 0.38 M_\odot$, $a_{bin} = 18.5$ AU and $e_{bin} = 0.36$, which translates into a binary period of $P = 56.7$ years. We assume that the primary star is surrounded by a flat circumstellar disk, where the binary orbit and the disk are coplanar. In a numerical hydrodynamical model of the system, the fact that the disk's vertical thickness $H(r)$ at a given distance r from the primary is typically small with respect to the radius ($H/r \ll 1$) is used to perform restricted two-dimensional (2D) simulations and neglect the vertical extent altogether. Here, we present such 2D hydrodynamical simulations of a circumstellar disk which is perturbed by the secondary. We assume that the effects of the intrinsic turbulence of the disk can be described approximately through the viscous Navier–Stokes equations, which are solved by a finite volume method that is second order in space and time. To substantiate our results, we utilize two different codes `RH2D` (Kley 1989, 1999) and `NIRVANA` (Nelson et al. 2000; Ziegler and Yorke 1997).

The planet evolution and the orbital motion of the binary proceed on time scales much longer than the typical dynamical timescale of the disk, and consequently very many time steps need to be calculated just in one simulation. To allow for parameter studies, we have increased the performance of the runs by implementing the `FARGO`-algorithm in our codes, which is particularly designed to model differentially rotating flows (Masset 2000) efficiently. For our chosen radial range and grid resolution, we find a speed-up factor of about 7.5 over the standard case. Then, applying a Courant number of 0.75 still about 160,000 time steps are required for only 10 binary orbits for our radial range and a standard resolution of 300 × 300. We checked the accuracy of the FARGO implementation against several test cases.

The results of these investigations of the γ Cep system are presented in more detail in Kley and Nelson (2008).

6.2.1.1 Numerical Setup

As the disk is orbiting only one star, we utilize an adapted cylindrical coordinate system (r, φ) which is centered on the primary. It extends radially from $r_{min} = 0.5$ AU to $r_{max} = 8$ AU and in azimuth around a whole annulus ($\varphi_{min} = 0$, $\varphi_{max} = 2\pi$). Within this domain, at the beginning of the simulations ($t = 0$) an axisymmetric disk (with respect to the primary) is initialized with a surface density profile $\Sigma(r) = \Sigma_0 r^{-1/2}$. The reference density Σ_0 is chosen such that the total mass in the computational domain (within r_{min} and r_{max}) equals $1.75 \times 10^{-3} M_\odot$ which implies $\Sigma_0 = 1.89 \times 10^{-5} M_\odot/\mathrm{AU}^2$. The temperature profile is fixed here and given by $T(r) \propto r^{-1}$, which follows from the assumed constancy of the scale height $h = H/r$, which is fixed to $h = 0.05$. For the viscosity we assume an α-type prescription where the coefficient of the kinematic viscosity is given by $\nu = \alpha c_s H$ with $\alpha = 0.005$, and the sound speed $c_s(r) = h\, v_{kep}(r)$. The binary is initialized at apocenter using the above orbital parameter.

The boundary conditions are chosen such that material may escape through the radial boundaries. At the outer boundary (r_{max}) we impose a so called zero-gradient outflow condition. During periastron when large spirals may extend beyond r_{max} this condition will allow material to leave the system and not create numerical artifacts. At the inner boundary we set a viscous outflow condition where the material may flow through r_{min} with the local (azimuthally averaged) viscous inflow referring to an accretion disk in equilibrium. No matter is allowed to flow back into the system and the mass of the disk will slowly decline. To ensure a uniform setup for the planets we rescale the disk mass when inserting them.

6.2.1.2 The Structure of the Disk

The presence of an eccentric secondary star leads to a strong periodic disturbance of the disk whenever the secondary is at periastron. This effect is illustrated in Fig. 6.1 where we display the surface density Σ of the disk in gray scale at two different times in the early evolution of the disk [see also Nelson (2000)]. As shown here, two strong spiral arms (shock waves) are created in the disk which carry material beyond the outer boundary of the computational domain. In between the periapses the disk settles down and becomes more circular again. Figure 6.2 shows the radial surface density and eccentricity of primary's circumstellar disk. Already the very first approaches with the binary lead to a truncation of the disk as visible in left panel of Fig. 6.2 for the curve at $t = 10$ binary orbits. Slowly the whole disk structure rearranges and equilibrates at around $t = 50$ where it shows a much steeper density slope than in the initial state. The timescale for this equilibration process depends on the magnitude of the disk viscosity. The eccentricity of the disk in the final quasi-equilibrium state of the simulations (after around 60–70 orbits) varies approximately between 0.1 and 0.16 depending on the position of the binary in its orbit as shown in the left panel of Fig. 6.3. The disk eccentricity $e_{disk}(r)$ has been obtained by calculating the eccentricity of each disk element, as if in a two-body motion with the primary star, and then averaged over the respective annulus. At the same time the disk as a whole precesses as is shown in the right panel of Fig. 6.3. This coherent slow retrograde precession with a pattern speed much smaller than the orbital period of the disk material around the star is caused by the non-negligible pressure forces operating in the disk. Similar behaviour has been demonstrated for disks with free eccentricities (Papaloizou 2005). This dynamical origin of an eccentric and precessing disk in a binary system is highly interesting and complex in itself, a fact which has been pointed out recently by Paardekooper et al. (2008).

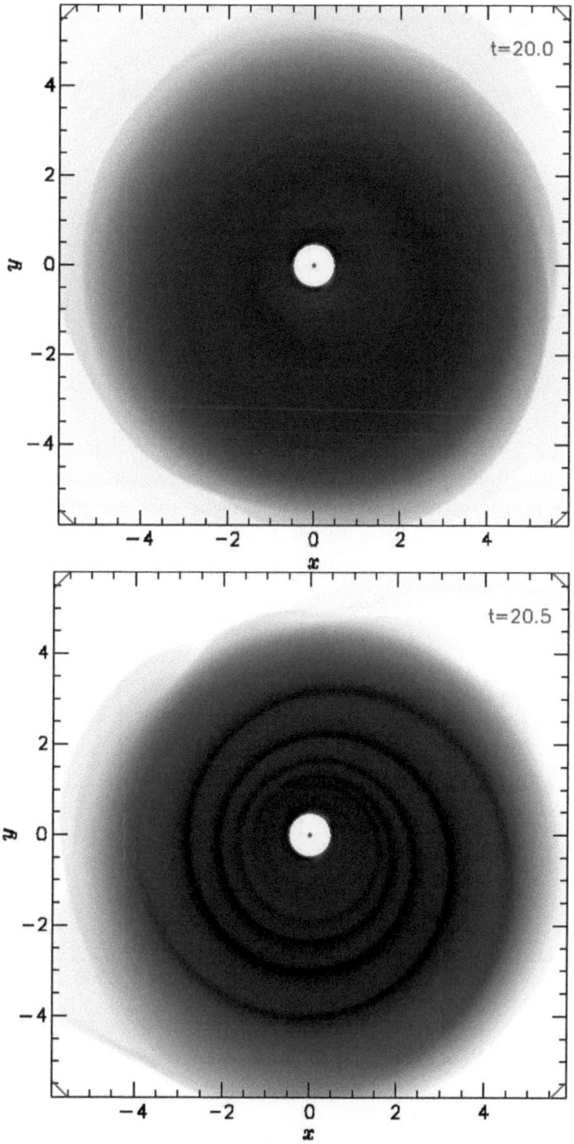

Fig. 6.1 Grayscale plot of the two-dimensional density distribution of the circumstellar disk around the primary at two different orbital phases of the binary. *Top* shortly after apocenter at about 20 binary orbits, and *Bottom* shortly after closest approach (pericenter)

6.2.1.3 The Orbital Elements of the Binary

In the previous section we have seen that the gravitational torques of the binary lead to a truncation of the disk and re-arrangement of the material within. In turn, we expect a change in the orbital elements of the binary.

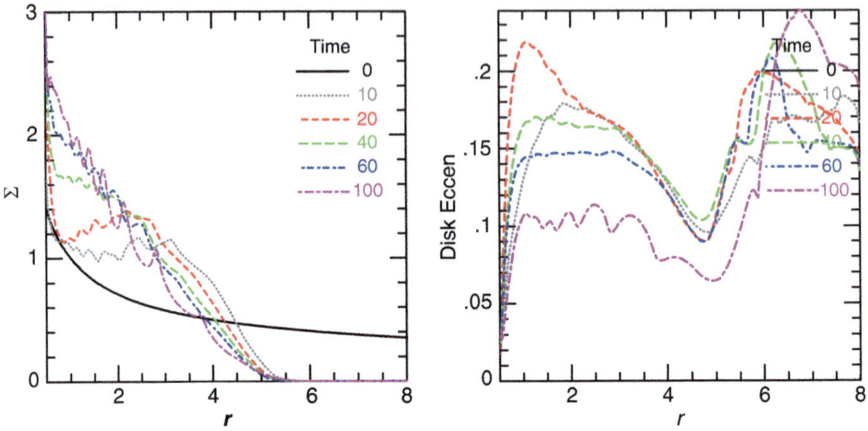

Fig. 6.2 The radial surface density distribution (*left*) and the eccentricity (*right*) of the circumstellar disk around the primary in the presence of the secondary. Time is given in units of the binary orbit, radial distance is in AU, and the density is in dimensionless units

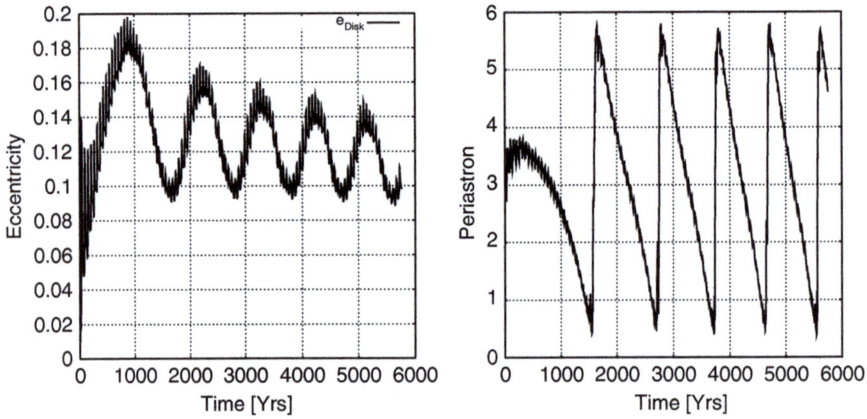

Fig. 6.3 The evolution of the global mass averaged disk eccentricity (*left*) and the position angle of the disk's periapse (*right*)

To estimate theoretically the magnitude of the back reaction a circumstellar disk has on the orbital elements of the binary during the initial phase of readjustment, we assume an idealized system consisting of a binary system and a ring-like mass distribution with mass m_{ring} orbiting the primary star at a distance (δ-function) of r_{ring}. The energy E_{bin} and angular momentum L_{bin} of the binary are given by

$$E_{bin} = -\frac{GM\mu}{2a_{bin}}, \tag{6.1}$$

and

$$L_{bin} = \mu \left[GM a_{bin} (1 - e_{bin}^2)\right]^{1/2}, \tag{6.2}$$

where $M = M_1 + M_2$ is the total mass of the two stars and $\mu = M_1 M_2/M$ is the reduced mass. The corresponding quantities of the ring are

$$E_{ring} = -\frac{GM_1 m_{disk}}{2r_{ring}} \tag{6.3}$$

and

$$L_{ring} = m_{ring} \left(GM_1 r_{ring}\right)^{1/2}, \tag{6.4}$$

Respectively. Now, suppose that the ring is shifted from its initial position r_{ring}^{α} to a smaller radius r_{ring}^{β} keeping all its mass. This change of radius mimics the initial truncation of disk by the binary. Through this process the ring's energy and angular momentum are reduced from E_{ring}^{α} and L_{ring}^{α} to E_{ring}^{β} and L_{ring}^{β}. From the conservation of total energy and angular momentum,

$$E = E_{ring} + E_{bin} \tag{6.5}$$

and

$$L = L_{ring} + L_{bin}, \tag{6.6}$$

we can calculate the corresponding change in the orbital elements of the binary from E_{bin}^{α} and L_{bin}^{α} to E_{bin}^{β} and L_{bin}^{β}. For the binary masses of $M_1 = 1.6M_\odot$ and $M_2 = 0.4M_\odot$, and with initial orbital elements of $a_{bin}^{\alpha} = 18.5$ AU and $e_{bin}^{\alpha} = 0.36$, we find that for the shift of a ring with $m_{ring} = 4 \times 10^{-3} M_\odot$ and initial radius $r_{ring}^{\alpha} = 4.0$ AU to a final radius of $r_{ring}^{\beta} = 2.0$ AU, the binary elements change to $a_{bin}^{\beta} = 19.4$ AU and $e_{bin}^{\beta} = 0.41$ – A quite substantial change considering the smallness of the ring's mass in comparison to the stellar masses. But the closeness to the primary allows to gain a substantial amount of binding energy from the ring. The calculation is approximate in the sense that the energy and angular momentum of the ring are calculated with respect to the non-inertial coordinate frame centered on the primary.

We can now compare this estimate with the previous hydrodynamical simulations, and plot in Fig. 6.4 the evolution of a_{bin} and e_{bin} for about the first 100 binary periods with no planet included. As demonstrated in this figure, the binary expands as it gains energy from the compressed disk and increases its eccentricity. The increase in e_{bin} does not lead to a decrease in the angular momentum however, since its separation has also increased, (see Eqs. 6.1 and 6.2). Whenever the secondary is near periastron, the gravitational interaction with the disk is maximal, which results in the strong periodic spikes in the binary elements. The changes in the orbital elements of the binary is somewhat smaller than the estimated values because (i) the mass of the disk is smaller in the hydrodynamic calculation and (ii) disk mass and angular momentum are stripped off by the secondary and are lost through the outer boundary of the computational domain. The loss through the (open) inner boundary

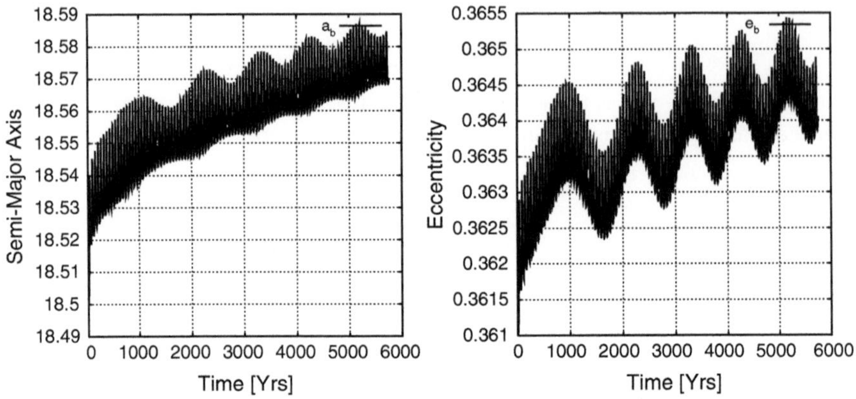

Fig. 6.4 The evolution of the binary elements due to the interaction with the circumstellar disk around the primary star without an embedded planet. One binary orbit refers to approximately 57 years. *Left*: $a_{bin}(t)$; *Right*: $e_{bin}(t)$

of the disk is only marginal. The changes in the orbital elements of the binary, as displayed in Fig. 6.4, is initially large during the equilibration phase of the disk and then slows down such that the semimajor axis and eccentricity change only very slowly. As long as a disk is present, the binary will evolve as well and only if the disk has dissipated, its evolution will come to a halt. This slow evolution, however, does not affect our conclusions about the planet formation process as discussed in this work.

6.2.1.4 The Behaviour of an Embedded Planet

In the previous section we have seen that the gravitational torques of the binary lead to a truncation of the disk and a rearrangement of the disk material. To study the influence of the companion on the evolution of small protoplanets, we embed, after an equilibration time of 100 binary orbits (nearly 6,000 years), a $30 M_{Earth}$ planet in the disk and follow its subsequent evolution. This rather time consuming procedure to generate the initial state is necessary to obtain realistic initial conditions for the growing protoplanet. At the time of insertion of the planet, the remaining disk mass is rescaled to contain 3 M_{Jup} within the computational domain.

As a first sample case we follow the planet's orbital evolution while keeping its mass constant, i.e., the planet is not allowed to accrete mass from its environment. This model will serve as a reference for the subsequent cases which will allow for planetary mass growth. The planet is released at $a_p = 2.5$ AU on a circular orbit. After insertion of the planet its orbital elements will change due to gravitational interaction with the disk and the binary. The planet migrates inward due to the torques of the disk, with a rate of 0.1 AU in about 2,800 years. While the overall migration is approximately linear over this time, it is modulated by the binary companion and the precessing, eccentric disk (see left panel of Fig. 6.5). At the same time, the

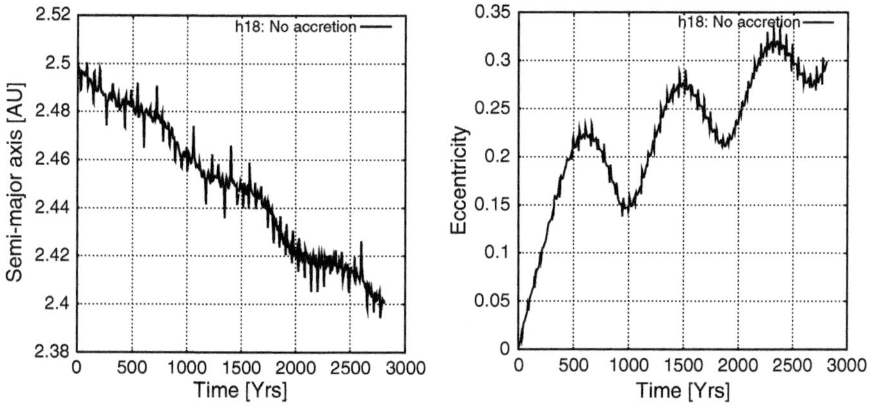

Fig. 6.5 The evolution of the semimajor (*left*) axis and eccentricity (*right*) of an embedded planet in the circumstellar accretion disk around the primary. Here, the planet is not allowed to accrete material from the disk and remains at 30 M_{Earth}. The planet is inserted after 100 orbital binary periods, and the time is reset to zero

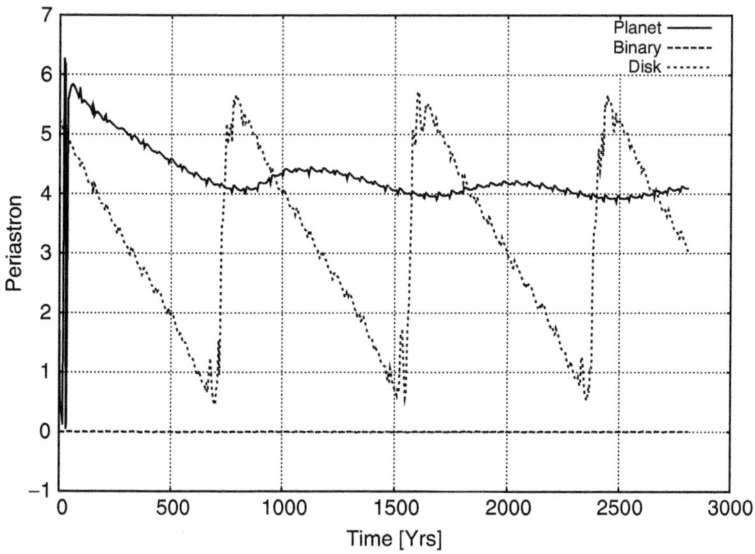

Fig. 6.6 The evolution of the argument of pericenter of the disk, the planet, and the binary after insertion of a 30 M_{Earth} object

planetary eccentricity e_p increases to about 0.3, with the eccentric disk yielding the prime contribution to its growth. The oscillatory behaviour originates from the changing degree of apsidal alignment between eccentric disk and planet as they undergo relative precession.

The evolution of the argument of pericenter of the disk, the planet and the binary are displayed in Fig. 6.6. While the disk continues its retrograde precession and the

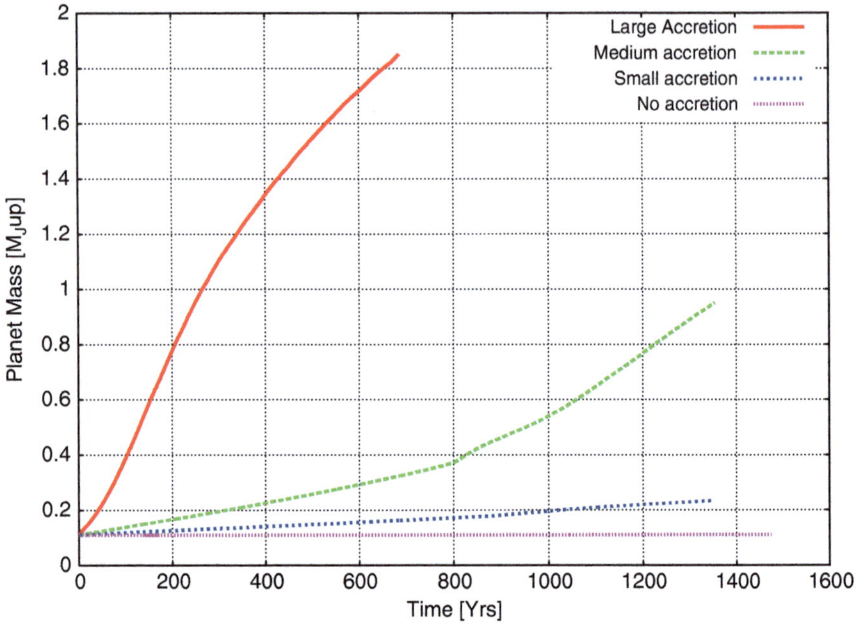

Fig. 6.7 The evolution of the mass of the planet for different accretion rates. Initially the planet had a mass of 30 M_{Earth}

binary remains unchanged, the planet undergoes initially a retrograde precession and then settles to an approximately constant value with oscillations whose frequency is given by the precession frequency of the whole disk in which it is embedded.

To study more realistic cases, we now allow the planet to grow in mass by accreting material from the disk during its motion through it. The accretion process is modeled numerically in a simple manner. At each time step, a certain fraction of the material within the Roche lobe of the planet is taken out of the computational domain and added to planet's mass. In Fig. 6.7 we show the evolution of the mass of the planet for different accretion rates. For the largest accretion rates, the planet acquires over $1.8M_{Jup}$ within the first 700 years of its evolution, a value that is unrealistically high. This model is used to set the limiting case for the others. The model with the small accretion only doubles the mass of the planet from 30 to 60 M_{Earth} during the first 1,000 years, which gives a more realistic accretion rate. The no-accreting case is given by the horizontal line.

More interesting is now the different orbital behaviour of each accreting planet, which is displayed in Fig. 6.8. The planet with the constant mass has the slowest migration, and the larger the accretion rate the larger is the migration speed. This is consistent with the estimated migration rates for different masses (D'Angelo et al. 2003). The planet with the maximum accretion rate grows rapidly in mass and approaches 1 M_{Jup} limit after 280 years, when its migration rate slows down

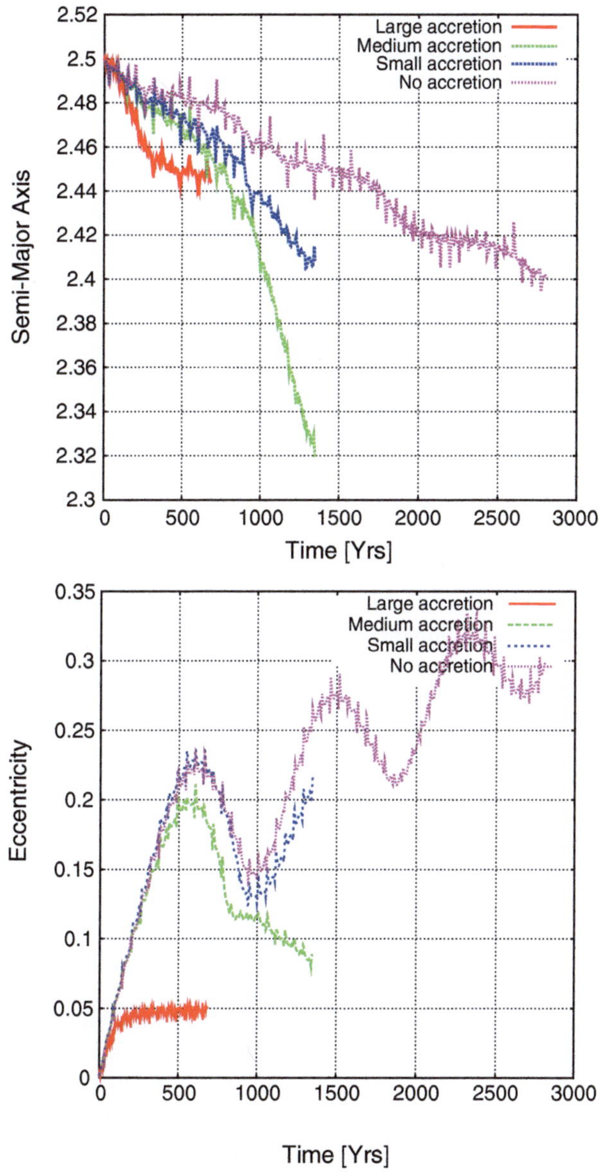

Fig. 6.8 The evolution of the semimajor axis (*top*) and eccentricity (*bottom*) of embedded planets in the circumstellar accretion disk. The planets all started with the same mass but accreted at different rates from the accretion disk (see Fig. 6.7). The planets are inserted after 100 orbital binary periods, and the time is reset to zero

and levels off as the mass in the disk decreases and the driving agent disappears. The intermediate cases migrate initially with the same speed as the non-accreting model but accelerate as the planetary mass increases.

Concerning the eccentricity evolution, the lightest planet experiences the largest growth. For the large accretion rate the eccentricity soon levels off to a value of $e_p = 0.05$.

6.2.1.5 Comparison with γ Cep

The most up to date observational data suggest the following parameters for the planet in the γ Cep system: $a_p \simeq 2.044$, $e_p \simeq 0.115$ and $m_p \sin i \simeq 1.60 M_{Jup}$ (Neuhäuser et al. 2007). If this planet formed according to the core accretion model, then our simulations raise a number of important questions that should be addressed in the future. Simulations which cover a wider parameter set of initial conditions are presented in Kley and Nelson (2008).

First, a low mass, non accreting planet embedded in an eccentric disk experiences substantial growth in eccentricity (see Fig. 6.5). This has clear implications for the accretion of planetesimals because their velocity dispersion may become very large due to this effect. Thébault et al. (2004) examined the evolution of planetesimal orbits under the influence of the binary companion and aerodynamical gas-drag. They concluded that accretion of planetesimals would occur in the shear dominated regime because orbital alignment was maintained due to the gas-drag. This work, however, did not include the effects of an eccentric disk, and so it remains unclear whether planetesimal orbits will remain aligned. We will discuss the effects of including the full dynamics of the disk when calculating the orbital evolution of planetesimals in the γ Cep system in the next section.

A second issue is related to the type I migration of the giant planet core that must survive before gas accretion occurs. Figure 6.5 shows the non-accreting, low-mass planet undergoing quite rapid inward migration. The migration, however, is modulated by the eccentricity of the planet, such that at high eccentricity phases the migration rate decreases. It is possible that longer run times will show an essential stalling of this migration if the planet eccentricity grows beyond its final value of $e_p \simeq 0.3$.

Once gas accretion is switched on, our calculations have shown that a disk mass of about three Jupiter masses, where the outer disk radius is tidally truncated at $r \simeq 5$ AU, will be sufficient to grow a planet that is close to the minimum observed mass of $m_p \sin i \simeq 2.044 M_{Jup}$. It is also clear that we can construct a model in which a low-mass planet growing from an initially circular orbit can achieve a final mass of $m_p \simeq 2 M_{Jup}$, and have a final eccentricity of $e_p \simeq 0.1$. Calculations to see if a planetary core on an initially eccentric orbit (as expected from Fig. 6.5), will circularize as it accretes gas from the disk, such as in a self consistent model that fits the observation, will be interesting to construct.

A final comment regarding to the final mass of the planet. Our simulations suggest that a disk mass of about three Jupiter-masses will be enough to form a gas-giant of the required minimum mass. A future test of the models by which the planet in γ Cep may have formed (gravitational instability versus core accretion) will be the determination of its actual mass. We suspect that a disk that is massive enough to form a planet through gravitational instability will lead to a planet whose final mass is substantially larger than the minimum value observed.

6.3 Evolution of Planetesimals in a Circumstellar Disk with a Companion

We now describe preliminary results from simulations of planetesimals embedded in circumstellar disks with a companion star. We take as our basic model the disk and binary system described in Section 6.2. As in the models in which low mass protoplanets were considered, we evolve the system for 100 binary orbits prior to inserting 100 planetesimals of various sizes. At the point when the planetesimals are inserted, the disk mass is augmented so that it contains three Jupiter masses in total. The planetesimals are randomly distributed initially between orbital radii of 1.5 and 2.5 AU on circular Keplerian orbits. We consider here planetesimals whose physical radii are 100 m and 1 km. A broader range of sizes will be discussed in Kley and Nelson (2008). The planetesimals experience aerodynamic gas-drag using the standard formulae found in Weidenschilling (1977), and also experience the gravitational force due to the disk, central star and companion star. Although the simulations we describe here are two dimensional, we assume that the planetesimals lie in the disk midplane and calculate the volumetric density from the surface density by assuming that the vertical density profile is Gaussian with scale height $H = 0.05\,r$, where r is the orbital radius. We use linear interpolation to calculate the gas density and velocity at the planetesimal positions for use in the gas-drag formula. The evolution of the semimajor axes, eccentricities and longitudes of pericenter for five representative 100 m-sized planetesimals are shown in Fig. 6.9. We see that the planetesimals migrate inward on the expected timescale due to the aerodynamic gas-drag, and are also excited onto orbits with high eccentricity ($e \geq 0.12$). The eccentricity is driven upward primarily by gravitational interaction with the eccentric gas disk, and not because of direct interaction with the binary companion. As the planetesimals drift inward their eccentricity decays slightly but still remains significant. We see also that the eccentric orbits remain highly-aligned, suggesting that encounter velocities between these same-sized planetesimals will remain dominated by Keplerian shear, and therefore will remain small. Similar results were found for the 1 km-sized bodies.

The picture changes, however, when we consider planetesimals of different sizes. In Fig. 6.10 we show the evolution of two representative planetesimals, one being of size 100 m, the other being 1 km in radius. The top left panel shows the semimajor axes, and we see that the 100 m sized body (which is initially further out in the disc) crosses the orbit of the 1 km sized planetesimal after $\sim$700 orbits, since it migrates in at a higher speed due to aerodynamic drag. The top right panel of Fig. 6.10 shows the eccentricities, and the lower left panel shows the longitudes of

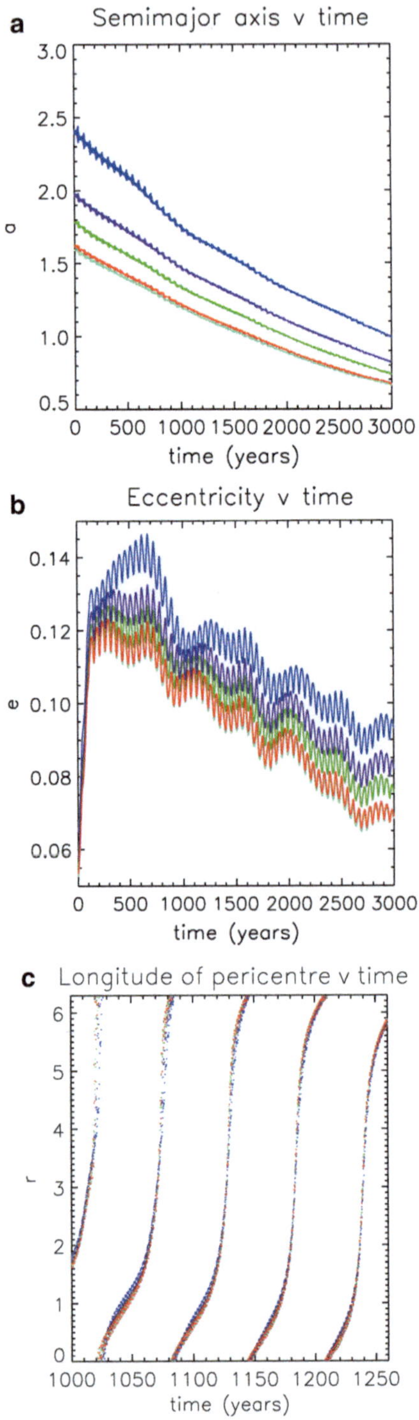

Fig. 6.9 The evolution of the semi-major axes (**a**), eccentricities (**b**), and longitudes of pericenter of embedded 100m-sized planetesimals in the circumstellar accretion disk (**c**)

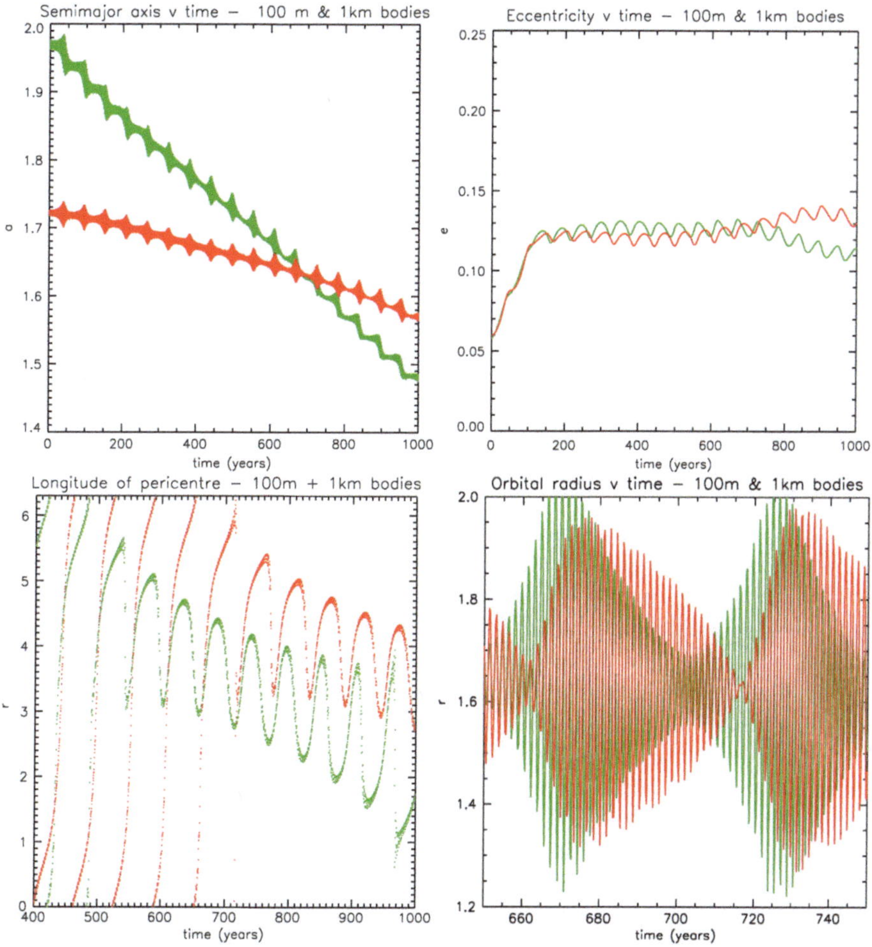

Fig. 6.10 The evolution of the semimajor axes (*top left panel*), eccentricities (*top right panel*), longitudes of pericenter (*lower left panel*) and orbital radii (*lower right panel*) for the 100 m and 1 km sized planetesimals described in the text. The 100 m sized planetesimal is shown by the green line, and the 1 km sized body is shown in red

pericenter. This latter panel shows that at the point where the two orbits cross, the eccentric orbits are significantly misaligned, resulting in the orbit crossing which is shown in the lower right panel, where the orbital radii are plotted. The implication is that large collisional velocities are highly likely between planetesimals of different sizes. Although we do not have sufficient numbers of particles to make statistically significant measures of planetesimal collision velocities, we note that eccentricities of $e \simeq 0.1$ at orbital radii $\simeq$1.5 AU imply encounter velocities $\simeq$2 km s^{-1}. Recent results presented by Paardekooper et al. (2008) who considered a similar model, suggest encounter velocities between 1 km and 5 km-sized bodies of $\simeq$500 m s^{-1}.

Simulations of colliding icy bodies with radii $\simeq$100 m performed by Benz and Asphaug (1999) suggest that disruption occurs for impact velocities $\simeq$15 m s^{-1}. Clearly the results presented here suggest that it will be difficult to build up large planetesimals and planetary embryos through planetesimal accretion in protoplanetary discs in close binary systems. This raises questions about the applicability of the standard core accretion model when applied to close binary systems such as γ Cep, as it would appear that impacts between planetesimals will be destructive rather than accretional.

6.4 Evolution of Planets in Circumbinary Disks

In this section we present the results of simulations that examine the evolution of both low and high mass protoplanets which form in circumbinary disks. A more expansive discussion of the work relating to low mass planets is presented in Pierens and Nelson (2007), and a detailed description of the simulations relating to high mass planets is presented in Nelson (2003).

We consider the interaction between a coplanar binary and protoplanet system and a two-dimensional, gaseous, viscous, circumbinary disk within which protoplanets are assumed to form. We do not address the formation process itself, but rather assume that circumbinary protoplanets can form, and examine its dynamical consequences.

Each of the stellar components and each protoplanet experience the gravitational force of the other two, as well as that due to the disk. For simplicity, we refer to a protoplanet as a planet in the rest of this section. The planet and binary orbits are evolved using a fifth-order Runge–Kutta scheme (Press et al. 1992). The force of the protoplanet on the disk, and of the disk on the planet, is softened using a gravitational softening parameter $b = 0.5a_p(H/r)$, where a_p is the semimajor axis of the planet, and H/r is the disk aspect ratio. We assume that the mass of the planet is fixed, and disk models have effective aspect ratio $H/r = 0.05$.

6.4.1 *Low Mass Circumbinary Planets*

The simulation described below was performed using the hydrodynamics code `GENESIS` (Pierens et al. 2005; de Val-Borro et al. 2006). The Shakura–Sunyaev viscosity parameter $\alpha = 2 \times 10^{-4}$, and the disk was initialized to have a mass of 0.04 $M_\odot$ within a radius of 40 AU. The interaction between gas disc and the circumbinary planet is expected to be very dominant compared to the interaction between the planet and any planetesimals that may be embedded in the circumbinary disc. Therefore dynamical friction due to planetesimals is not included. An expanded version of the following discussion is presented in Pierens and Nelson (2007).

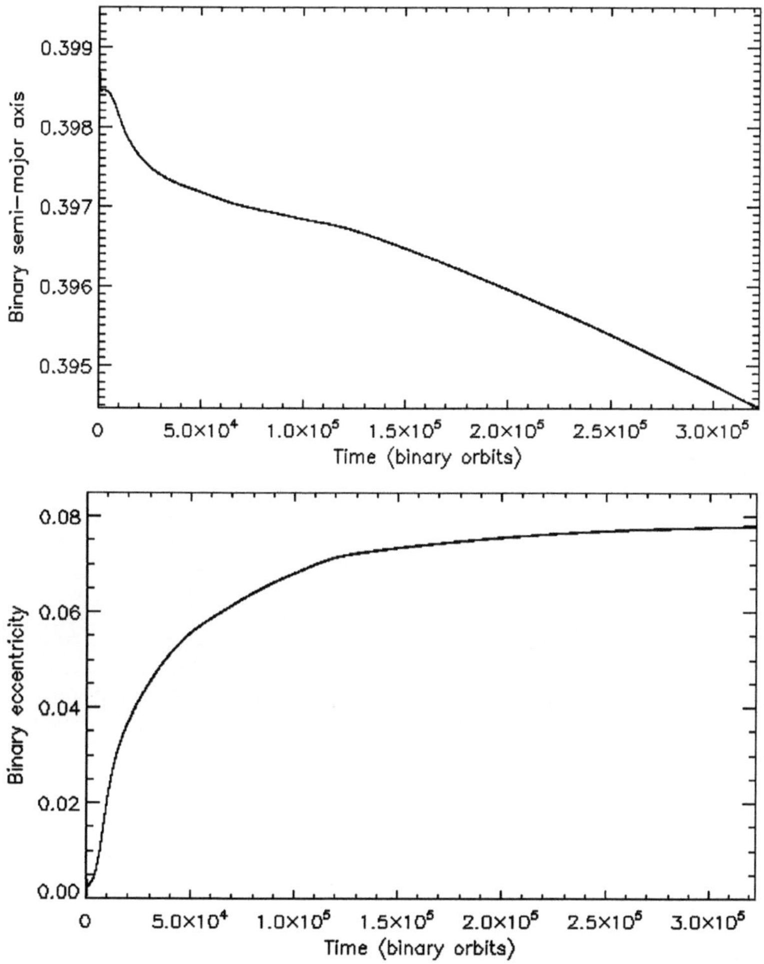

Fig. 6.11 The evolution of the binary elements due to interaction with the circumbinary disk. The *top panel* shows the semimajor axis evolution over time (expressed in binary orbits), and the *bottom panel* shows the eccentricity evolution. The binary orbital period is ~92 days

The simulation was initialized with a binary star system on a circular orbit surrounded by an unperturbed circumbinary disk. The stellar masses were $M_1 = (1/11)M_\odot$ and $M_2 = (1/110)M_\odot$ (i.e., the mass ratio was $q = 0.1$), and the semimajor axis $a_{bin} = 0.4$ AU. The top panel of Fig. 6.11 shows the slow decline of the binary semimajor axis over a timescale of about 80,000 years (the binary orbital period is approximately 92 days) and the bottom panel shows the growth and saturation of the binary eccentricity. As expected, interaction with the disk drives the growth of binary eccentricity through the action of eccentric Lindblad resonances (e.g., Papaloizou, Nelson and Masset 2001), and the eccentricity eventually reaches a steady value of $e_{bin} \simeq 0.08$, at which point additional resonances have presumably become active further out in the disc which contribute to eccentricity damping.

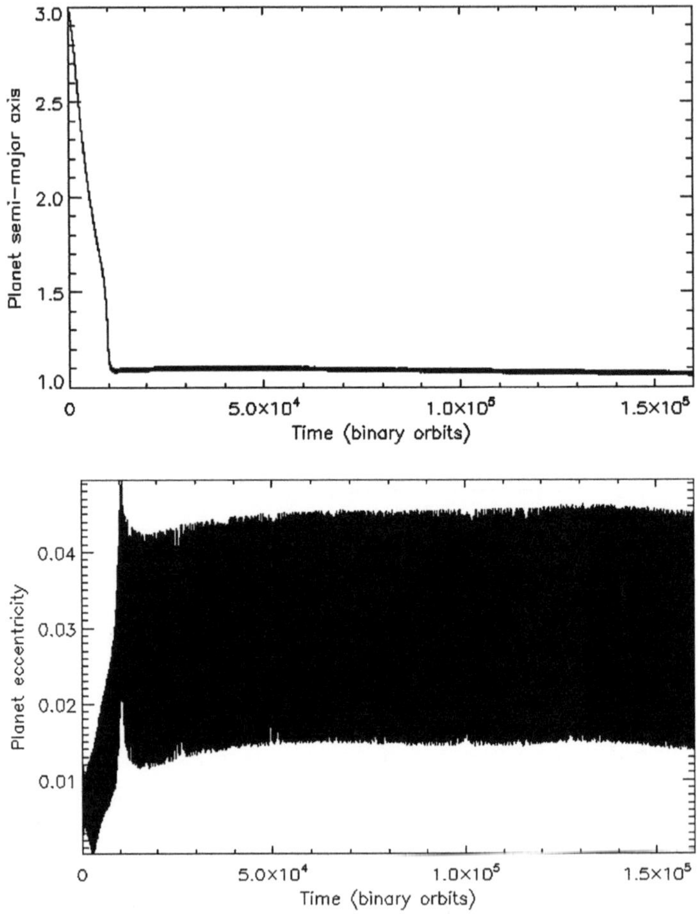

Fig. 6.12 The evolution of the planet elements due to interaction with the circumbinary disk. The *top panel* shows the semimajor axis evolution over time in years, and the *bottom panel* shows the eccentricity evolution

Once the binary eccentricity reaches a constant value, a low-mass protoplanet ($m_p = 50\ M_\oplus$) was inserted in the disk on a circular orbit with semimajor axis $a_p = 3$ AU and allowed to evolve. The planet migrates inward due to interaction with the disk, as shown in Fig. 6.12, which also shows the planet eccentricity evolution. As the planet semimajor axis reaches a value of $a_p \simeq 1.1$ AU, we see that migration suddenly stalls. This halting of migration appears to be robust, and occurs for planets whose masses are too small for gap formation in the gas disk to occur (Pierens and Nelson 2007). We ascribe this behavior to an increase in the corotation torque as the planet enters the inner cavity that is cleared by the tidal torques of the binary. A similar effect has been described by Masset et al. (2006) who show that planet migration can be halted due to the action of corotation torques at surface density transitions. As such, we expect this stalling of migration for low-mass planets to be a generic feature within circumbinary disks, and to occur near the edge of the tidally truncated cavity generated by the binary. We note here that the stability analysis presented by Holman and Wiegert (1999) suggests that the critical semimajor axis for stability of a planet orbiting a binary system with $e_b \simeq 0.08$ and mass ratio of 0.1 is $\simeq 2.3\ a_b$. The planets in our simulation stall at with semimajor axes $\simeq 3.0\ a_b$, suggesting that they will be stable on long time scales. The top panel of Fig. 6.13 shows the azimuthally averaged surface density in the disk as a function of radius

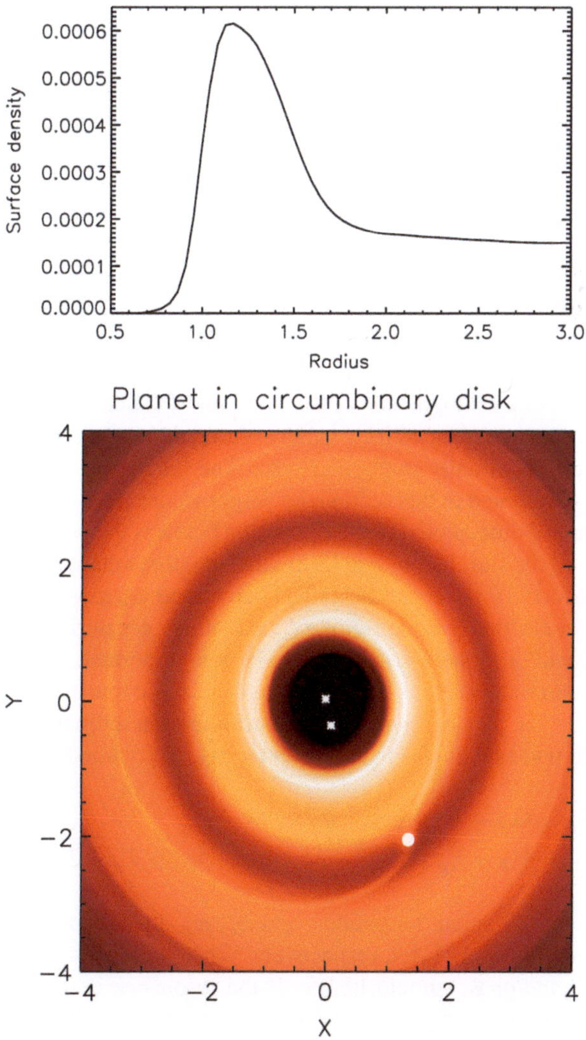

Fig. 6.13 The *top panel* shows the azimuthally averaged surface density profile at the end of the simulation. The *bottom panel* shows an image of the disk along with the planet and binary system. This image corresponds to an earlier time during which the planet is migrating inward toward the central binary system

at the end of the simulation, and illustrates the point that the planet stalls within the inner cavity due to corotation torques. The bottom panel shows an image of the binary, planet and circumbinary disk at the end of the simulation.

6.4.2 High Mass Circumbinary Planets

The simulations described below were evolved using the hydrodynamics code NIRVANA (Ziegler and Yorke 1997). The viscosity parameter is $\alpha = 5 \times 10^{-3}$, and the surface density is normalized such that the disk contains about four Jupiter masses interior to the initial planet semimajor axis (Nelson 2003). The total mass of the binary plus protoplanet system is assumed to be 1 $M_{\odot}$. We use units in which the gravitational constant $G = 1$, and the unit of length is approximately 3.6 AU. The initial binary semimajor axis is $a_{bin} = 0.4$ in our computational units, and the initial planet semimajor axis is $a_p = 1.4$. The latter corresponds to 5 AU in physical units. Thus the planet lies just outside the 6:1 mean motion resonance with the binary. According to the stability criterion obtained by Holman and Wiegert (1999), planets orbiting at this semimajor axis should be stable for all binary eccentricities considered here, though it is worth noting that Holman and Wiegert (1999) found islands of instability at the locations of exterior mean motion resonances in their study. The effect of a protoplanetary disc, however, was not included in their study.

Simulations were performed for a variety of initial binary eccentricities, e_{bin}, and the protoplanet was initially in circular orbit. The binary mass-ratio $q_{bin} = 0.1$ for all simulations presented in this section, but larger values were considered in Nelson (2003). The unit of time quoted in the discussion below is the orbital period at $r = 1$.

The results of the simulations can be divided into three main categories (Mode 1, Mode 2, and Mode 3), which are described below, and are most strongly correlated with changes in the binary mass-ratio, q_{bin}, and binary eccentricity e_{bin}. Changes to the disk mass and/or planet mass appear to be less important. Here we present the results of just three simulations that illustrate these basic modes of evolution.

In some runs, the planet entered the 4:1 mean motion resonance with the binary. The associated resonant angles in the coplanar case are defined by:

$$\begin{aligned} \psi_1 &= 4\lambda_s - \lambda_p - 3\omega_s \quad \psi_2 = 4\lambda_s - \lambda_p - 3\omega_p \\ \psi_3 &= 4\lambda_s - \lambda_p - 2\omega_s - \omega_p \quad \psi_4 = 4\lambda_s - \lambda_p - 2\omega_p - \omega_s \end{aligned} \tag{6.7}$$

where λ_s, λ_p are the mean longitudes, and ω_s, ω_p are the longitudes of pericenters of the secondary and planet, respectively. When the planet is in resonance with the binary, ψ_3 and/or ψ_4 should librate. If the resonance is such that ψ_3 and ψ_4 both librate, then ψ_1 and ψ_2 will also librate because they can be expressed as linear combinations of ψ_3 and ψ_4. In principle the planet is able to enter higher order resonances than 4:1, such as 5:1 or 6:1, since its initial location lies beyond these resonance locations. However, none of the simulations presented here resulted in such a capture. Test calculations indicate that capture into higher order resonances requires slower planetary migration rates than those that arise in these simulations. For significantly faster migration rates the planet may pass through the 4:1 resonance (Nelson 2003).

6.4.2.1 Mode 1 – Planetary Scattering

A number of simulations resulted in a close encounter between the planet and binary system, leading to gravitational scattering of the planet to larger radii, or into an unbound state. We label this mode of evolution as 'Mode 1'. Typically the initial scattering causes the eccentricity of the planet to grow to values $e_p \simeq 0.9$, and the semimajor axis to increase to $a_p \simeq 6-8$ in code units (this corresponds to an actual distance of 15–24 AU since a code unit corresponds to 2.5 AU. Note that the code unit in these simulations are different from the simulations of Section 6.4.2). In runs that were continued for significant times after this initial scattering, ejection of the planet could occur after subsequent close encounters.

We illustrate this mode of evolution using a simulation with $m_p = 3$ Jupiter-masses and $q_{bin} = 0.1$. A series of snapshots of the simulation are shown in Fig. 6.14. Mode 1 evolution proceeds as follows. The planet migrates in toward the central binary due to interaction with the circumbinary disk, and temporarily enters the 4:1 mean motion resonance with the binary. The migration and eccentricity evolution is shown in the top panel of Fig. 6.15, and the resonance angles are shown in the bottom panel. The resonant angle ψ_3 librates with low amplitude, indicating that the planet is strongly locked in the resonance. The resonance drives the eccentricity of the planet upward until it has a close encounter with the secondary star during or close to periapse. At this state, the planet is scattered out of the resonance into a high eccentricity orbit with significantly larger semimajor axis. We note that the existence of a resonance normally helps to maintain the stability of two objects orbiting about a central mass. However, when one of the objects is a star, the large perturbations experienced by the planet can cause the resonance to break when the eccentricities are significant. Once out of resonance, the chances of a close encounter and subsequent scattering are greatly increased. This provides a method of forming 'free-floating planets'.

6.4.2.2 Mode 2 – Near-Resonant Protoplanet

A mode of evolution was found in some of the simulations leading to the planet orbiting stably just outside of 4:1 resonance. We label this mode as Mode 2 and illustrate it by a simulation corresponds to $m_p = 1$, $q_{bin} = 0.1$, and $e_{bin} = 0.1$. The evolution of the orbital elements at this mode are shown in Fig. 6.16. As shown in this figure, the planet migrates inward and becomes *weakly* locked into the 4:1 resonance, with the resonant angle ψ_3 librating with large amplitude. The resonance becomes undefined and breaks when the planet's eccentricity momentarily approaches zero during its high amplitude oscillations that accompany the libration

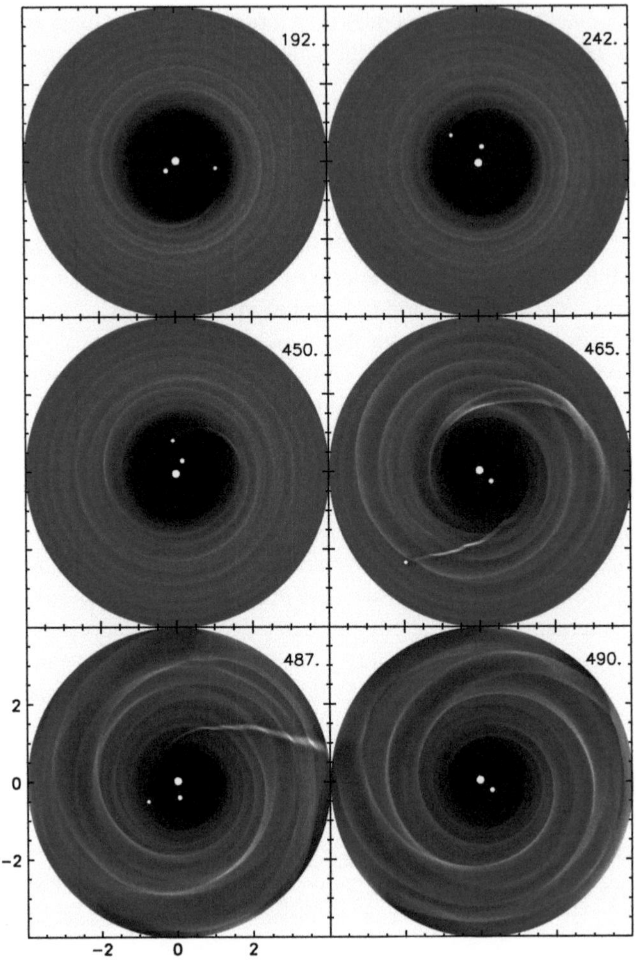

Fig. 6.14 This figure shows surface density contours for run in which the planet is ejected by the binary

of ψ_3. At this stage, the planet undergoes a period of outward migration through interaction with the disk by virtue of the eccentricity having attained values of $e_p \simeq 0.17$ once the resonance is broken. Unpublished simulations show that gap-forming planets orbiting in tidally truncated disks undergo outward migration if they are given eccentricities of this magnitude impulsively. The outward migration moves the planet to a safer distance away from the binary, helping to avoid instability. We note that the stability criterion derived by Holman and Wiegert (1999) for binaries with mass-ratios of 0.1 and eccentricities 0.1 suggests that planets orbiting with semimajor axes $a_p = 2.4066\, a_{bin}$ are on the cusp of instability. The planet shown in Fig. 6.16 orbits beyond $a_p = 2.52\ a_{bin}$, and is therefore expected to be stable in the long-term.

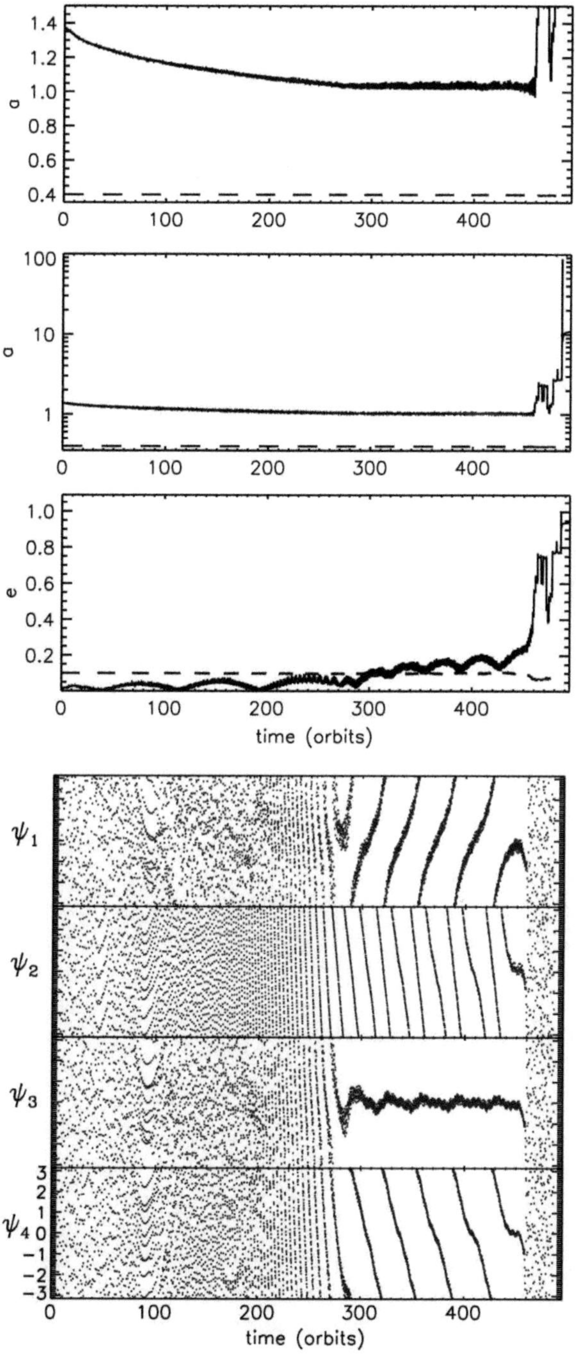

Fig. 6.15 The *top panel* shows the semimajor axes and eccentricities for a run in which planet is scattered by the binary. The *bottom panel* shows the resonant angles for the 4:1 resonance, indicating capture into this resonance prior to scattering

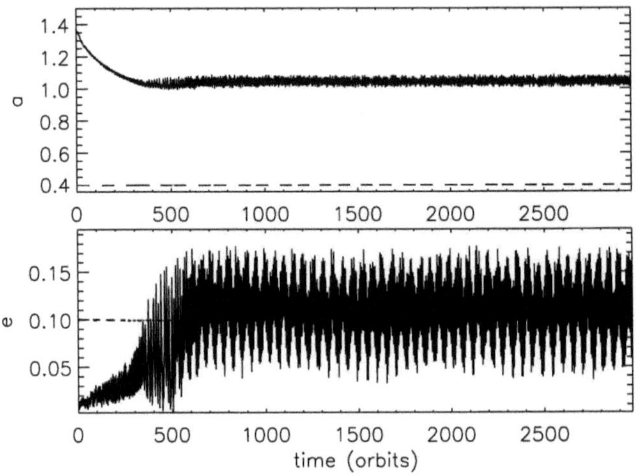

Fig. 6.16 This figure shows semimajor axes and eccentricities for the Mode 2 run described in the text

Once the planet has migrated to just beyond the 4:1 resonance, the outward migration halts since the planet's eccentricity reduces slightly. Planet, then, remains there for the duration of the simulation and the system achieves a balance between eccentricity damping by the disk and eccentricity excitation by the binary, maintaining a mean value of $e_p \simeq 0.12$ (Nelson 2003). The torque exerted by the disk on the protoplanet is significantly weakened by virtue of the finite eccentricity (Nelson 2003), preventing the planet from migrating back toward the binary.

Continuation of this run in the absence of the disk indicates that the planet remains stable for over 6×10^6 orbits. This is in good agreement with the stability criteria obtained by Holman and Wiegert (1999) since the protoplanet lies just outside of the zone of instability found by their study.

6.4.2.3 Mode 3 – Eccentric Disk

A mode of evolution was found in which the planetary migration was halted before the protoplanet could approach the central binary and reach the 4:1 resonance. This only occurred when the central binary had an initial eccentricity of $e_{bin} \geq 0.2$. The migration stalls because the circumbinary disk becomes eccentric. We label this mode of evolution as Mode 3, and illustrate it using a simulation with $m_p = 1\ M_{Jup}$, $q_{bin} = 0.1$, and $e_{bin} = 0.2$. The top panel of Fig. 6.17 shows snapshots of the surface density at different times during the simulation, with the disk becoming noticeably eccentric. Interaction between the planet and the eccentric disk leads to a dramatic reduction or even reversal of the time-averaged torque driving the migration. This is because the disk–planet interaction becomes dominated by the $m = 1$ surface density perturbation in the disk rather than by the usual interaction at Lindblad

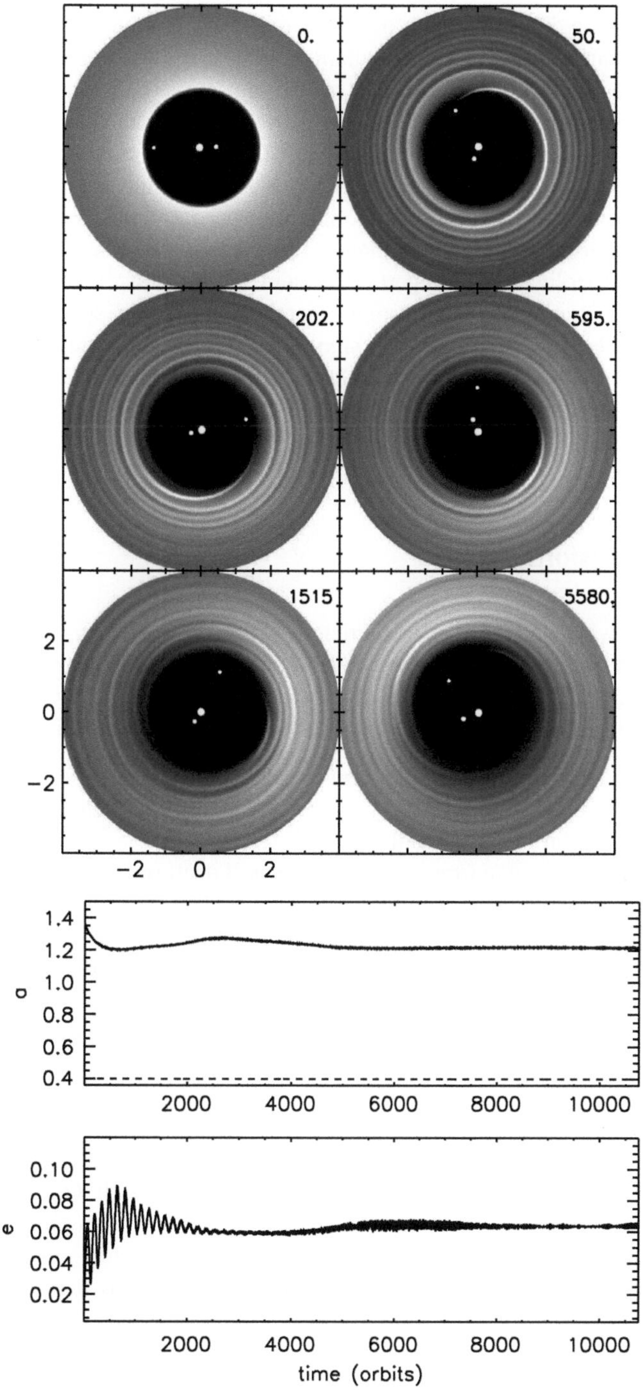

Fig. 6.17 The *top panel* shows contours of surface density for the Mode 3 run described in the text. The *bottom panel* shows the resulting changes to the semimajor axis and eccentricity of the protoplanet

resonances in the disk. Here m is the azimuthal wave number. The value of $m = 1$ refers to eccentric disks. Linear calculations of planets orbiting in eccentric disks also show the possibility of outward or stalled migration (Papaloizou 2002).

The bottom panel of Fig. 6.17 shows the evolution of the semimajor axis and eccentricity of the planet, illustrating the stalled migration. Simulations of this type can be run for many thousands of planetary orbits without any significant net inward migration occurring. Such systems are likely to be stable long after the circumbinary disk has dispersed, since the planets remain in the region of stability defined by the work of Holman and Wiegert (1999) and are probably the best candidates for finding stable circumbinary extrasolar planets. Interestingly, spectroscopic binary systems with large eccentricities are significantly more numerous than those with lower eccentricities (Duquennoy and Mayor 1991; Mathieu et al. 2000), suggesting that circumbinary planets may be common if planets are able to form in circumbinary disks.

6.5 Conclusions

Much of the work presented in this article is presently under active research. More details can be found in (Pierens and Nelson 2007; Paardekooper et al. 2008; Kley and Nelson 2008). The conclusions about planet formation and evolution in binary systems via the core-accretion scenario that we are able to draw thus far are:

- In systems such as γ Cep, the nascent circumstellar disk is expected to be tidally truncated at a radius of $\simeq$4 AU, and to be driven into an eccentric and precessing state by the binary gravitational potential.
- A low-mass planet that forms in such a disk will itself become eccentric, and will migrate inward on a fairly rapid time scale.
- Gas accretion onto such a planet is likely to be highly efficient because of the induced orbital eccentricity, such that a large fraction of the disk gas will accrete onto the planet. Simulations indicate that a gas disk containing $\simeq$3 Jupiter-masses will form a planet of $\simeq$2 Jupiter-masses, as required to fit the minimum mass of the planet detected in the γ Cep system.
- Simulations of different-sized planetesimals orbiting in a tidally truncated and eccentric protoplanetary disk indicate that high velocity collisions are likely, in agreement with Paardekooper et al. (2008). Such collisions will probably lead to fragmentation of the planetesimals rather than their growth. Our simulations of planetesimals with the same size show that the orbits remain aligned, in agreement with the results Thébault et al. (2004), suggesting that collisions between equal sized bodies will not be destructive. Further work is required to confirm this picture.
- Low-mass planets in circumbinary disk migrate inward until they reach the gap edge, where they appear to stall due to the action of corotation torques.

- Should these low-mass planets grow to become gas giants, a range of outcomes seem likely. These include stalled migration leading to the formation of stable circumbinary giant planets, and inward migration followed by scattering and ejection by the central binary.

References

Armitage, P. J., Clarke, C. J., and Tout, C. A. 1999, MNRAS, 304, 425
Artymowicz, P. and Lubow, S. H. 1994, ApJ, 421, 651
D'Angelo, G., Kley, W., and Henning, T. 2003, ApJ, 586, 540
de Val-Borro, M., Edgar, R. G., Artymowicz, P., et al. 2006, MNRAS, 370, 529
Duquennoy, A. and Mayor, M. 1991, A&A, 248, 485
Dvorak, R., Pilat-Lohinger, E., Bois, E., et al. 2004, in Revista Mexicana de Astronomia y Astrofisica Conference Series, 222–226
Eggenberger, A., Udry, S., and Mayor, M. 2004, A&A, 417, 353
Haghighipour, N. 2006, ApJ, 644, 543
Haghighipour, N. 2007, ApJ, 666, 436
Hatzes, A. P., Cochran, W. D., Endl, M., et al. 2003, ApJ, 599, 1383
Holman, M. J. and Wiegert, P. A. 1999, AJ, 117, 621
Innanen, K. A., Zheng, J. Q., Mikkola, S., and Valtonen, M. J. 1997, AJ, 113, 1915
Kley, W. 1989, A&A, 208, 98
Kley, W. 1999, MNRAS, 303, 696
Kely, W. 2000, in Birth and Evolution of Binary Stars, IAU Symposium No. 200 on The Formation of Binary Stars, Eds. B. Reipurth and H. Zinnecker, 211
Kley, W. and Burkert, A. 2000, in ASP Conference Series, 219: Disks, Planetesimals, and Planets, 189
Kley, W. and Nelson, R. 2008, A&A, 486, 617
Larwood, J. D., Nelson, R. P., Papaloizou, J. C. B., and Terquem, C. 1996, MNRAS, 282, 597
Lissauer, J. J., Quintana, E. V., Chambers, J. E., Duncan, M. J., and Adams, F. C. 2004, in Revista Mexicana de Astronomia y Astrofisica Conference Series, 99–103
Masset, F. 2000, A&AS, 141, 165
Masset, F. S., Morbidelli, A., Crida, A., and Ferreira, J. 2006, ApJ, 642, 478
Mathieu, R. D., Ghez, A. M., Jensen, E. L. N., and Simon, M. 2000, Protostars and Planets IV, 703
Monin, J.-L., Clarke, C. J., Prato, L., and McCabe, C. 2007, *in Protostars and Planets V*, ed. B. Reipurth, D. Jewitt, and K. Keil, 395–409
Mugrauer, M., Neuhaeuser, R., and Mazeh, T. 2007, å, 469, 755
Mugrauer, M. and Neuhäuser, R. 2005, MNRAS, 361, L15
Nelson, A. F. 2000, ApJ, 537, L65
Nelson, R. P. 2003, MNRAS, 345, 233
Nelson, R. P., Papaloizou, J. C. B., Masset, F. S., and Kley, W. 2000, MNRAS, 318, 18
Neuhäuser, R., Mugrauer, M., Fukagawa, M., Torres, G., and Schmidt, T. 2007, A&A, 462, 777
Paardekooper, S., Thebault, P., and Mellema, G. 2008, MNRAS, 386, 973
Papaloizou, J. C. B. 2002, A&A, 388, 615
Papaloizou, J. C. B. 2005, A&A, 432, 757
Pierens, A., Dutrey, A., Guilloteau, S., and Huré, J.-M. 2005, in SF2A-2005: Semaine de l'Astrophysique Francaise, ed. F. Casoli, T. Contini, J. M. Hameury, and L. Pagani, 733
Pierens, A. and Nelson, R. P. 2007, A&A, 472, 993
Queloz, D., Mayor, M., Weber, L., et al. 2000, A&A, 354, 99
Quintana, E. V., Adams, F. C., Lissauer, J. J., and Chambers, J. E. 2007, ApJ, 660, 807
Thébault, P., Marzari, F., Scholl, H., Turrini, D., and Barbieri, M. 2004, A&A, 427, 1097

Torres, G. 2007, ApJ, 654, 1095
Turrini, D., Barbieri, M., Marzari, F., Thebault, P., and Tricarico, P. 2005, Memorie della Societa Astronomica Italiana Supplement, 6, 172
Turrini, D., Barbieri, M., Marzari, F., and Tricarico, P. 2004, Memorie della Societa Astronomica Italiana Supplement, 5, 127
Verrier, P. E. and Evans, N. W. 2006, MNRAS, 368, 1599
Weidenschilling, S. J. 1977, MNRAS, 180, 57
Ziegler, U. and Yorke, H. 1997, Computer Physics Communications, 101, 54

Chapter 7
Dynamics and Planet Formation in/Around Binaries

Francesco Marzari, Philippe Thébault, Steve Kortenkamp, and Hans Scholl

7.1 Introduction

7.1.1 The "Standard" Planet Formation Scenario

It is believed that extrasolar planets, detected in orbits around an increasing number of nearby stars may have formed either by core accretion or by disk instability (e.g., Boss 1997, 2007). While core accretion requires a few million years to form a gas-giant planet (Alibert et al. 2004), in the competing model, self-gravitating clumps formed by disk instability will contract to planetary densities in times of a few hundred years (Boss 2004, 2006). The two formation mechanisms are both viable and, at the moment, there is no reason to definitively exclude one in favor of the other. However, it is believed that extrasolar planets around metal-rich stars preferentially formed by core accretion because of a significant parent-star metallicity correlation (Santos et al. 2004). The recent finding of an exoplanet with a massive core of about 70 $M_{\oplus}$ (Sato et al. 2005) seems to confirm this view. In this chapter we will concentrate on planetesimal accretion as a common mechanism for forming either terrestrial planets or the core of gaseous giants.

A fundamental step in the "standard" scenario for the formation of planets, either terrestrial or gaseous giant, is the collisional accumulation of kilometer-sized

F. Marzari (✉)
Dipartimento di Fisica, Via Marzolo 8, 35131 Paduva, Italy
e-mail: marzari@pd.infn.it

P. Thébault (✉)
Observatoire de Paris, Section de Meudon, F-92195 Meudon, Principal Cedex, France
e-mail: philippe.Thebault@obspm.fr

S. Kortenkamp
Planetary Science Institute, 1700 East Fort Lowell Road, Tucson, AZ 85719-2395
e-mail: kortenka@psi.edu

H. Scholl
Observatoire de la Côte d'Azur, 06304 Nice, Cedex 4, France
e-mail: scholl@obs-nice.fr

N. Haghighipour (ed.), *Planets in Binary Star Systems*, Astrophysics and Space Science Library 366, DOI 10.1007/978-90-481-8687-7_7,

planetesimals in successively larger bodies. It is widely accepted that planetesimals emerge from the coagulation of silicate and ice grains in the midplane of protoplanetary disks on timescales of the order of 10^4 years, even though at present the details of this coagulation process are not well understood. Colliding with each other, planetesimals grow into large planetary embryos (Lunar- to Mars-size) through a phase of "runaway" and "oligarchic" growth on a timescale of the order of 10^4–10^5 years (Greenberg et al. 1978; Wetherill and Stewart 1989; Barge and Pellat 1993; Lissauer 1993; Kokubo and Ida 1998, 2000; Rafikov 2003, 2004). This is a robust finding corroborated by distinct numerical methods (see review by Kortenkamp et al. 2000). For terrestrial planets, the formation process reaches completion after a final phase of giant impacts and mutual accretion of planetary embryos resulting in 10^{27}–10^{28} g full-size bodies in about 10^7–10^8 years on stable separated orbits (Chambers and Wetherill 1998). In the case of giant planets, protoplanets accumulate into a solid core that, once sufficiently massive, begins to accrete a gaseous envelope. Initially, this envelope is in hydrostatic equilibrium sustained by the luminosity provided by the accreting planetesimals and is increasing at a slow rate compared to that of the solid core. However, at a critical core mass (in between 5 and 15 $M_\oplus$) a phase of runaway gas accretion occurs, the planet mass increases rapidly and it reaches its final value on a timescale estimated to be only a few 10^3 years. For a given density of solids and gas in the disk, hydrodynamical calculations can determine for which mass of the core the envelope is able to remain static or collapse, and the rate at which the mass flows onto the planet (see, e.g., Wuchterl et al. 2000). At the end of the gas rapid infall, the planet evolves into the isolated stage during which it cannot grow any further because of the exhaustion of the gas supply either because of its removal by effects of the star or as a result of gap formation around the planet orbit triggered by disk-planet tidal interaction (Papaloizou and Lin 1984). The planet finally cools down and contracts to its present size. This scenario for giant planet formation, called core accretion (Pollack et al. 1996), is based on the interior models for Jupiter and Saturn which strongly suggest the presence of a dense central core surrounded by an envelope composed mainly by hydrogen and helium for both planets. There are large uncertainties on the mass of the core and detailed calculations performed by e.g., Guillot (1999) and Guillot (2004) predict an upper limit of 10 $M_\oplus$ for Jupiter's core while for Saturn the core mass ranges from 6 to 17 $M_\oplus$.

In a given protoplanetary disk, the potentiality of a planetesimal swarm to accumulate into larger bodies is measured by two critical parameters.

- *The Surface Density of Solid Material* The concentration of condensible material determines the final mass in the case of a terrestrial planet, while it must lead to a reasonable formation time for the core of a giant planet that must be compatible with the lifetime of the gaseous disk. Disk ages are estimated to range from 0.1 to 10 Myr (Strom et al. 1993; Haisch et al. 2001; Chen and Kamp 2004), while 3 Myr is the age at which half of the stars show evidence of disks. Simulations of the accumulation of Jovian planets (Bodenheimer and Lin 2002) show that if the surface density of solids in the disk is assumed to be at least three times larger than that of the minimum-mass solar nebula (Weidenschilling 1977), then the formation time for Jupiter mass planets is within 2–3 Myr at 5 AU from the star. If planetary migration is included in the simulations, the timescale for forming Jupiter is reduced to 1–2 Myr even with solid surface density closer to the minimum mass solar nebula (Rice and Armitage 2003).

- *The Relative Encounter Velocities Between Planetesimals* The random encounter velocities in a planetesimal swarm determine whether mutual impacts will result in accretion or on the contrary, in cratering or fragmentation. The key mechanisms controlling the impact velocities are: (1) mutual gravitational encounters between planetesimals, (2) physical collisions, (3) gas-drag, in presence of the gaseous component of the disk. Several numerical studies have shown that in an unperturbed swarm of kilometer-sized planetesimals around a single star, the average random encounter velocity $\langle \Delta v \rangle$ is low enough to favor the collisional accumulation of bodies (e.g., Greenberg et al. 1978; Wetherill and Stewart 1989, 1993; Weidenschilling et al.1991, 1997). At the early stages of accretion, a low $\langle \Delta v \rangle$ is also a fundamental requirement for permitting a phase of rapid "runaway" growth which significantly shortens the timescale for the formation of planetary embryos. Indeed, under this condition, the ratio $v_{esc(R)}/\langle \Delta v \rangle$ between the escape speed of the largest bodies and the average impact velocities is $\gg$1, thus making their gravitational cross-section quite larger than the geometrical cross-section. This gravitational focusing enhances the growth-rate of the larger planetesimals which grow much faster than the rest of the population (e.g., Lissauer 1993). A direct consequence of this crucial $\langle \Delta v \rangle$ dependence is that planetesimal growth is very sensitive to any increase of the encounter velocities. It could thus be significantly slowed down or even stopped if the gravitational influence of a massive body external to the swarm, like a pre-existing giant planets (e.g., Jupiter in our solar system), or a companion star in a binary system is able to excite large impact velocities. Thébault et al. (2002) have shown that the formation of terrestrial planets in those exoplanetary systems hosting a giant planet on an external orbit may have been indeed halted. They find that in both the ϵ Eridani and 47 Uma planetary systems, the relative velocity between planetesimals in the terrestrial planet region may have quickly exceeded the critical threshold value v_{ero} beyond which erosion dominates before the completion of the runaway growth if the massive outer planet rapidly reaches its present mass. Of course, the threshold velocity that preferentially result in disruption rather than accretion is a sensitive function of the internal strength of planetesimals, their bulk density, and the fraction of impact energy partitioned into the fragments after the impact. These physical parameters strongly depend on the initial radial density and temperature profile in the disk midplane.

7.1.2 Planets in Binary Star Systems

In this chapter we will focus on how the presence of a companion star affects the planetesimal accretion process either in the formation of terrestrial planets or the core of a giant planet. While the existence of extrasolar Earth-like planets has yet to be assessed, some of the giant exoplanets (more than 40 at the time of the writing of this chapter) have been detected in the so-called S-type orbits that encircle one component of a binary system (e.g., Raghavan et al. 2006; Desidera and Barbieri 2007). Most of these are widely separated pairs where the stellar companion probably does not have a significant influence on the planetesimal accretion around the other star. However, in three of the systems, γ Cephei, GL 86, and HD41004, the companion is within 20–25 AU from the star hosting the planet. The γ Cephei system consists of a central star with an estimated mass of $1.40 \pm 0.12 M_\odot$, a planet orbiting at 2.13 ± 0.05 AU with a minimum mass of $1.46 \pm 0.32\ M_J$ and a companion star on an eccentric orbit ($e = 0.4085 \pm 0.0065$) with semimajor axis of 20.18 ± 0.66 AU and mass of about $0.4 M_\odot$ (see also the chapter by Hatzes et al. in this volume Hatzes et al. 2003; Torres 2007; Neuhäuser et al. 2007). GL 86 is a dwarf star somewhat less massive than the Sun (about $0.7 M_\odot$, e.g., Santos et al. 2004 with a white dwarf companion of $0.55 M_\odot$ Mugrauer and Neuhauser 2005 on an orbit having a semimajor axis of 18.42 AU and an eccentricity of e = 0.3974 (Lagrange et al. 2006). A massive Jovian planet ($M \sin(i) = 3.91 M_J$) moves around the primary star on an almost circular orbit ($e = 0.0416$) with a semimajor axis of $a = 0.113$ AU (Eggenberger et al. 2003). HD41004 A is a $0.7 M_\odot$ star with a dwarf M4V companion orbiting at $\simeq$22 AU. The primary star has a $M \sin(i) = 2.3 M_J$ planetary companion on a very eccentric $e = 0.39$ orbit at $a = 1.31$ AU (Zucker et al. 2004; Raghavan et al. 2006). Note that this system is a hierarchical quadruple, since the stellar companion is itself orbited at $a = 0.016$ AU by a brown dwarf of minimum mass 18.4 M_J (Zucker et al. 2004).

The observed characteristics of these systems, e.g., the proximity of the companion star and its eccentric orbit, strongly suggest that the presence of the secondary star must have had an influence on the formation of the detected planets. The gravitational pull of the companion star acts on the $\langle \Delta v \rangle$ of the planetesimal disk around the primary star. As previously mentioned, this can modify the course of the accretion process which strongly depends on the ratio $v_{esc}/\langle \Delta v \rangle$, where v_{esc} is the surface escape velocity of the biggest accreting bodies. Timing is an important issue since the perturbations will be effective in altering the planetesimal formation process only if t_{sec}, the timescale required by the secular perturbations of the star to induce a significant Δv increase, is shorter than t_{grow}, the timescale for the runaway/oligarchic formation of embryos which is typically of the order of 10^4 to 10^5 years. We can outline three possible different evolutions for a perturbed planetesimal swarm:

- If $\langle \Delta v \rangle$ remains smaller than v_{esc} during the growth phase, planetary formation proceeds almost unaffected by the companion perturbations and the "classical" planetesimal accumulation scenario holds.
- For values of $\langle \Delta v \rangle$ larger than v_{esc} but still smaller than v_{ero}, the threshold velocity for erosion to dominate over accretion, planetary accretion will still be possible but runaway growth will either not occur or will occur only after large enough bodies have formed such that $v_{esc} > \langle \Delta v \rangle$ (the so-called Type II runaway growth described in Section 7.6 below).

- If $\langle \Delta v \rangle$ is increased beyond v_{ero}, in the majority of collisions cratering and fragmentation will overcome accretion and the planetesimal population will be slowly ground down to dust, failing to form a full size planet.

In the following, we will focus on the determination of $\langle \Delta v \rangle$ for different parameters of the binary star system with the goal of evaluating when planetesimal accretion is possible. We have performed a series of deterministic numerical simulations, following the dynamical evolution of a swarm of test planetesimals, under the influence of the companion star's gravitational pull and friction by the gas in which the planetesimals are imbedded. Crucial parameters, such as the companion star's mass (relative to the mass of the primary, i.e., the binary mass-ratio) m_b, semimajor axis a_b and orbital eccentricity e_b are explored as free parameters.

We use a code initially developed for the study of planetesimal systems perturbed by giant planets (Thébault and Brahic 1998; Thébault and Beust 2001; Thébault et al. 2002) and adapted to the circumprimary case (Marzari and Scholl 2000; Thébault et al. 2004, 2006). This 3-D code has a close encounter tracking algorithm which enables precise determination of the Δv within the system.

In a first step we present a detailed study of the planetesimal secular dynamics in presence of a companion star (Section 7.2). We will then proceed by introducing the effects of gas-drag and by estimating how the combined action of the friction force with the secular perturbations affect the value of $\langle \Delta v \rangle$ (Section 7.3). We then compare the derived $\langle \Delta v \rangle$ with an approximate estimate of v_{ero} to derive a yes-or-no criterion for planetary formation as a function of the binary orbital parameters (Section 7.4). Finally, in Section 7.5, we present and describe one possible accretion-growth mode for planetesimals in perturbed binary star systems, the so-called Type II runaway growth mechanism.

7.2 Planetesimal Dynamics in a Binary Star System: the Secular Approximation

Hereinafter, we will assume that the orbit of the secondary star is coplanar to the planetesimal disk so that we can neglect the inclination of the bodies. This is a reasonable assumption when the star is close to the primary because of the short relaxation time of the disk into the plane of the binary orbit.

Heppenheimer (1978) developed a simplified theory for the evolution of the planetesimal eccentricity with time based on an expansion of the disturbing function in a power series of the ratio of the semimajor axes a/a_b where a is the semimajor axis of the planetesimal and a_b that of the companion star. The approximation holds if a is small compared to a_b and it stays within the critical semimajor axis a_c for dynamical stability derived using direct numerical integrations by Holman and Wiegert (1999). By truncating the disturbing function of the secondary star at the second order in the eccentricity of the planetesimal, the secular equations simplify and lead to the following expressions for the forced eccentricity e_F and the frequency u of circulation of the perihelion ϖ:

$$e_F = \frac{5}{4}\frac{a}{a_b}\frac{e_b}{1-e_b^2}, \tag{7.1}$$

$$u = \frac{3}{4}\pi\frac{1}{\left(1-e_b^2\right)^{3/2}}m_b\frac{a^{3/2}}{a_b^3}, \tag{7.2}$$

with

$$\tan(\varpi(t)) = -\frac{\sin(2ut)}{1-\cos(2ut)}. \tag{7.3}$$

Note that the units chosen in these equations are such that all masses are renormalized to the mass of the primary star (which has been assumed to be 1 solar-mass), all distances to 1 AU, and all times to 1 year, i.e., the orbital period at 1 AU from a solar-mass primary. In these units, the quantity m_b, the mass-ratio of the binary, will also represent the mass of the secondary star when the primary is solar-mass.

Thebault et al. (2006) compared the predictions of the above equations with direct numerical simulations and found significant discrepancies, especially for the oscillation frequency u. They empirically derived a revised expression for the frequency u which reads:

$$u = \frac{3}{4}\pi\frac{1}{\left(1-e_b^2\right)^{3/2}}m_b\frac{a^{3/2}}{a_b^3}\left[1+32\frac{m_b}{\left(1-e_b^2\right)^3}\left(\frac{a}{a_b}\right)^2\right]. \tag{7.4}$$

The discrepancy between the numerical and analytical estimate of u is reduced from about 70%, in most of the cases with high e_b and e, to less than 5%.

To illustrate the dynamical evolution of a planetesimal swarm under the secular perturbations of a companion star, we have performed a "pedagogical" numerical simulation where we integrate the orbits of massless test bodies, initially on circular orbits, for about 10^5 years. In Fig. 7.1 we plot the location of each particle in the (a,e) plane at different times. The eccentricity of each single body oscillates with a frequency u that depends on the semimajor axis of the body. If we freeze the dynamical system at subsequent times, the slow de-phasing of the eccentricity of nearby orbits will translate into a wavy pattern whose density grows with time.

7.3 Effects of the Secular Perturbations on $\langle \Delta v \rangle$

If the orientations of the perihelia were fully randomized in the planetesimal swarm, the effect of the secular perturbations of the companion star on $\langle \Delta v \rangle$ would be easy to derive through the standard relation (Lissauer and Stewart, 1993):

$$\langle \Delta v \rangle \simeq (5/4)^{1/2} e_F v_K \tag{7.5}$$

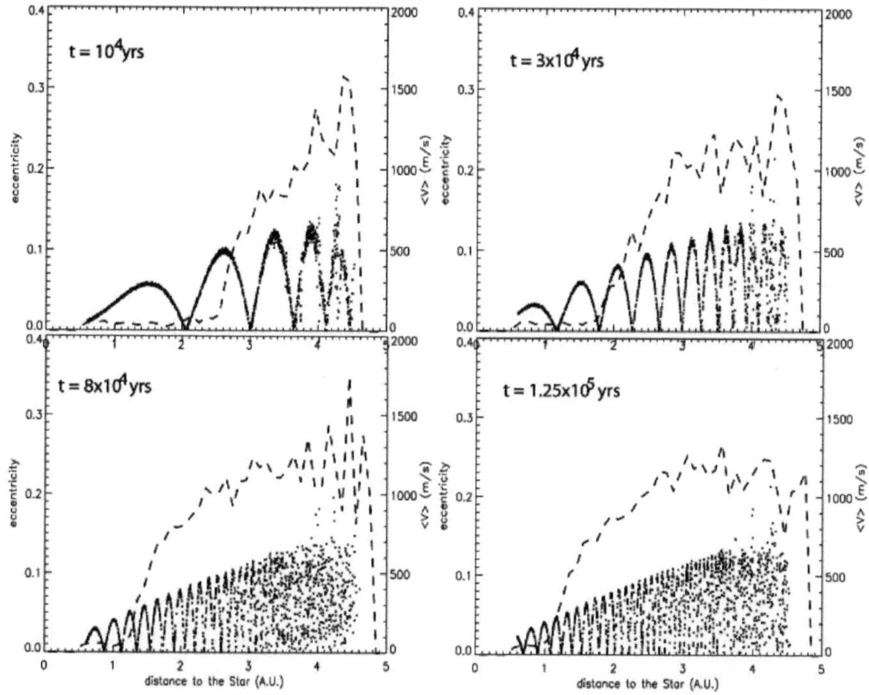

Fig. 7.1 Evolution of a test particle population perturbed by a stellar companion having approximately the characteristics of γ Cephei with $m_b = 0.25$, $a_b = 20$ AU, and $e_b = 0.3$. The *dotted line* represents the distribution of average encounter velocities between the bodies of the system

where v_K is the local Keplerian velocity. According to this equation, even very low values of forced eccentricity would halt accretion. In the γ Cephei binary system, for example, $\langle \Delta v \rangle$ would be approximately 1.8×10^3 m s^{-1} at 2 AU, the distance from the star where the planet has been found. This large relative velocity would certainly prevent mutual accretion in any swarm of kilometer-sized planetesimals (which have escape velocities of the order of a few m s^{-1}). In addition, a proper eccentricity component e_p has to be added to the planetesimal eccentricities, causing oscillations of the eccentricity around e_F with an amplitude equal to e_p, which leads to even higher impact velocities. If this was the scenario in the early phases of planetesimal accumulation in close binary systems, secular perturbations would pose a serious threat to planetary formation. However, the predictions of Eq. 7.5 strongly depend on the assumption that planetesimal trajectories have a randomized distribution of orbital angles. This is not true at the beginning of the dynamical evolution, because secular perturbations force a strong *phasing* of the orbits. The large eccentricity oscillations induced by the companion star are in-phase and associated to a very efficient periastron alignment between neighboring orbits. This configuration leads to planetesimal encounters on almost tangential trajectories and, notwithstanding the very high values of forced eccentricity, to low and accretion friendly *relative velocities*, dominated by the Keplerian shear.

At subsequent times, however, the dependence of the precession frequency u of ϖ on $a^{3/2}$ causes a progressive de-phasing of nearby orbits. At some point neighboring orbits eventually cross [see the detailed discussion in Thébault et al. (2006)], leading to a sudden and very sharp increase of the encounter velocities. This is illustrated in Fig. 7.1 where we have superimposed to the (e, a) graph the corresponding evolution of the average relative velocity $\langle \Delta v \rangle$. The transition between the inner regions, where the orbital phasing is still strong by preventing orbital crossing, and the outer regions, where orbits already cross is very abrupt: $\langle \Delta v \rangle$ changes from a few m s^{-1} to about 1 km s^{-1}, an accretion-inhibiting value for kilometer-sized planetesimals, within $\simeq$0.1 AU. Thébault et al. (2006) have derived an analytical estimate of the location in semimajor axis a_{cross} where the orbital crossing occurs as a function of time:

$$C_1 a_{cross}^{1/2} \left[\left(\frac{a_{cross}}{a_b} \right)^2 + C_2 \left(\frac{a_{cross}}{a_b} \right)^4 \right] = \frac{1}{t}, \tag{7.6}$$

where

$$C_1 = \frac{45\pi}{16} \frac{e_b}{\left(1 - e_b^2\right)^{5/2}} m_b \frac{1}{a_b^2}; \quad C_2 = \frac{224}{3} \frac{m_b}{\left(1 - e_b^2\right)^3}. \tag{7.7}$$

Conversely, for a given location a in the planetesimal swarm, the time t_{cr} required by the wavefront of high impact speeds to reach a_{cross} is (Thébault et al. 2006):

$$t_{cr} = 6.98 \times 10^2 \frac{\left(1 - e_b^2\right)^3}{e_b} \left(\frac{\mu_b}{M_\odot} \right)^{1.1} \left(\frac{a_b}{10\text{AU}} \right)^{4.3} \left(\frac{a_{cross}}{1\text{AU}} \right)^{-2.8} \text{year}. \tag{7.8}$$

These analytical formulae hold for $a_b > 10$ AU and for $0.05 < e_b < 0.8$. In Fig. 7.2 we show the time required by the high speed regime to reach 1 AU for different values of a_b and e_b and for $m_b = 0.5$. Note that for low e_b and large a_b, t_{cr} is longer than 10^5 years, giving a chance to runaway growth to build up large protoplanets before an erosion regime is established.

An implicit assumption of this scenario is that the *initial* encounter velocities in the planetesimal swarm are very low, i.e., that the initial orbits are either circular or fully phased. This assumption is commonly adopted when addressing planetesimal accretion in binaries (e.g., Heppenheimer 1978; Whitmire et al. 1998; Thébault et al. 2004), but should however be handled with care. Its validity depends on how and how fast planetesimals decouple from the gas of the protoplanetary disk to follow their own Keplerian orbits around the main star. If the decoupling occurs almost at the same time for all planetesimals, and if they are at that moment on almost circular orbits, their proper eccentricities e_p will be approximately all equal to e_F, and their ϖ's will be grouped around 0 with consequent low relative velocities. Basically, it all goes down to how the "initial" kilometer-sized planetesimals are being formed. Unfortunately, so far there are no clues on how planetesimals form in a disk perturbed by the gravity of the secondary star and how these perturbations may affect their trajectories as they emerge from the gas of the disk. As a matter of fact,

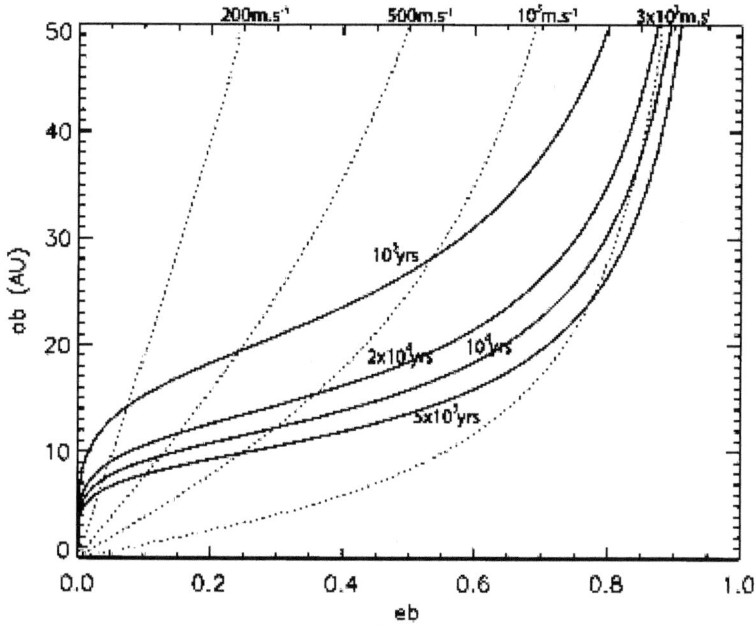

Fig. 7.2 Value of the minimum companion semimajor axis, leading to orbital crossing of planetesimals at 1 AU, for different values of the crossing time t_{cr}, as a function of the companion eccentricity. The companion's mass is fixed with $m_b = 0.5$. The *dotted lines* represent constant values of $\langle \Delta v \rangle$ at orbital crossing, as given by Eq. 7.5

there is not even full agreement on how planetesimals form around "normal" quiet single-stars. There are schematically two competing models for planetesimal formation in a quiescent protoplanetary disk: gravitational instability (e.g., Goldreich and Ward 1973; Youdin and Shu 2002; Youdin and Chiang 2004), where kilometer-sized bodies directly form from small solids in a dense unstable solid grain layer in the midplane of the protoplanetary disk, and collisional-chemical sticking (e.g., Weidenschilling 1980; Dominik and Tielens 1997; Dullemond and Dominik 2005; Johansen et al. 2006, 2007; Johansen and Youdin 2007), in which planetesimals are the outcome of a progressive mutual grain sticking. If gravitational instability works out also for disks in binary systems, then the planetesimal formation timescale is likely to be negligible compared to the typical timescales for runaway growth and for the onset of fully de-phased secular perturbations by the companion star. In this case, it is reasonable to expect an initial orbital alignment within the swarm. The decoupling from the gas occurs at a slower rate in the chemical sticking scenario where planetesimals grow progressively. They might begin to feel the companion star perturbations when they are not yet fully detached from gas streamlines. However, numerical simulations indicate that the whole growth process from grains to kilometer-sized bodies might not exceed a few 10^3 years (Weidenschilling 2000), i.e., still shorter than both runaway growth and secular perturbations timescales.

7.4 Role of Gas Drag

So far, all effects other than the gravitational perturbations of the companion star have been neglected. However, in the standard planetary formation scenario, the initial stages of planetesimal accumulation are believed to take place in presence of significant amounts of primordial gas. Marzari and Scholl (2000) have shown that gas-drag damps planetesimal eccentricities and also restores a strong perihelion alignment for equal-size bodies. If the density of the gas is large enough all the ϖs of planetesimals halt their circulations and tend towards a fixed value ϖ_{al} equal to 270°. This tends to cancel out the large eccentricity oscillations forced by the companion. In Fig. 7.3 we show this phenomenon for 5 km planetesimals embedded in a protostellar disk similar to the minimum-mass solar nebula and perturbed by an outer star. The parameters of the binary system recall those of α Centauri. In this simulation and in all the following ones, the drag force is modeled in laminar gas approximation as

$$\ddot{\mathbf{r}} = -K v_{rel} \mathbf{v_{rel}}, \tag{7.9}$$

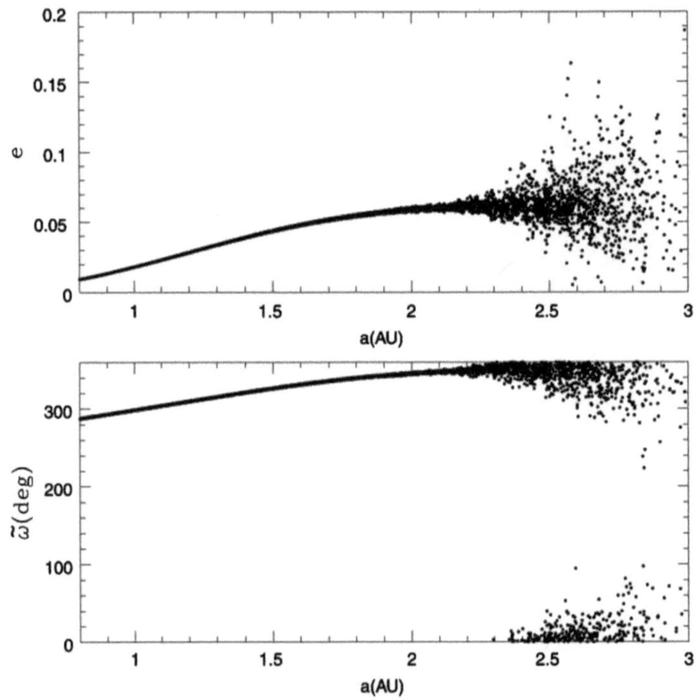

Fig. 7.3 Periastron and eccentricity distributions of a swarm of 5 km diameter planetesimals surrounding the main star of the Alpha Centauri system. After 10^4 years from the beginning of the simulation, the planetesimals orbits were strongly aligned because of the coupling between gas-drag and secular perturbations of the companion star

where K is the drag parameter given by $K = 3\rho_g C_d / 8\rho_{pl} R$ (Kary et al. 1993), R is the radius of a planetesimal, ρ_{pl} its mass density, ρ_g is the density of the gas in the protoplanetary disk, and C_d is a dimensionless drag coefficient related to the shape of the body (~0.4 for spherical bodies).[1]

The orbits of the planetesimals are all aligned for an indefinite time up to about 2 AU from the star and the $\langle \Delta v \rangle$ is of the order of a few meters per second. Under this condition, planetesimal collisions lead to accretion all the time.

Unfortunately, this alignment depends on the balancing between the drag force and the strength of the secular perturbations. For a lower gas density or for a larger size of the planetesimals, the perihelia moves away from ϖ_{al} towards larger values and the alignment is perturbed by a wavy pattern that forewarns the onset of a regime dominated by secular perturbations. This behavior is illustrated in Fig. 7.4 for a bimodal planetesimal population characterized by two different sizes.

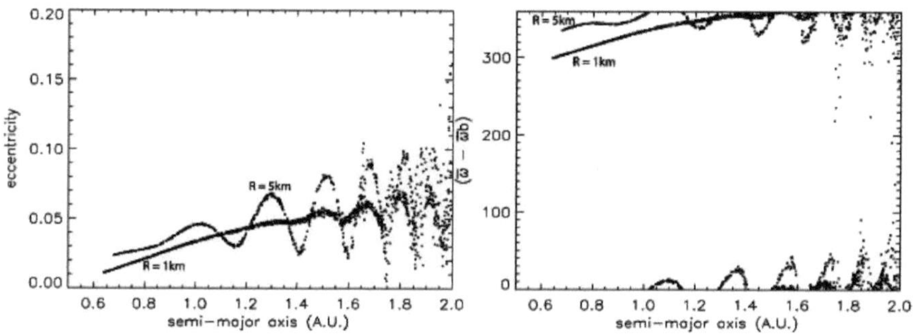

Fig. 7.4 Example gas-drag run. Snapshots (at $t = 3 \times 10^3$ years) of the (e, a) and $(\varpi - \varpi_b, a)$ distributions for two planetesimal populations of sizes $R_1 = 1$ km and $R_2 = 5$ km. The quantity $\varpi - \varpi_b$ is the difference, in angular degrees, between the particles and the companion star's longitude of periastron. The companion star orbital parameters are $a_b = 10$ AU, $e_b = 0.3$, and $m_b = 0.5 M_\odot$

[1] Our gas-drag model is a simplified one where the gas disk is assumed to be fully axisymmetric and follows a classical (Hayashi 1981) power law distribution. It is however more than likely that in reality the gas disk should depart from this simplified view because it would also "feel" the companion star's perturbations. Several numerical studies have investigated the complex behavior of gaseous disks in binary systems. They all show that pronounced spiral structures rapidly form within the disk (e.g., Artymowicz and Lubow 1994; Savonije et al. 1994) and that gas streamlines exhibit radial velocities. To follow the dynamical behavior of planetesimals in such non-axisymmetric gas profiles would require a study of the *coupled* evolution of both gas and planetesimal populations, which have to rely on hydrodynamic-code modeling of the gas in addition to N-body type simulations for the planetesimals. Such an all-encompassing gas + planetesimals modeling is clearly the next step in binary disk studies and have already been started by several teams. It is interesting to note that preliminary results seem to show that planetesimal behaviors in systems with "realistic" gas disk modeling do not seem to drastically depart from the behavior in the axisymmetric case. There is in particular no phase alignment between eccentric planetesimal orbits and gas streamlines, so that gas friction on planetesimals is still very high (S.J. Paardekooper, private communication).

The different alignment of planetesimals according to their size has a crucial impact on the accretion process. While equal-size planetesimals keep a low $\langle \Delta v \rangle$, since their orbits are oriented towards the same angle, when different-size planetesimals encounter each other, their relative velocities are large most of the time because of their forced unpaired orbital alignment. The high eccentricities of these objects,induced by the secular perturbations, introduce large non-tangential components in the relative velocities that easily overcome the Keplerian shear and cause high velocity impacts. In Tables 7.1 and 7.2 we report the average encounter velocities $< \Delta v(R_1, R_2) >$ between planetesimals of size R_1 and R_2 for two different binary star orbital configurations. The assumed density of the gas corresponds to that of the minimum-mass solar nebulae derived by (Hayashi 1981). The velocity values have been obtained with numerical integrations of the planetesimal trajectories over

Table 7.1 Average encounter velocities in m s^{-1} at 1 AU from the primary within a population of "small" planetesimals ($1 < R < 10$ km) for a gas-drag simulation with the companion star with parameters $m_b = 0.5$, $a_b = 10$ AU and $e_b = 0.3$. The values of the quantity $\Delta v_{R1,R2}$ are averaged over the time interval $0 < t < 2 \times 10^4$ years. Initial starting encounter velocities are such that $\Delta v_0 \simeq 10$ m s^{-1}. The values of Δv in bold correspond to accreting impacts for all tested collision outcome prescriptions. Underlined values are those for which we obtained different accretion vs. erosion balance depending on the tested prescription. Values in classical roman characters correspond to cases for which all tested models agreed on a net erosive outcome

Sizes (km)	1	2	3	4	5	6	7	8	9	10
1	**10**	154	233	285	327	360	391	426	452	458
2	172	**10**	94	133	187	223	262	287	316	334
3	238	84	**11**	54	99	137	177	200	230	254
4	289	144	63	**12**	<u>40</u>	80	115	149	171	198
5	325	188	103	<u>43</u>	**12**	<u>32</u>	<u>70</u>	100	122	154
6	373	228	144	83	<u>32</u>	**11**	**36**	<u>56</u>	84	104
7	400	261	182	113	<u>68</u>	**36**	**12**	**35**	<u>48</u>	<u>76</u>
8	428	298	212	147	98	<u>56</u>	**35**	**12**	**36**	<u>45</u>
9	450	310	238	168	123	83	<u>48</u>	**36**	**13**	**31**
10	453	338	263	196	152	107	<u>73</u>	<u>48</u>	**31**	**13**

Table 7.2 Same as Table 7.1, but with companion star parameters $m_b = 0.5$, $a_b = 20$ AU and $e_b = 0.4$

Sizes (km)	1	2	3	4	5	6	7	8	9	10
1	**11**	127	204	255	298	342	368	390	417	442
2	126	**10**	84	139	185	227	258	290	317	340
3	200	91	**11**	68	108	158	186	218	246	272
4	258	146	61	**9**	<u>48</u>	88	120	154	186	209
5	301	192	113	54	**12**	<u>44</u>	<u>75</u>	111	136	164
6	339	232	152	99	<u>43</u>	**10**	**31**	<u>66</u>	92	119
7	361	262	181	126	<u>77</u>	**28**	**13**	**26**	<u>56</u>	87
8	395	295	219	159	112	<u>68</u>	**26**	**11**	**28**	<u>48</u>
9	425	320	246	190	136	92	<u>55</u>	**25**	**11**	**23**
10	446	346	266	208	163	122	82	<u>49</u>	**22**	**12**

2×10^4 years. For the case displayed in Table 7.1, it is worth noticing that, in absence of the gaseous component, the parameters of the binary system would lead to orbital crossing of nearby planetesimal orbits in less than $\simeq 5\times10^3$ years (see Fig. 7.2). The swarm would stop growing and erosion would take place. When we include the effects of gas-drag, the $\langle \Delta v \rangle$ remains always low for equal-size planetesimals. However, for bodies of different sizes, the different phasing caused by the coupling between gas-drag and secular perturbations, pump ups the relative velocities.

7.5 Dependence of the Accretion Process on the Binary Parameters

Once we determine the $\langle \Delta v_{(R_1,R_2)} \rangle$ for any pair of planetesimal sizes, we must compare it with the threshold velocity v_{ero} for that pair of projectile-target to test whether either erosion or accretion will be the result of the collision. As pointed out in Thébault et al. (2006), one can find in the literature many prescriptions of the collisional outcome between two planetesimals given their impact velocity. Adopting the approach outlined in Thébault et al. (2006), three different values of v_{ero} are computed according to Holsapple (1994), Marzari et al. (1995) and Benz and Asphaug (1999). Erosion (or accumulation) is assumed to occur when $\langle \Delta v_{(R_1,R_2)} \rangle$ is higher (or lower) than all the three values of v_{ero}. In the intermediate cases we prefer not to draw any definitive conclusion letting them as undefined cases. In Tables 7.1 and 7.2 a different text font is used to indicate when the computed $\langle \Delta v_{(R_1,R_2)} \rangle$ leads to accretion, erosion, or to an undefined situation. An inspection of these tables show that anytime we depart from the diagonal stripe $R_1 = R_2$, there will be a sharp increase in $\langle \Delta v \rangle$, almost always exceeding the threshold value for accretion, at least for the binary parameters adopted in the tables.

The knowledge of the relative velocities between different-size planetesimals is the first step towards imposing constraints on a dynamical environment that allows planetary formation. These velocities should be used in the context of a full numerical model for planetesimal accretion that also takes into account the frequency with which different or equal-size planetesimals collide. If the size distribution of the swarm privileges impacts between equal-size planetesimals, it comes out clearly from the data in the tables that accretion is also possible for very close and eccentric binary systems. On the other hand, if collisions between different-size planetesimals dominate the evolution of the planetesimal swarm, then the different alignment of perihelia might halt the accretion process. Unfortunately, the size distribution of a planetesimal swarm is poorly constrained at the beginning of the accretion process and it may also be a misleading concept (Wetherill and Inaba 2000).

The frequently adopted assumption that "initial" planetesimals have a given size R is numerically convenient but certainly an oversimplification. Regardless of the details of the planetesimal formation process, there has to be a dispersion ΔR in planetesimal sizes. The amplitude of ΔR is difficult to estimate, but it seems likely that it spans over at least one (if not several) order(s) of magnitude (see for instance

Figure 3 of Weidenschilling 2000). If this is the case, then in the binary stars cases explored here, encounters leading to erosion ("NA" type encounters) might largely dominate over those leading to accretion (type "A"). We have checked this by performing a simplified test where we assume for the planetesimal size distribution a Gaussian function centered on $R = 5\,\text{km}$ of variance ΔR^2. By counting the number of A-type encounters vs. NA-types using the values of $\langle \Delta v \rangle_{(R_1,R_2)}$ reported in Tables 7.1 and 7.2 we find that accretion dominates over erosion only when $\Delta R < 0.8\,\text{km}$. This value of ΔR appears to be small, possibly too small for any realistic initial size distribution. As a consequence, gas-drag tends to prevent, or at least slow down, accretion in the initial stages of planetesimal evolution. As accretion proceeds, larger bodies are formed and impacts occur more and more frequently between different size bodies. The different phasing caused by gas-drag for different-size planetesimals increases the number of NA-type encounters and erosion dominates, at least for the binary parameters given in the tables.

The limit of a purely dynamical approach that estimates the planetesimals' relative velocities but neglects their size distribution can be partly overcome by taking into account that on average the projectiles delivering the maximum kinetic energy to a target of size R_2 are roughly those of size $R_1 \simeq 1/2\, R_2$ (see Thébault et al. 2006). With this assumption we can probe the feasibility of accretion by testing $\langle \Delta v \rangle$ against v_{ero} only for a limited number of different-size body pairs. By exploiting this simplification, Thébault et al. (2006) have tested the chances of planetesimal accretion for 120 different binary systems with semimajor axes ranging from 10 to 50 AU and eccentricity from 0.05 to 0.9. The mass-ratios of these binaries were kept constant and equal to $\mu_b = 0.5$. In all simulations only two size pairs were considered, $(R_1, R_2) = (2.5, 5)$ km, representative of small planetesimals, and $(R_1, R_2) = (15, 50)$ km, representative of intermediate-size bodies in the swarm. In Figs. 7.5 and 7.6 different color coding are used to outline the planetesimal collisional fate, depending on the binary parameters:

- *Dark Green* The evolution of a planetesimal disk in binaries belonging to these regions is not significantly affected by the companion star perturbations. The quantity $\langle \Delta v \rangle$ remains low and accretion should proceed as in the standard model for single stars.
- *Light Green* Within these regions, $\langle \Delta v \rangle$ is increased by the gas-drag coupling to the secular perturbations but remains below v_{ero}. Accretion is still possible but the amplitude of the gravitational focusing factor $(v_{esc(R_2)}/\Delta v)^2$ is significantly reduced. This slows down the accretion rate compared to the single-star unperturbed case.
- *Yellow* Here $\langle \Delta v \rangle$ is comparable to the erosion velocities computed with the different prescriptions adopted at the beginning of the study for v_{ero}. This is a limiting situation where the preference of the system towards accretion or erosion depends on the details of the still not well-defined collisional physics. We prefer not to draw any definite conclusion in this case.
- *Orange and Red* In these regions, $\langle \Delta v \rangle$ exceeds v_{ero} for all three different collision outcome prescriptions considered. Collisions always result in a net mass-loss and mutual accretion is impossible.

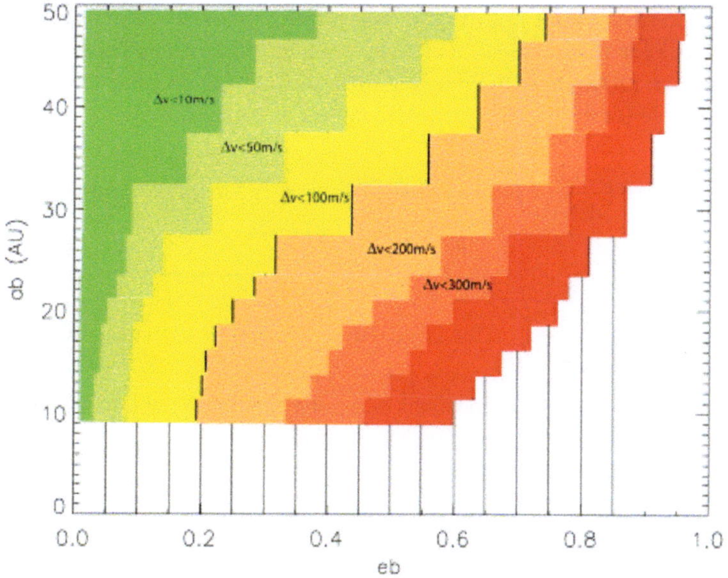

Fig. 7.5 Encounter velocities between two objects with sizes $R_1 = 2.5$ km and $R_2 = 5$ km at 1 AU from the primary star, averaged over the time interval of $0 < t < 2 \times 10^4$ years, for different values of the semimajor axis and eccentricity of the stellar companion. The *short black vertical segments* mark the limit beyond which the values of $\langle \Delta v_{(R1,R2)} \rangle$ correspond to eroding impacts for all tested collision outcome prescriptions

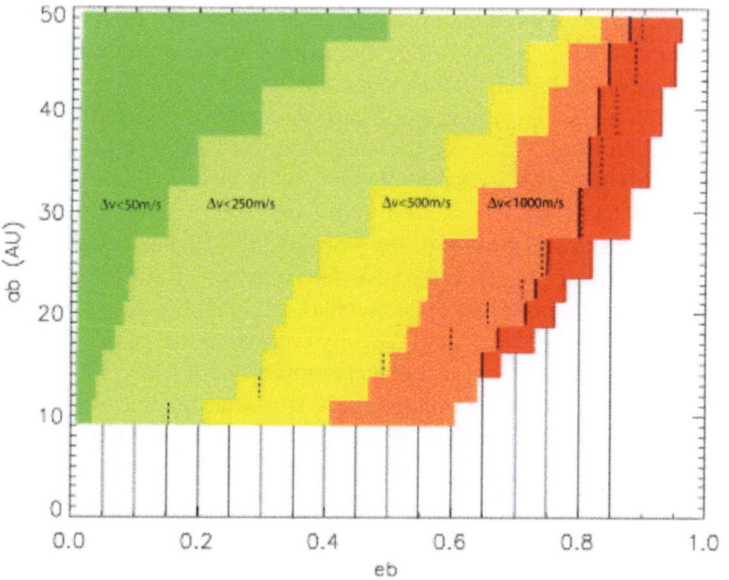

Fig. 7.6 Same as in Fig. 7.5, but for $R_1 = 15$ km and $R_2 = 50$ km bodies. The *short black vertical segments* mark the limit beyond which $\langle \Delta v_{(R1,R2)} \rangle$ corresponds to eroding impacts for all tested collision outcome prescriptions. The *short dashed vertical segments* mark the position beyond which, despite the effects of gas-drag, orbital crossing occurs for 50 km bodies

As expected, systems with higher eccentricity or lower binary separation are more critical for planetesimal accretion. An empirical fit has been performed in Thébault et al. (2006) to outline the region in the (a_b, e_b) phase space where accretion is possible for the two different-size ranges of planetesimals:

$$e_b \simeq 0.013 \left(\frac{a_b}{10\,\mathrm{AU}}\right)^2, \text{ for small } 1 < \mathrm{R} < 10\,\mathrm{km} \text{ planetesimals,} \tag{7.10}$$

and

$$e_b \simeq 0.018 \left(\frac{a_b}{10\,\mathrm{AU}}\right)^2, \text{ for large } 10 < \mathrm{R} < 50\,\mathrm{km} \text{ planetesimals.} \tag{7.11}$$

By extrapolating the fit to large values of a_b one can figure out that for binary separations $a_b \geq 90$AU the planetesimal accretion process is not significantly perturbed by the companion star gravity.

The results shown in Figs. 7.5 and 7.6 may be easily extended to different mass-ratios. In addition, focusing on a single system makes it possible to improve the N-body model and explore in more details the accretion versus fragmentation trend for planetesimal populations with any given size distribution, and to investigate the influence of the gas disk density, radial profile, and dissipation timescale (Xie and Zhou 2008) on the process. In recent studies (Thebault et al. 2008, 2009), we concentrated on the planetesimals-to-embryos accretion phase around both Alpha Cen A and B, our closest neighbors in space. In Fig. 7.7, we illustrate how the accumulation process depends on the initial size distribution of planetesimals. The fraction of impacts leading to accretion is computed at different distances from the primary star (Alpha Cen A) assuming a Maxwellian distribution for the planetesimal sizes peaking at a median value of 5 km for the planetesimal radius, a collisional-'equilibrium' power-law with a −3.5 exponent, and Gaussians centered on a radius of 5 km and with various variances. Apart for the unrealistic case with an almost Dirac-like Gaussian distribution with variance 0.25 km^2, all models predict that planetesimal accumulation is halted for $r > 0.5$ AU from Alpha Cen A. A similar study for different density values and profiles of the gaseous disk (Thebault et al. 2008) shows that the accretion limit moves to 0.75 AU for a gas density equal to 10 times that of the MMSN (Minimum Mass Solar Nebula), while it shrinks for decreasing gas density down to 0.2 AU when the gas density is 1/100 of that of the MMSN.

One intriguing possibility is that planets formed in the Alpha Cen system when the stars had a different orbital configuration. Most stars are born in clusters where close stellar encounters are frequent. A binary system may have had different orbital parameters at the time of planet formation compared to the present ones. A present-day binary could even be what is left of an initially triple (or more) system (Marzari and Barbieri 2007a,b). As a consequence, there might be an important difference between the presently observed orbit of a binary system and the one it had when planetary-formation was undergoing. A crucial question is, which initial orbital configurations of the Alpha Cen system could have allowed planetesimal accretion to occur in the Habitable Zone (HZ) of the stars? We have explored this issue

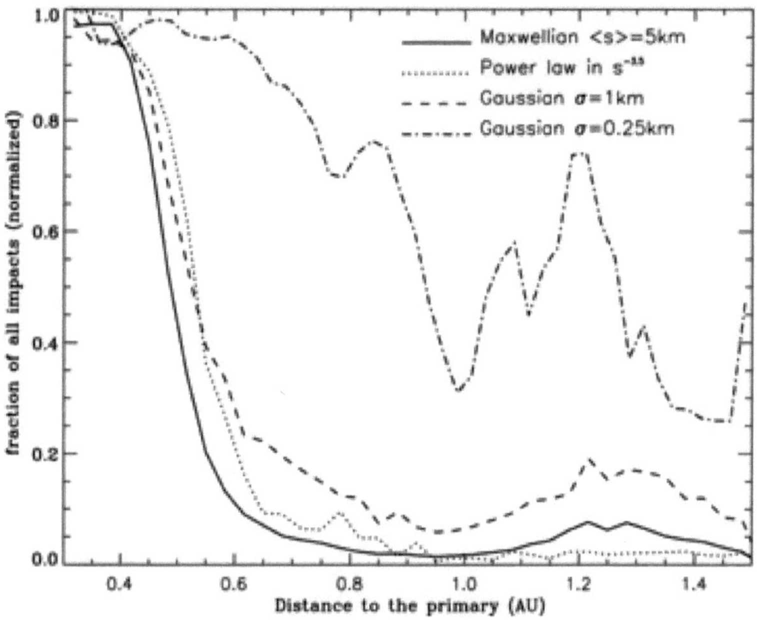

Fig. 7.7 Fraction of accreting impacts for different prescriptions of the initial planetesimal size-distribution. The curve corresponding to the extremely narrow Gaussian is possibly unrealistic. Planetesimal accretion occurs in spite of the binary perturbations when the fraction of impacts leading to accumulation is larger than 0.5. The gas density is assumed to be equal to that of the standard MMSN with $\rho_{g0} = 1.4 \times 10^{-9}\,\mathrm{g\,cm^{-3}}$ at 1 AU, and declining as $\rho_g = \rho_{g0} r^{-2.75}$

for Alpha Cen B finding that in order to have the HZ fully accretion-friendly, one needs an initial semimajor axis for the system of $a_b \geq 37$ AU. More favorable is a scenario in which the initial binary eccentricity e_b was lower. As an example, for $e_b = 0.26$ (half of the present-day value), the entire HZ becomes accretion-friendly when $a_b \geq 26$ AU (see Fig. 7.8).

7.6 Inclined Binaries

A systematic study by Hale (1994) on binary systems with solar-type components suggests that in those systems, the spin of the two stars is aligned only for binary separations of 30–40 AU or less. Beyond these values, the primary's equator, and thus a putative planetesimal disk surrounding it, would be randomly inclined with respect to the binary orbit. As a consequence, planetary accretion may occur in a significantly inclined disk when $a_b > 30$–40 AU. In this scenario, planetesimals would most likely decouple from the gaseous disk because of the forced inclination due to the companion star and also because of the randomization of the nodal longitudes. For inclinations larger than approximately 10°, the evolution of planetesimals would

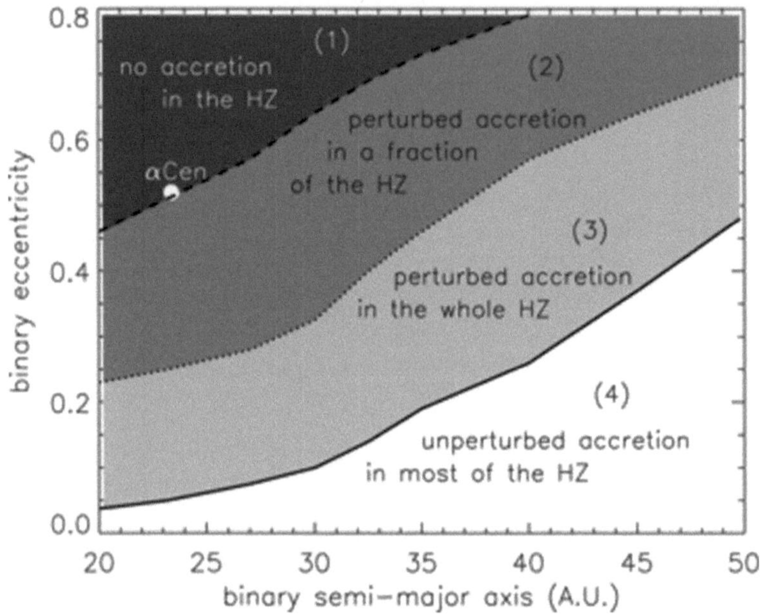

Fig. 7.8 Dominant collisional outcome (erosion, perturbed accretion, unperturbed accretion) in a planetesimal disk located in the HZ (0.50.9 AU) around Alpha Cen B for different possible primordial orbital configurations of the Alpha Cen binary system

occur as in a gas-free environment. The gaseous disk, in fact, either remains a long-lived coherent entity in spite of the binary perturbations, and planetesimals would cross it only at the nodal lines of their orbits, or it is warped and finally disrupted by the binary perturbations losing coherence. Even in this second case, planetesimals would not be affected by gas-drag during the accumulation process. A third possibility is that, for very large viscosities, the disk relaxes to the binary plane on a short timescale and planet accretion would occur as in initially coplanar systems.

In the two more likely scenarios in which planetesimals evolve free from gas friction, accretion would strongly depend on the balance between the timescale for node randomization and that for planetesimal accumulation into bigger bodies. The nodal randomization has in fact two negative effects on accretion: (1) it introduces an additional out-of-plane component in the relative velocity between the bodies favoring cratering and fragmentation, and (2) it leads to a sparser distribution of planetesimals in space causing a lower impact rate. In spite of these adverse effects, planet formation can still occur if the relative impact velocity is lower than the erosion limit for a time long enough for the planetesimals to grow big enough to sustain high energy collisions. As it can be seen in Fig. 7.9, in a binary configuration with $a_b = 50$ AU, $e_b = 0$ and $i_b = 30°$, the whole system is accretion-friendly within 4 AU from the star for a population of 10 km size bodies for 10^5 years, a conservative timescale for runaway growth. We recall here that in binaries, runaway growth (when it occurs) is necessarily slower compared to the case of single stars because of the companion perturbations.

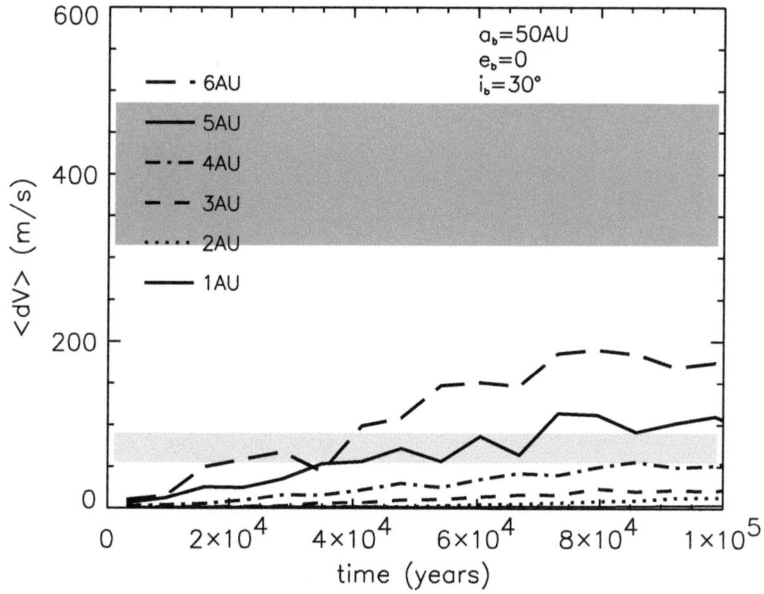

Fig. 7.9 Average impact rate in a planetesimal swarm in an inclined binary system as a function of time at different radial distances from the primary star. The grey bands show the erosion limit for planetesimals 1–10 km in size (*lower band*) and 10–50 km (*upper band*). The initial inclination between the planetesimal disk and the binary orbit is $i_b = 30°$

Marzari et al. (2009) have computed the distance from the primary star within which accretion is possible for different binary orbital parameters and the outcome of this analysis is shown in the form of 2-dimensional maps in Fig. 7.10.

For small values of the binary semimajor axis, there is a significant dependence of the limiting semimajor axis beyond which planetesimal accretion is possible (a_l) on i_b showing that the cause of the shrinking of the planet formation zone is the randomization of the nodes. For binary semimajor axes larger than 50 AU, the effect of the secular perturbations of the inclined binaries are weaker except for $i_b \geq 40°$ where Kozai oscillations dominate the planetesimal dynamics. In this regime, the impact velocities halt the planetesimal accretion close to the star also for large binary semimajor axes.

7.7 Type II Runaway Growth

One interesting feature of Figs. 7.5 and 7.6 is the extent of the "light-green" region where planetesimal accretion is not halted, but possibly slowed down since the gravitational focusing factor is reduced. As a consequence, "classical" runaway growth is not possible and might be replaced by the alternative, possibly slower type II runaway mode identified by Kortenkamp et al. (2001).

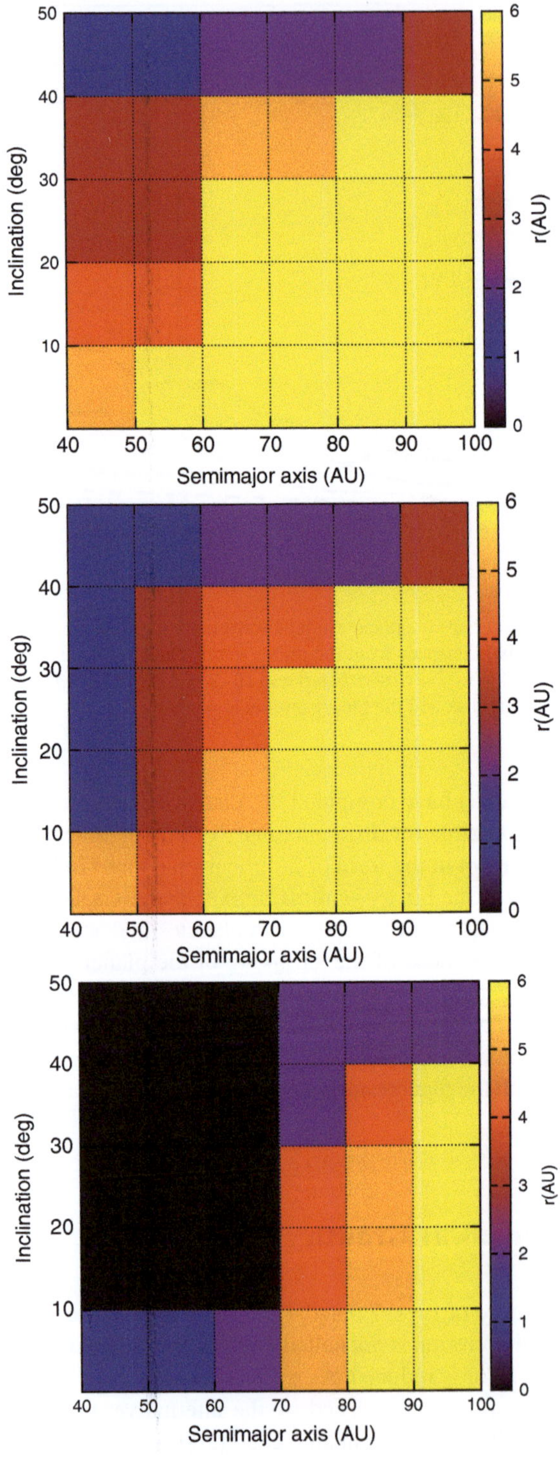
Inclination (deg)
Semimajor axis (AU)
r(AU)
0
1
2
3
4
5
6
10
20
30
40
50
60
70
80
90
100

In this section we describe details of planetesimal accretion simulations in the Sun–Jupiter "binary" system (Kortenkamp and Wetherill 2000a; Kortenkamp et al. 2001). This system has traditionally been used for initial characterizations of dynamical effects in binary-star systems (e.g., Heppenheimer 1978; Whitmire et al. 1998). In the models described here we use units of binary separation D rather than AU. Note also that the modeling included Saturn as well (a triple system) but we omit Saturn from most of the discussion, referring simply instead to the Sun–Jupiter "binary."

Initial orbits for the planetesimals were circular, coplanar with the nebular midplane, and uniformly distributed from 10–50% of D. As the system evolves, the combination of gas-drag and secular perturbations from the companion Jupiter (and Saturn) leads to a size-dependence in the planetesimal eccentricity and inclination as well as in the phasing of the orbital orientation angles, as described earlier in this chapter. Note that the system is fully three dimensional. The companion orbits are inclined with respect to each other and the protoplanetary disk.

As described above, the distribution of encounter velocities is a critical factor which determines how a population of planetesimals will accumulate into larger bodies. High encounter velocities can actually lead to impact erosion of the target rather than growth. On the other hand, low encounter velocities can gravitationally enhance the collision cross-section of the target (Öpik 1951). The effective cross-section becomes

$$\sigma_{\text{eff}} = \sigma \left(1 + \frac{V_{esc}^2}{v_{enc}^2} \right), \tag{7.12}$$

where σ is the target's geometric cross-section, v_{enc} is the projectile's encounter velocity, and V_{esc} is the combined surface escape velocity of the target and projectile.

The size-dependent phasing of orbital elements leads to low encounter velocities between similar size bodies and high encounter velocities between bodies of markedly different sizes. For the Sun–Jupiter model described above "binary" encounter velocities were calculated for all intersecting orbits over the entire range of planetesimal sizes. Figure 7.11 shows examples of these encounter velocities for various target bodies and equal or lesser size projectile bodies with orbits near 33% D, where, again, D is the Sun–Jupiter separation distance.

Modeling the growth of planetesimals in this scenario is considerably more difficult than tracking their orbital evolution. Full-scale N-body simulations of planetesimal growth that include mutual gravitational perturbations, secular perturbations, and gas-drag are beyond the reach of current techniques. Theoretically

Fig. 7.10 Maps showing the limiting values for accretion a_l as a function of the binary orbital parameters (a_b, i_b). The *top plot* refers to the case with $e_b = 0.0$, the middle plot to $e_b = 0.2$, and the lower plot to $e_b = 0.4$. The color coding gives different values of a_l (in AU), the limiting semimajor axis beyond which planetesimal accretion is possible. Each *square* of the map refers to the lower value of the labels in the axes. The cases for $i_b = 0°$ do not include gas-drag, so they are only indicative

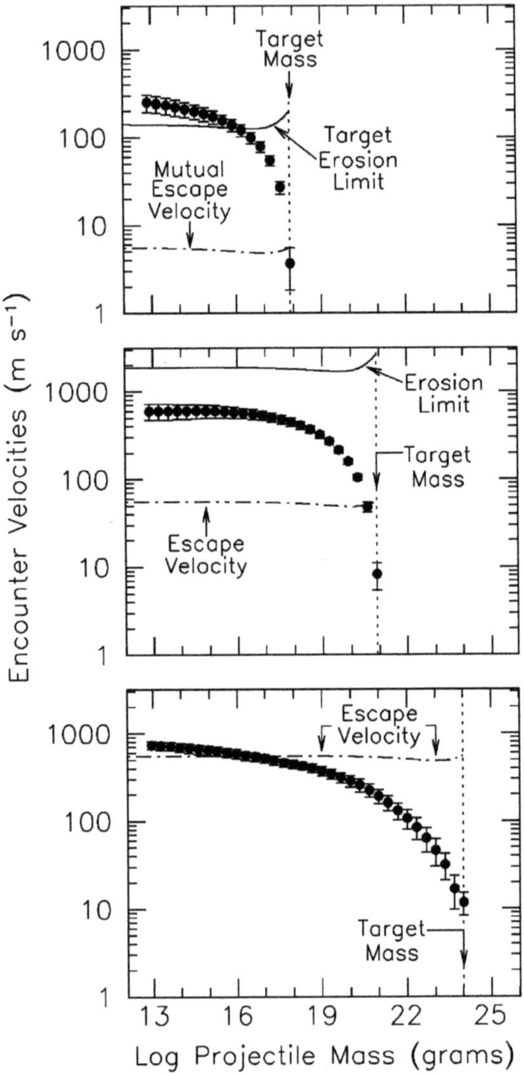

Fig. 7.11 Mean encounter velocities ($\pm\sigma$) are shown for various size planetesimals with intersecting orbits near 33% D, where D is the Sun–Jupiter separation distance. The size-dependent orbital phasing of bodies on intersecting orbits leads to high encounter velocities between bodies of markedly different sizes. The surface escape velocity of the combined target/projectile is shown by the horizontal dashed line. In the *top two panels* the erosion limit is indicated by the horizontal solid line. For encounter velocities above the erosion limit impacting projectiles crush and eject more than their own mass (crushing strength was assumed to be 10^8 ergs/g). *Top*: Encounter velocities for 10^{18} g targets and projectiles of equal or lessor mass. *Middle*: Targets with masses of 10^{21} g are large enough that impacts of all projectiles lead to growth. *Bottom*: Planetesimals capable of growing to 10^{24} g have surface escape velocities near 500 m s^{-1}. Encounter velocities between these large planetesimals and bodies as much as 8–10 orders of magnitude less massive are below the escape velocity, allowing for significant gravitational focusing

one would need to include $\sim 10^{12}$ small ($\sim 10^{14}$ g) planetesimals to form a single 10^{26} g terrestrial planet embryo. Direct N-body integrations of mutually perturbing planetesimals cannot even remotely approach this figure, treating only about $\sim 10^4$ bodies over the time scale required. On the other hand, existing statistical simulations of planetary growth are not limited by the number of bodies. However, these simulations assume that the orbits are completely randomized so they cannot easily accommodate the size-dependent orbital evolution described above. To overcome these shortcomings, Kortenkamp et al. (2001) developed a hybrid approach that capitalizes on the strengths of each technique. They used N-body integration to map the size-dependent velocity distributions. These velocity distributions are then used in modified statistical accretion simulations to model the collisional accumulation of the planetesimals.

This approach led to the identification of an alternative form of runaway growth that is facilitated by the secular perturbations of massive companions (Kortenkamp et al. 2001). Figure 7.12 shows results from these growth simulations.

Four separate simulations were performed, centered on 15%, 25%, 33% and 40% of D. For comparison, the habitable zone extends from about 10–25% of D (crudely defined as the Venus–Mars region). In all four regions the growth is characterized as "orderly" (non-runaway, see Safronov 1969) until the distribution reaches approximately 10^{24} g, or about the size of the largest asteroid – (1) Ceres. The planetesimal size distribution then becomes bi-modal, transitioning to runaway growth and producing Mars-size embryos. All planetesimals were assumed to be uniformly distributed across a region and any bodies separated by more than 10 mutual Hill radii were considered dynamically isolated (Chambers et al. 1996) and not allowed to collide. This resulted in multiple runaway embryos emerging in each region.

Note that mutual perturbations between planetesimals were not included in these calculations. This new form of runaway growth arises when secular perturbations and gas-drag act together to establish size-dependent encounter velocities that remain low when colliding bodies are of similar size. Collisions between bodies of significantly different sizes are at high velocity and can lead to cratering and erosion, but these simulations show that growth overcomes erosion (Fig. 7.12). This alternative mode of growth, which was eloquently dubbed "Type II" runaway growth, allowed for the formation of terrestrial planet embryos throughout the habitable zone of the Sun–Jupiter "binary." This general result should apply regardless of whether the perturbations are from Jupiter-like companions formed earlier by disk-instability (e.g., Boss 1997; Mayer et al. 2002), stellar-mass objects in multiple-star systems, or some other source, as long as the source of the dynamical perturbations pre-dates the formation of terrestrial planet embryos. It is conceivable to expect that in binary star systems, "Type II" runaway growth leads to the formation not only of terrestrial planets but also of giant planets by favoring the growth on a short timescale of a massive core, the first step of the core-accretion model.

Classical "Type I" runaway growth occurs in a self-gravitating population of planetesimals. Random orbital kinetic energy is exchanged during gravitational encounters between large and small bodies and the population trends towards energy equipartition, a process dubbed "dynamical friction" (Stewart and Kaula 1980).

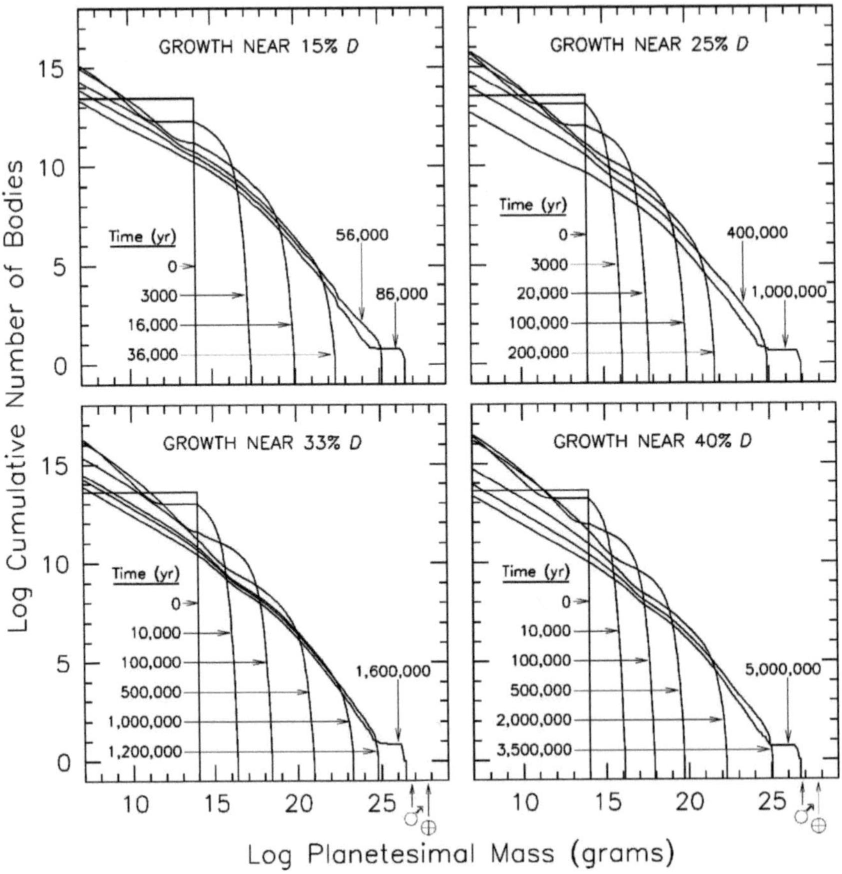

Fig. 7.12 Growth of planetesimals in a circumsolar disk perturbed by the "companion" Jupiter at a distance *D* from the sun. The habitable zone, simply defined here as the Venus–Mars region, stretches from about 10–25% of *D*. Initially all planetesimals have identical masses of 10^{14} g. The initial surface density of planetesimals at 15% of *D* is 10 g cm^{-2}, scaling as $r^{-3/2}$ with heliocentric distance r. This is roughly consistent with the so-called "minimum mass" surface density. Collision fragments smaller than 10^7 g ($\sim$1m) are presumed lost via nebular gas-drag and therefore removed from the simulation. Growth is calculated in four different regions centered on 15%, 25%, 33%, and 40% of *D*. For comparison, the masses of Earth and Mars are indicated by their respective symbols. The *top two panels* represent growth in the habitable zone and show initial orderly growth transitioning to Type II runaway growth of Mars-size planetary embryos in 10^5 to 10^6 years. Beyond the habitable zone (*bottom panels*) growth is slower but eventually also produces Mars-size embryos. Figure is adapted from Kortenkamp et al. (2001)

Dynamical friction lowers the encounter velocities of the larger bodies with respect to each other, enhancing their effective collision cross-sections and increasing the rate at which they accumulate each other. Under these conditions nearly the entire growth period up to embryo-size is in type I runaway mode. In our simulations, which are not self-gravitating, the size-dependent phasing of orbital elements holds encounter velocities low between all similar-size bodies (typically 1–10 $\mathrm{m\,s^{-1}}$, see Fig. 7.11). Initially these encounter velocities exceed the planetesimal escape velocities so there is no enhancement of collision cross-sections and growth is orderly. As larger and larger bodies grow their escape velocities approach and then exceed the relatively low encounter velocities, causing the transition from orderly growth to type II runaway growth. In this way, the effects of dynamical friction are mimicked by the size-dependent phasing of orbital elements.

7.8 Conclusions

The process of planetary formation around the primary star of a binary system is complicated in all its stages by the gravitational perturbations of the companion star. However, the existence of some gas giant planets in binary systems with separation of a few tens of AU suggests that these perturbations are not always strong enough to prevent the formation of a planet. We have analyzed within the standard model of planet growth how the secular perturbations of the companion star affect the various stages of planetesimal accretion. Ours studies show that when we include also the effects of gas-drag on the motion of small planetesimals, their eccentricities and perihelion longitudes become aligned (Marzari and Scholl 2000). At a first sight, this alignment favors fast accretion keeping low the encounter velocities between the bodies in spite of the large values of eccentricity forced by the secular perturbations. However, the angle towards which the perihelia align depends on the size of the planetesimals and on their distance from the star (Thébault et al. 2004, 2006). In the earlier stages of planetesimal accretion, all the bodies have roughly the same size and the encounter velocities are low because of the local strong orbital alignment. However, as soon as larger bodies emerge from the population by coagulation, their perihelia are no longer aligned to those of smaller planetesimals leading to larger impact velocities. This may prevent the onset of runaway growth or even cause erosion of the bodies, inhibiting the formation of a planet. Through a full numerical approach, we have analyzed to what extent the different alignment influences the random planetesimal velocities. We have also mapped the values of the binary orbital parameters (semimajor axis and eccentricity) for which accumulation dominates over erosion despite of the secular perturbations, or vice-versa. Our modeling shows that for binaries of separations $a_b \leq 40$ AU, only very low e_b allow planetesimal accretion to proceed as in the standard single-star case. On the contrary, only relatively high e_b values (of at least 0.2 in the closest $a_b = 10$AU separation explored and at least $\sim$0.7 for $a_b = 40$AU) lead to a complete stop of planetesimal accretion. In most cases, when the perturbations of the massive companion on the planetesimal disk is significant but not strong enough to halt accretion, runaway growth can still occur, but in a different way with respect to the classical "Type I" runaway growth typical of planetesimal populations around single stars. This new type of growth termed "Type II" runaway growth, allows planet formation to occur in binary-star systems with much tighter orbits than previously suggested (Heppenheimer 1978; Whitmire et al. 1998). However, there is (at least) one caveat. As already noted, the growth simulations described above and represented in Fig. 7.12 did not include mutual perturbations of the planetesimals themselves. Crude attempts at including self-gravitating planetesimals (Kortenkamp and Wetherill 2000b) indicated that when the size distribution reaches 10^{24} to 10^{25} g the mutual perturbations are beginning to become important, although they are still dominated by the size-dependent phasing of secular perturbations. This suggests that just as type II runaway growth is getting underway accretion may either stall or perhaps transition to the classical type I runaway growth. To explore these possibilities we are working to modify the multi-zone planetesimal accretion code of Weidenschilling et al. (1997) to include secular perturbations from massive companions as well as mutual perturbations from planetesimals and nebular gas drag (Kortenkamp et al. 2006).

Acnowledgments S. Kortenkamp acknowledges support from NASA for some of this work under grants NNG04GP56G and NNG04GI14G.

References

Adachi, I., Hayashi, C., Nakagawa, K., 1976, The gas drag effect on the elliptical motion of a solid body in the primordial solar nebula, Prog. Theor. Phys., 56, 1756

Alibert, Y., Mordasini, C., Benz, W., 2004, Migration and giant planet formation, A&A, 417, L25–L28

Artymowicz, P., Lubow, S. H., 1994, Dynamics of binary-disk interaction. 1: Resonances and disk gap sizes, ApJ, 421, 651

Barge, P., Pellat, R., 1993, Mass spectrum and velocity dispersions during planetesimal accumulation, Icarus, 104, 79

Benz, W., Asphaug, E., 1999, Catastrophic disruptions revisited, Icarus, 142, 5

Bodenheimer, P., Lin, D. N. C., 2002, Implications of extrasolar planets for understanding planet formation, Annual Review of Earth and Planetary Sciences, 30, 113–148

Boss, A. P., 1997, Giant planet formation by gravitational instability, Science, 276, 1836–1839

Boss, A. P., 2007, Testing disk instability models for giant planet formation, ApJ, 661, L73–L76

Boss, A. P., 2004, Convective cooling of protoplanetary disks and rapid giant planet formation, ApJ, 610, 456–463

Boss, A. P., 2006, Gas giant protoplanets formed by disk instability in binary star systems, ApJ, 641, 1148–1161

Chambers, J. E., Wetherill, G. W., 1998, Making the terrestrial planets: N-body integrations of planetary embryos in three dimensions, Icarus, 136, 304–327

Chambers J.E., Wetherill G.W., Boss A.P., 1996, The stability of multi-planet systems, Icarus, 119, 261

Chen, C. H., Kamp, I., 2004, ApJ, 602, 985–992

Cuzzi, J., Weidenschilling, S., Particle-gas dynamics and primary accretion, 2006, in Meteorites and the Early Solar System II, D. S. Lauretta and H. Y. McSween Jr. (eds.), University of Arizona Press, Tucson, 943 pp., p. 353–381

Desidera, S., Barbieri, M., 2007, A&A, 462, 345–353

Dominik, C., Tielens, A., The physics of dust coagulation and the structure of dust aggregates in space, 1997, ApJ, 480, 647

Dullemond, C., Dominik, C., Dust coagulation in protoplanetary disks: A rapid depletion of small grains, 2005, A&A, 434, 971

Duquennoy, A., Mayor, M., 1991, Multiplicity among solar-type stars in the solar neighborhood. II-Distribution of the orbital elements in an unbiased sample, A&A, 248, 485
Eggenberger, A., Udry, S., Mayor, M., 2003, in ASP Conf. Ser. 294, Scientific Frontiers in Research on Extrasolar Planets, ed. D. Denning & S. Seager, 43
Goldreich, P., Ward, W., 1973, The formation of planetesimals, ApJ, 183, 1051
Greenberg, R., Hartmann, W. K., Chapman, C. R., Wacker, J. F., 1978, Planetesimals to planets – Numerical simulation of collisional evolution, Icarus, 35, 1
Guillot, T., 1999, A comparison of the interiors of Jupiter and Saturn, Planetary and Space Science, 47, 1183–1200
Guillot, T., 2004, The interiors of giant planets: Models and Outstanding Questions, Annual Review of Earth and Planetary Sciences, 33, 493–530
Haisch, K. E., Lada, E. A., Lada, C. J., 2001, ApJ, 553, L153–L156
Hale, A., 1994, AJ, 107, 306
Hatzes, A. P., Cochran, W. D., Endl, M., McArthur, B., Paulson, D. B., Walker, G. A. H., Campbell, B., Yang, S., 2003, A planetary companion to gamma Cephei A, ApJ, 599, 1383–1394
Hayashi, C., 1981, Structure of the solar nebula, growth and decay of magnetic fields and effects of magnetic and turbulent viscosities on the nebula, PthPS, 70, 35
Heppenheimer, T., 1978, On the formation of planets in binary star systems, A&A, 65, 421
Holman, M. J., Wiegert, P. A., 1999, Long-term stability of planets in binary systems, AJ, 117, 621–628
Holsapple, K. A., 1994, Catastrophic disruptions and cratering of solar system bodies: A review and new results, P&SS, 42, 1067
Johansen, A., Oishi, J. S., Low, M. M., Klahr, H., Henning, T., Youdin, A., 2007, Rapid planetesimal formation in turbulent circumstellar disks, Nature, 448, 1022–1025
Johansen, A., Youdin, A., 2007, Protoplanetary disk turbulence driven by the streaming instability: Nonlinear saturation and particle concentration, ApJ, 662, 627–641
Johansen, A., Klahr, H., Henning, T., 2006, Gravoturbulent formation of planetesimals, ApJ, 636, 1121–1134
Kary, D. M., Lissauer, J. J., Greenzweig, Y., 1993, Nebular gas drag and planetary accretion, Icarus, 106, 288–307
Kokubo, E., Ida, S., 1998, Oligarchic growth of protoplanets, Icarus, 131, 171
Kokubo, E., Ida, S., 2000, Formation of protoplanets from planetesimals in the solar nebula, Icarus, 143, 15
Kortenkamp S. J., Wetherill G. W., 2000a, Terrestrial planet and asteroid formation in the presence of giant planets I. Relative velocities of planetesimals subject to Jupiter and Saturn perturbations, Icarus, 143, 60
Kortenkamp S. J., Wetherill G. W., 2000b, Formation of the asteroid belt, Lunar Plan. Sci. Conf., 31, abstract #1813
Kortenkamp S. J., Kokubo E., Weidenschilling S.J., 2000, Formation of planetary embryos, in Origin of the Earth and Moon, R. M. Canup and K. Righter (eds.), University of Arizona Press, Tucson, pp. 85–100
Kortenkamp, S., Wetherill, G., Inaba, S., 2001, Runaway growth of planetary embryos facilitated by massive bodies in a protoplanetary disk, Science, 293, 1127
Kortenkamp S. J., Weidenschilling S. J., Marzari F., 2006, A new code for modeling planetesimal accretion in protoplanetary disks perturbed by massive companions 38th DPS Meeting, Pasadena, CA, abstract #63.03
Lagrange, A.-M., Beust, H., Udry, S., Chauvin, G., Mayor, M., 2006, New constrains on Gliese 86 B - VLT near infrared coronographic imaging survey of planetary hosts, A&A, 459, 955
Lissauer, J.J., 1993, Planet formation, ARA&A, 31, 129
Lissauer J., Stewart G., 1993, Growth of planets from planetesimals, in *Protostars and Planets III*, the Univ. of Arizona Press, Tucson, 1061
Marzari, F., Davis, D., Vanzani, V., 1995, Collisional evolution of asteroid families, Icarus, 113, 168
Marzari F., Scholl H., 1998, Capture of Trojans by a growing proto-jupiter, Icarus, 131, 41

Marzari F., Scholl H., 2000, Planetesimal accretion in binary star systems, ApJ, 543, 328
Marzari, F., Barbieri, M., 2007a, A&A, 467, 347
Marzari, F., Barbieri, M., 2007b, A&A, 472, 643
Marzari, F., Thébault, P., Scholl, H., 2009, A&A, 507, 505
Mayer L., Quinn T., Wadsley J., Stadel J., 2002, Formation of giant planets by fragmentation of protoplanetary disks, Science, 298, 1756
Mugrauer, M., Neuhauser, R., 2005, Gl 86B: A white dwarf orbits an exoplanet host star, MNRAS, 361, L15
Nelson, A., 2000, Planet formation is unlikely in equal-mass binary systems with $a \simeq 50$ AU, ApJ, 537, 65
Neuhäuser, R., Mugrauer, M., Fukagawa, M., Torres G., Schmidt, T., 2007, Direct detection of exoplanet host star companion gamma Cep B and revised masses for both stars and the substellar object, A&A, 462, 777
Öpik E. J., 1951, Collision probabilities with the planets and the distribution of interplanetary matter, Proc. Irish Acad., 54, 165
Papaloizou, J., Lin, D. N. C., 1984, On the tidal interaction between protoplanets and the primordial solar nebula. I – Linear calculation of the role of angular momentum exchange, ApJ, 285, 818–834
Pollack, J. B., Hubickyj, O., Bodenheimer, P., Lissauer, J. J., Podolak, M., Greenzweig, Y., 1996, Formation of the giant planets by concurrent accretion of solids and gas, Icarus, 124, 62–85
Pourbaix, D., Neuforge-Verheecke, C., Noels, A., 1999, Revised masses of alpha Centauri, A&A, 344, 172–176
Rafikov, R., 2003, The growth of planetary embryos: Orderly, runaway, or oligarchic?, AJ, 125, 942
Rafikov, R., 2004, Fast Accretion of Small Planetesimals by Protoplanetary Cores, AJ, 128, 1348
Raghavan, D., Henry, T. J., Mason, B. D., Subasavage, J. P., Jao, W.-C., Beaulieu, T. D., Hambly, N.C., 2006, Two suns in the sky: Stellar multiplicity in exoplanet systems, ApJ, 646, 523
Rice, W. K. M., Armitage, P. J., 2003, On the formation timescale and core masses of gas giant planets, ApJ, 598, L55–L58
Safronov, V. S., 1969, Evolution of the protoplanetary cloud and formation of the earth and the planets. Israel program for scientific translation, TT-F 677
Santos, N. C., Israelian, G., Mayor, M., 2004, Spectroscopic [Fe/H] for 98 extra-solar planet-host stars. Exploring the probability of planet formation, A&A, 415, 1153–1166
Sato, B., Fischer, D. A., Henry, G. W., Laughlin, G., Butler, R. P., Marcy, G. W., Vogt, S. S., Bodenheimer, P., Ida, S., Toyota, E., Wolf, A., Valenti, J. A., Boyd, L. J., Johnson, J. A., Wright, J. T., Ammons, M., Robinson, S., Strader, J., McCarthy, C., Tah, K. L., Minniti, D., 2005, The N2K consortium. II. A transiting hot saturn around HD 149026 with a large dense core, ApJ, 633, 465–473
Savonije, G. J., Papaloizou, J. C. B., Lin, D., 1994, On tidally induced shocks in accretion disks in close binary systems, MNRAS, 268, 13
Stewart G. R., Kaula W. M., 1980, Gravitational kinetic theory for planetesimals, Icarus, 44, 154
Strom, S. E., Edwards, S., Skrutskie, M. F., Evolutionary time scales for circumstellar disks associated with intermediate- and solar-type stars. In Protostars and Planets III, University of Arizona Press, Tucson, pp. 837–866
Thébault, P., Marzari, F., Scholl, H., 2002, Terrestrial planet formation in exoplanetary systems with a giant planet on an external orbit, A&A, 384, 594
Thébault, P., Brahic, A., 1998, Dynamical influence of a proto-Jupiter on a disc of colliding planetesimals, P&SS, 47, 233
Thébault, P., Beust, H., 2001, Falling evaporating bodies in the Ŏ3b2 Pictoris system. Resonance refilling and long term duration of the phenomenon, A&A, 376, 621
Thébault P., Augereau, J.-C., Beust, H., 2003, Dust production from collisions in extrasolar planetary systems. The inner beta Pictoris disc, A&A, 408, 775
Thébault, P., Marzari, F., Scholl, H., Turrini, D., Barbieri, M., 2004, Planetary formation in the γ Cephei system, A&A, 427, 1097

Thébault, P., Marzari, F., Scholl, H., 2006, Relative velocities among accreting planetesimals in binary systems: The circumprimary case, Icarus, 183, 193
Thebault, P., Marzari, F., Scholl, H., 2008, MNRAS, 388, 1528
Thebault, P., Marzari, F., Scholl, H., 2009, MNRAS, 393, L21–L25
Torres, G., 2007, The planet host star γ Cephei: Physical properties, the binary orbit, and the mass of the substellar companion, ApJ, 654, 1095
Weidenschilling, S., 1977, The distribution of mass in the planetary system and solar nebula, Astrophysics and Space Science, 51, 153–158
Weidenschilling, S., 1980, Dust to planetesimals – Settling and coagulation in the solar nebula, Icarus, 44, 172
Weidenschilling, S. J., 2000, Formation of planetesimals and accretion of the terrestrial planets, SSRv, 92, 295
Weidenschilling, S. J., Davis, D. R., 1985, Orbital resonances in the solar nebula – Implications for planetary accretion, Icarus, 62, 16
Weidenschilling S. J., Spaute D., Davis D. R., Marzari F., Ohtsuki K., 1997, Accretional evolution of a planetesimal swarm II: The terrestrial zone, Icarus, 128, 429
Spaute, D., Weidenschilling, S. J., Davis, D. R., Marzari, F., 1991, Accretional evolution of a planetesimal swarm. I - A new simulation, Icarus, 92, 147–164
Wetherill, G. W., Stewart, G. R., 1989, Accumulation of a swarm of small planetesimals, Icarus, 77, 330
Wetherill, G. W., Stewart, G. R., 1993, Formation of planetary embryos – Effects of fragmentation, low relative velocity, and independent variation of eccentricity and inclination, Icarus, 106, 190
Wetherill, G. W., Inaba, S., 2000, Planetary accumulation with a continuous supply of planetesimals, SSRv, 92, 311
Whitmire, D., Matese, J., Criswell, L., 1998, Habitable planet formation in binary star systems, Icarus, 132, 196
Wuchterl, G., Guillot, T., Lissauer, J. J., 2000, Giant planet formation, in Protostars and Planets IV, V. Mannings, A. P. Boss, S. S. Russell, (eds.), University of Arizona Press, Tucson, p. 1081
Xie, J.-W., Zhou J.-L., 2008, ApJ, 686, 570
Youdin, A., Shu, F., 2002, Planetesimal formation by gravitational instability, ApJ, 580, 494
Youdin, A., Chiang, E., 2004, Particle pileups and planetesimal formation, ApJ, 601, 1109
Zucker, S., Mazeh, T., Santos, N. C., Udry, S., Mayor, M., 2004, Multi-order TODCOR: Application to observations taken with the CORALIE echelle spectrograph. II. A planet in the system HD 41004, 2004, A&A, 426, 695

Chapter 8
Gravitational Instability in Binary Protoplanetary Disks

Lucio Mayer, Alan Boss, and Andrew F. Nelson

8.1 Introduction

Gravitational instabilities (GIs) can occur in any region of a gas disk that becomes sufficiently cool or develops a high enough surface density. In the nonlinear regime, GIs can produce local and global spiral waves, self-gravitating turbulence, mass and angular momentum transport, and disk fragmentation into dense clumps and substructure. It has been quite some time since the idea was first suggested by Kuiper (1951) and Cameron (1978), and revived by Boss (1997, 1998) stating that the dense clumps in a disk fragmented by GIs may become self-gravitating precursors to gas giant planets. This particular idea for gas giant planet formation has come to be known as the *disk instability* theory. The idea is appealing since gravitational instability develops on very short timescales compared to the accumulation of planetesimals by gravity and the subsequent accretion of gas by a rocky core, the conventional two-stage giant planet formation theory known as *core accretion* (see the chapter by Marzari et al.).

The particular emphasis of this review chapter is on how gravitational instability develops when a protoplanetary disk is not isolated but is a member of a binary or multiple star system. Indeed such a configuration is likely to be the most common in the Galaxy: the majority of solar-type stars in the Galaxy belong to double or multiple stellar systems (Duquennoy and Mayor 1991; Eggenberger et al. 2004). Radial velocity surveys have shown that planets exist in binary or multiple

L. Mayer (✉)
Institute for Theoretical Physics, University of Zurich, and Institute for Astronomy, ETH Zurich, Zurich, Switzerland
e-mail: lucio@phys.ethz.ch

A. Boss (✉)
Carnegie Institution of Washington, Washington, USA
e-mail: boss@dtm.ciw.edu

A.F. Nelson
Los Alamos National Laboratory, USA
e-mail: andy.nelson@lanl.gov

N. Haghighipour (ed.), *Planets in Binary Star Systems*, Astrophysics and Space Science Library 366, DOI 10.1007/978-90-481-8687-7_8,

stellar systems where the stars have separations from ~20 to several thousand AU (Eggenberger et al. 2004; see the chapter by Eggenberger and Udry). Although the samples are still small, attempts have been made to compare properties of planets in single and multiple stellar systems (Patience et al. 2002; Udry et al. 2004). Adaptive optics surveys designed to quantify the relative frequency of planets in single and multiple systems are underway (Udry et al. 2004; Chauvin 2006). At least 30% of extrasolar planetary systems appear to occur in binary or multiple star systems (Raghavan et al. 2006). These surveys could offer a new way to test theories of giant planet formation, provided that different formation models yield different predictions about the effects of a stellar companion.

8.1.1 Gravitational Instabilities

The stability analysis of density perturbations in self-gravitating gaseous disks can be studied analytically only in a local sense and only in the case of infinitesimally thin disks. This analysis is conducted by seeking perturbative solutions to the fluid and Poisson equations treating the disk as an ideal gas. A major outcome of such analysis is a parameter that determines whether such a thin gas disk is stable or unstable to local axisymmetric perturbations (one can show that such a disk will be always stable to local non-axisymmetric perturbations, see Binney and Tremaine 1987) is

$$Q = c\kappa/\pi G\Sigma, \tag{8.1}$$

where c is the sound speed, κ is the epicyclic frequency at which a fluid element oscillates when perturbed from circular motion, G is the gravitational constant, and Σ is the surface density. In a nearly Keplerian disk, κ is approximately equal to the rotational angular speed Ω. For axisymmetric (ring-like) disturbances, disks are stable when $Q > 1$ (Toomre 1964). At high Q-values, pressure, represented by c in equation (8.1), stabilizes short wavelengths, and rotation, represented by κ, stabilizes long wavelengths. The most unstable wavelength when $Q < 1$ is given by $\lambda_m \approx 2\pi^2 G\Sigma/\kappa^2$, and it is known as Toomre wavelength.

The study of global stability to both axisymmetric and non-axisymmetric perturbations can be carried out with the aid of numerical simulations. With three-dimensional simulations, this analysis can be extended further to realistic disks with a finite thickness. For axisymmetric perturbations, $Q = 1$ represents the neutral stability regime exactly as in the local case. For non-axisymmetric perturbations, instead, several works, beginning with Papaloizou and Savonije (1991), show that these become unstable for $Q < \sim 1.5$. They manifest themselves as multi-armed spirals in the disk, grow exponentially on the timescale of a rotation period $P_{\rm rot} = 2\pi/\Omega$, have a predominantly trailing pattern, and several modes can appear simultaneously (Boss 1998; Laughlin et al. 1997; Nelson et al. 1998; Pickett et al. 1998). Spiral waves grow, produce shocks, and thus strong localized heating (Pickett et al. 1998, 2000; Nelson 2000). Gas is also heated by compression on larger scales and through net mass transport due to gravitational torques.

The subsequent evolution of the spiral modes, in particular whether their nonlinear amplitude saturates or continues to grow until the gas becomes dense enough to collapse locally under the action of gravity, depends on mainly two factors, disk thermodynamics and nonlinear mode coupling. During the highly nonlinear phase, the vertical structure of the disk also plays a crucial role, both for cooling and for essential aspects of the dynamics (Pickett et al. 1998, 2000, 2003, Boley and Durisen 2006).

Laughlin et al. (1997, 1998) studied nonlinear mode coupling. Using second and third-order governing equations for spiral modes, and comparing their results with a full nonlinear hydrodynamics treatment, these authors showed that even if only a single mode becomes nonlinear, power is quickly distributed over modes with a wide variety of wavelengths and number of arms. At this point the disk enters in a turbulent regime. The turbulence being generated by self-gravity itself. As high densities are generated by the various modes all over the disk, the chances of fragmentation increase.

Ultimately, the balance between heating and the loss of disk thermal energy by radiative or convective cooling is what decides if fragmentation happens or not. In locally isothermal simulations, in which the disk instantaneously radiates away all the heat generated by self-gravity, local thin-disk calculations, as well as 3D hydro simulations, find fragmentation when $Q < 1.4$ (Johnson and Gammie 2003, Boss 2000; Nelson et al. 1998; Pickett et al. 2000, 2003; Mayer et al. 2002, 2004). Figure 8.1 shows a classic example of a fragmenting disk.

A thermal self-regulation regime of GIs can be reached according to numerical experiments of various authors, provided that the cooling time is moderate (Tomley et al. 1991; Pickett et al. 1998, 2000, 2003; Nelson et al. 2000 Gammie 2001; Rice et al. 2003b; Lodato and Rice 2004, 2005; Mejía et al. 2005; Cai et al. 2006a,b). During self-regulation, Q remains close to 1 but does not fall below unity. Instead, if the cooling rate is fast, no self-regulation occurs and Q drops below unity. In this case, the disk fragments. There have been various attempts to model radiative cooling in self-gravitating disks. For a recent overview of the different methods appearing in the literature and for a general discussion of gravitational instability in protoplanetary disks, we point the reader to Durisen et al. (2007). In this chapter, we will focus on the radiative cooling models that have been used in the few existing works on binary self-gravitating protoplanetary disks.

8.1.2 *Fragmentation and Survival of Clumps*

When the disk cools fast, on a timescale comparable or shorter than the local orbital time, fragmentation happens, both in local thin-disk calculations (Gammie 2001) and in global 3D simulations (Rice et al. 2003b; Mejía et al. 2005; Mayer et al. 2005). If we define the cooling time $t_{\mathrm{cool}} = U/\dot{U}$, where U is the internal energy of the gas, then the condition for fragmentation is $t_{\mathrm{cool}} \leq 3\Omega^{-1}$, or, equivalently, $t_{\mathrm{cool}} \leq P_{\mathrm{rot}}/2$. Finite thickness has a slight stabilizing influence (Rice et al. 2003b; Mayer et al. 2004). If radiative cooling is included, the cooling time varies as the

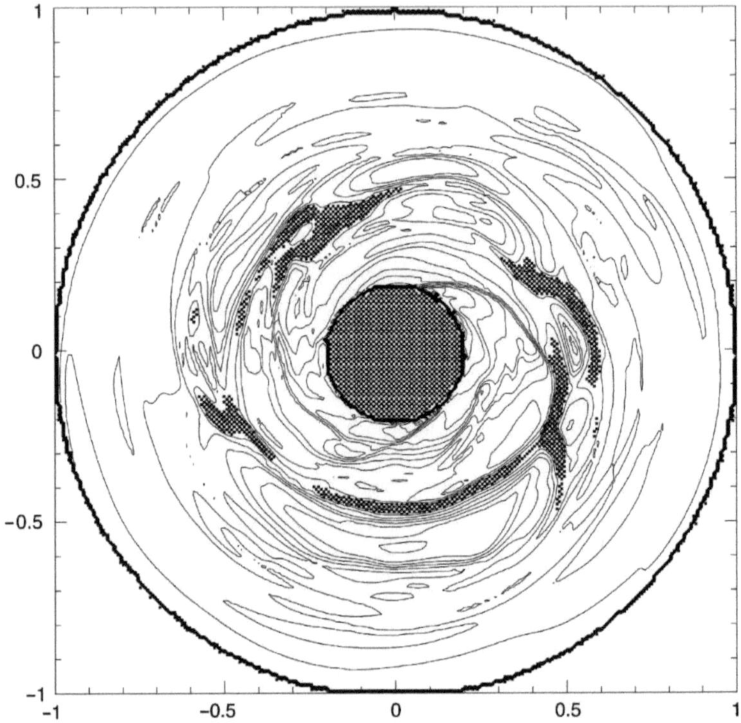

Fig. 8.1 Midplane density contours after 339 years of evolution of a $0.091 M_{\odot}$ disk in orbit around a single $1 M_{\odot}$ protostar, showing the formation of a self-gravitating clump of mass $1.4 M_{\mathrm{Jup}}$ at 4 o'clock (Boss 2001)

gravitational instability develops and grows (Johnson and Gammie 2003), and in particular it tends to increase in the highly nonlinear phase (a tendency which opposes fragmentation – see Cai et al. 2006a,b).

Rice et al. (2003b) found a dependence of the fragmentation criterion on disk mass, with the critical cooling time rising up to $t_{\mathrm{cool}} = 5\Omega^{-1}$ for their most massive disks. The same happens if the adiabatic index is decreased, typically from $\gamma = 5/3$ to $\gamma = 7/5$ (Rice et al. 2003b; Mayer et al. 2005). This latter estimate is in agreement with the timescales for cooling found in 3D models with diffusion approximation radiative transfer and convective-like motions that led to fragmentation into self-gravitating clumps (Boss 2001, 2004a).

Although there is general agreement on conditions for fragmentation, two important questions remain. Do real disks ever cool fast enough for fragmentation to occur, and do the fragments last long enough to contract into permanent protoplanets before being disrupted by tidal stresses, shear stresses, physical collisions, and shocks? Recent simulations have just begun to address the issue of the long term survival of clumps once they have been produced in a disk (Durisen 2007). None of these long-term simulations has explored the case of binary systems and therefore their results are not necessarily valid in that case (for example, they do not take into account the effect of eventual orbital resonances with the companion). However, except for clumps forming at the very periphery of one of the two disks, one would expect the tidal stresses to be dominated by the central star of their own disk, in which case the results of isolated disks are still relevant. In addition, clumps are unlikely to form near the outskirts of disks since the surface density should be too low there. High spatial resolution appears to be crucial for the survival of clumps. Pickett et al. (2003) found that 256 azimuthal cells were not enough to resolve self-gravity on a scale of a fraction of AU, leading to artificial dissolution of overdensities. An increased ability of clumps to persist and become gravitationally bound as the resolution is increased was also found by Boss (2001) and Mayer et al. (2004). High spatial resolution of the gravitational force is crucial, as is the accuracy of a gravity solver for a given resolution element. These features are extremely code-dependent and are briefly discussed below in Section 8.2.2.

8.2 Numerical Techniques and Assumptions

To date, only three papers have been published that consider the possibility of forming giant planets by disk instability in binary star systems: Nelson (2000); Mayer et al. (2005); and Boss (2006). In short, Nelson (2000) found that binarity prevented fragmentation from occurring, Boss (2006) found that binarity could enhance clump formation, and Mayer et al. (2005) found the binarity could discourage fragmentation in some cases, but permit it in other cases. These three papers are the focus of the remainder of this chapter, as we try to decipher the reasons for this apparent dispersion in outcomes.

8.2.1 Hydrodynamics Methods

Three codes have been used so far to follow the evolution of binary protoplanetary disks, two smoothed particle hydrodynamics (SPH) codes (Nelson 2000; Mayer et al. 2005) and one finite-difference grid-based code (Boss 2006; described in detail by Boss and Myhill 1992). The two SPH codes are, respectively, a modified version (Nelson 2000) of a code originally developed by Benz (1990) and GASOLINE (Wadsley et al. 2004; Mayer et al. 2005). We begin with a description of the codes.

Although the codes used, respectively, in Mayer et al. (2005) and Nelson (2000) are based on a similar implementation of SPH, there are some differences in the details of the method that need to be recalled. One major difference is that the version of the code used in Nelson (2000) is 2D while GASOLINE is 3D, as is the Boss (2006) code. Evidence has been accumulated recently that gravitational instability in a protoplanetary disk is an intrinsically three-dimensional phenomenon (Cai et al. 2006a,b). At the same time, at the resolution for which affordable simulations can be done, 3D codes resolve only very poorly the structure of the disk in the third dimension. Nelson (2006) has shown that if the vertical structure is not well resolved, both from a hydrodynamical standpoint and from a radiative transfer standpoint, serious errors in the evolution of the disks may develop. Therefore, even with all other things being equal, this difference alone could result in a different evolution of the spiral patterns, and thus of the outcome of gravitational instability. In what follows, we will highlight the most important features of the two SPH codes and we will explicitly indicate in what ways the two codes differ and what the expected outcome of such differences will be.

SPH is an approach to hydrodynamical modeling first developed by Lucy (1977) and Gingold and Monaghan (1977). It is a particle-based method that does not refer to grids for the calculation of hydrodynamical quantities: all forces and fluid properties are determined by moving particles, and the resulting calculation of smoothed physical variables is spatially adaptive in all modern implementations of the method. The Boss and Myhill (1992) code is an Eulerian code, with all quantities defined on a spherical coordinate grid. The code is second-order accurate in both space and time, a crucial factor for keeping numerical diffusion at a tolerable level.

The basis of the SPH method is the Lagrangian representation and evolution of smoothly varying fluid quantities whose value is only known at disordered discrete points in space occupied by particles. Particles are the fundamental resolution elements comparable to cells in a grid. SPH operates through local summation operations over particles weighted with a smoothing kernel, W, that approximates a local integral. The smoothing operation provides a basis from which to obtain derivatives. Thus, estimates of density-related physical quantities and gradients are generated.

Both GASOLINE and Nelson's codes use a fairly standard implementation of the hydrodynamical equations of motion for SPH (Monaghan 1992). The density at the location of each particle with index i is calculated from a sum over particle masses m_j

$$\rho_i = \sum_{j=1}^{n} m_j W_{ij}. \tag{8.2}$$

where j is an index running on the entire set of n particles and W_{ij} is the kernel function used to compute smoothed averages (see Monaghan 1992). The momentum equation is expressed as

$$\frac{d\boldsymbol{v}_i}{dt} = -\sum_{j=1}^{n} m_j \left(\frac{P_i}{\rho_i^2} + \frac{P_j}{\rho_j^2} + \Pi_{ij} \right) \nabla_i W_{ij}, \tag{8.3}$$

where P_j is pressure, $\boldsymbol{v}_i$ is velocity, and the artificial viscosity term is Π_{ij}. This form of the momentum equation has to be viewed as the usual Euler equation for an inviscid, ideal gas while the artificial viscosity term is purely motivated by numerical reasons (see Section 8.2.4) and is not intended to represent a physical viscosity.

The kernel is a standard B-spline with compact support in both codes (Hernquist and Katz 1989). The number of neighbors, or in other words, the number of particles around a given particle that are considered for smoothed sums, is fixed in GASOLINE at 32 and varies between 10 and 30 in the 2D version of Nelson's code, depending on the local flow.

The Boss and Myhill (1992) code solves the equations of hydrodynamics in spherical coordinates (r, θ, ϕ) in conservation law form. The continuity equation is

$$\frac{\partial \rho}{\partial t} + \nabla \cdot (\rho \mathbf{v}) = 0, \tag{8.4}$$

while the three momentum equations are

$$\frac{\partial(\rho v_r)}{\partial t} + \nabla \cdot (\rho v_r \mathbf{v}) = -\left(\rho \frac{\partial \Phi}{\partial r} + \frac{\partial p}{\partial r}\right) + \frac{\rho}{r}\left(v_\theta^2 + v_\phi^2\right) - \frac{4\pi}{3c}\frac{\partial J}{\partial r} - \frac{1}{r^3}\frac{\partial(r^3 Q_r^r)}{\partial r}, \tag{8.5}$$

$$\frac{\partial(\rho v_\theta)}{\partial t} + \nabla \cdot (\rho v_\theta \mathbf{v}) = -\frac{1}{r}\left(\rho \frac{\partial \Phi}{\partial \theta} + \frac{\partial p}{\partial \theta}\right) - \frac{\rho}{r}\left(v_r v_\theta - v_\phi^2 \cot\theta\right) - \frac{4\pi}{3cr}\frac{\partial J}{\partial \theta} - \frac{1}{r\sin\theta}\frac{\partial\left(\sin\theta Q_\theta^\theta\right)}{\partial \theta}, \tag{8.6}$$

$$\frac{\partial(\rho A)}{\partial t} + \nabla \cdot (\rho A \mathbf{v}) = -\left(\rho \frac{\partial \Phi}{\partial \phi} + \frac{\partial p}{\partial \phi}\right) - \frac{4\pi}{3c}\frac{\partial J}{\partial \phi} - \frac{\partial Q_\phi^\phi}{\partial \phi}, \tag{8.7}$$

where ρ is the mass density, $\mathbf{v} = (v_r, v_\theta, v_\phi)$ is the Eulerian fluid velocity, $A = r\sin\theta v_\phi$ is the specific angular momentum, p is the gas pressure, Φ is the gravitational potential, J is the mean intensity, c is the speed of light, and the Q terms are the diagonal terms of a tensor artificial viscosity. Quantities Φ, J, and the Q terms are discussed in further detail in the appropriate sections below.

8.2.2 Gravity Solvers

Clearly gravity solvers represent a crucial aspect of simulations of self-gravitating disks. They need to be both accurate and efficient, and possibly parallelized in order to take advantage of modern computer architectures and permitting very high resolution calculations to be performed. Both GASOLINE and Nelson's code compute gravity using a tree-based solver, which is fast, easily parallelizable and a natural choice for a particle-based hydrodynamical code such as SPH, since once a tree is built it can also be re-used as an efficient search method for hydrodynamical forces as well. Both codes use a modified versions of the binary tree described in Benz (1990) which approximates the gravity of groups of distant particles in a multipole expansion while calculating interactions of nearby particles explicitly. Gravitational forces due to neighbor particles are softened to avoid divergences as particles pass near each other. The force calculation in tree algorithms requires work proportional to $\mathcal{O}(N \log N)$, where N is the number of particles in the simulation, as opposed to work proportional to N^2 in "direct" N-Body algorithms where all gravitational forces between individual particles are computed directly. The drawback is that, except for very nearby particles whose interactions may be calculated as in direct summation codes, the forces are approximate rather than exact when using a tree. A particularly useful property of tree codes is the ability to efficiently calculate forces for a subset of the bodies. This is critical if there is a large range of timescales in a simulation and multiple independent timesteps are employed (see below). At the cost of force calculations no longer being synchronized among the particles, substantial gains in time-to-solution may be realized. Multiple timesteps are particularly important for applications where the primary interest and thus need for high spatial resolution tends to be focused on small regions within a larger simulated volume; a protoplanetary disk undergoing fragmentation locally is one such application. GASOLINE uses fourth (hexadecapole) rather than second (quadrupole) order multipole moments of the gravitational potential (as used by most tree codes, including Nelson's) to represent the gravitational force from the mass distribution organized in cells at each level of the tree. This results in less computation for the same level of accuracy, better pipelining, smaller interaction lists for each particle, and reduced communication demands in parallel. The current implementation in GASOLINE uses reduced moments that require only $n + 1$ terms to be stored for the nth moment. For a detailed discussion of the accuracy and efficiency of the tree algorithm as a function the order of the multipoles used, see Stadel (2001).

Relaxation effects compromise the attempt to model continuous fluids using particles. Both in GASOLINE and in Nelson's code the particle masses are effectively smoothed in space using the same spline kernels employed in the SPH calculation. This means that the gravitational forces vanish at zero separation and return to Newtonian $1/r^2$ at a separation of $\epsilon_i + \epsilon_j$ where ϵ_i is the gravitational softening applied to each particle. In this sense the gravitational forces are well matched to the SPH forces. However, in GASOLINE, gravitational softening is constant over time and fixed at the beginning of the simulation, while in Nelson's code it is time-dependent and always equal to the local SPH smoothing length.

The use of adaptive softening in Nelson's code ensures that gravitational and pressure forces are always represented with the same resolution. Bate and Burkert (1997) have shown that when an imbalance between pressure and gravitational forces occurs, artificial fragmentation or suppression of physical fragmentation can arise. Bate and Burkert (1997) found that such imbalance leads to spurious results when it occurs at a scale comparable with the local Jeans length of the system, where the Jeans length is defined as $\lambda_j = (\pi c^2/G\rho)^{1/2}$ More recently Nelson (2006) has shown that particle clumping was amplified for imbalances in softening/smoothing even when the Toomre wavelength (see Section 8.1.1) was well resolved, which is a much more limiting condition. When the gravitational softening is fixed over time, such as in GASOLINE, care has to be taken that this be comparable to the SPH smoothing length at the scale of the Jeans length. Mayer et al. (2005) choose the softening according to the latter prescription at the beginning of the simulation and set its value so that outside 10 AU the softening drops to ~1/2 the local smoothing length. An argument can be made that later in the evolution, when strong overdensities develop along the spiral arms, the SPH smoothing length drops significantly, becoming smaller than the gravitational softening, thereby degrading the propensity for numerical fragmentation. Both the initial softening/smoothing inequality and the later evolution of disks towards fragmentation were examined by Nelson (2006). Nelson (2006) found that when the softening is initially smaller than the smoothing length, fragmentation was enhanced, but after it began, the fragments did not continue to contract indefinitely, rather they remained comparable in size to the fixed softening value. Mayer and collaborators have not published systematic tests of the same kind yet, although Mayer et al. (2004) did compare simulations with increasing mass resolution for isolated disks (in the highest resolution simulations the smoothing length was always comparable or smaller than the softening length, even in the outer disk) and found fragmentation to be confirmed and even enhanced as the mass resolution, namely the number of SPH particles, was increased compared to the standard setup (i.e. that in which the softening is $\sim$1/2 the local smoothing length in the outer disk). In conclusion, a slight inequality of softening and smoothing length seems to be a problem only when insufficient mass resolution is used. Moreover, as the new study conducted within the Wengen code comparison is showing, using equal lengths would not guarantee that the numerical result is correct, but requires further validation by means of convergence tests with increasing mass resolution (see Mayer and Gawryszczak 2008). Adaptive softening codes, which guarantee that smoothing and softening lengths are always equal, have also other possible flaws; particles have a time-dependent potential energy, which induces fluctuations in the potential that can in principle increase force errors, eventually affecting the accuracy of the integration. This is known to be problematic for purely gravitational simulations such as those of cosmological structure formation, but its consequences have not been studied systematically for the case of self-gravitating fluids. Recent work of Price and Monaghan (2006) has shown how adaptive softening may be used while still conserving energy, but their technique has not yet seen wide adoption.

The Wengen code comparison represents the first effort to compare SPH and adaptive mesh refinement (AMR) codes for the case of self-gravitating protoplanetary disks (Mayer and Gawryszczak 2008, see also Durisen et al. 2007) and will provide an independent check of the reliability of fragmentation in SPH. Preliminary results show with the standard choice of numerical parameters used in Mayer et al. (2004, 2005) for the SPH code GASOLINE the results converge to those of AMR codes such as FLASH in the case of isothermal calculations (Mayer and Gawryszczak 2008). The only drawback of non-adaptive softening manifesting in those calculations is that, once the clumps form, their collapse is artificially slowed down at some stage, but the same problem is seen also in AMR codes once the maximum level of refinement is reached (the codes are thus comparable at similar spatial resolution). The only way to improve here is to continue increase the resolution of the calculations.

In the Boss and Myhill (1992) code, Poisson's equation

$$\nabla^2 \Phi = 4\pi G\rho, \tag{8.8}$$

is solved for the gravitational potential at each time step. This solution is achieved by using a spherical harmonic (Y_{lm}) expansion of the density and gravitational potential, with terms in the expansion up to $l, m = 32$ or 48 typically being used. Boss (2000, 2001) found that the number of terms in this expansion was just as important for robust clump formation as the spatial resolution. Because the computational effort involved scales as the number of terms squared, however, in practice this value cannot be increased much beyond 48 without having the Poisson solver dominate the effort. Boss (2005) showed that the introduction of point masses to represent very high density clumps led to better defined, more massive clumps, but the computational effort associated with introducing these point masses also led to an appreciable slowing the execution speeds.

8.2.3 Timestepping

Nelson's code uses the Runge–Kutta–Fahlberg method to evolve the equations of motion. It employs a global timestep for all particles in the disk, which is increased or decreased depending on the conditions in the simulations. The integrator is a first order accurate method with second order error control and the scheme provides limits on the second order error term in all of the various variables. GASOLINE incorporates the timestep scheme described as Kick-Drift-Kick (KDK) (see Wadsley et al. 2004). Without gas forces, this is a symplectic leap-frog integration, which ensures the conservation of total energy. This being not guaranteed with the Runge–Kutta method (Quinn et al. 1997; Tremaine 2003), though much better conservation can be had when a single time step for all particles is used, as was done in Nelson (2000). The leap-frog scheme requires only one force evaluation and minimum storage. GASOLINE uses multiple timesteps, hence at any given time, different particles in the disk can be evolved with different timesteps. The base (maximum) timestep is divided in a hierarchy of smaller steps (rungs), with different particles being assigned to different rungs. This allows the use of a much smaller step size when and where it is required, allowing the code to probe a much higher dynamic range and follow correctly the dynamics of regions with very high densities. The drawback is that the scheme is no longer strictly symplectic if particles change rungs during the integration which they generally must do to satisfy their individual timestep criteria. Adaptivity in the time integration, hence the ability to achieve smaller timesteps than the dynamics or hydrodynamics require, can be important to model correctly the formation and evolution of overdensities and clumps. In fact, if the step size Δt is not small enough, the acceleration inside the overdensities, which varies as $1/\Delta t^2$, can be underestimated. This is equivalent to underestimating the self-gravity of clumps and can in principle lead to their artificial dissolution, although no systematic tests have ever been performed. GASOLINE uses a standard timestep criterion based on the particle acceleration (see Wadsley et al. 2004) and for gas particles, the Courant condition and the expansion cooling rate. Nelson's code adopts the Courant condition as well for hydrodynamical forces and additionally a set of constraints based on the position and velocities of particles. If the latter are not met after a timestep, the scheme tries again with a smaller timestep.

The Boss and Myhill (1992) code uses a single-size timestep based on the Courant condition, and uses a predictor–corrector method to achieve second-order accuracy in time.

8.2.4 *Artificial Viscosity*

Most hydrodynamic methods, including SPH, need artificial viscosity to stabilize the flow by avoiding particle interpenetration and to resolve, in an approximate manner, the physical dissipation present in shocks. Both Mayer et al. (2005) and Nelson (2000) use bulk and von Neumann–Richtmyer (so called $\bar{\alpha}$ and β) viscosities to simulate viscous pressures, which are linear and quadratic in the velocity divergence. They both incorporate a switch (see Balsara 1995) that acts to reduce substantially the large undesirable shear component associated with the standard form.

In both GASOLINE and Nelson's code, the artificial viscosity term reads

$$\Pi_{ij} = \begin{cases} \dfrac{-\alpha\frac{1}{2}(c_i + c_j)\mu_{ij} + \beta\mu_{ij}^2}{\frac{1}{2}(\rho_i + \rho_j)} & ; \quad \text{for} \quad \boldsymbol{v}_{ij}\cdot\boldsymbol{r}_{ij} < 0 \\ 0 & ; \quad \text{otherwise} \end{cases} \tag{8.9}$$

where

$$\mu_{ij} = \frac{h(\boldsymbol{v}_{ij}\cdot\boldsymbol{r}_{ij})}{\boldsymbol{r}_{ij}^2 + 0.01(h_i + h_j)^2}\,. \tag{8.10}$$

In these equations, $\boldsymbol{r}_{ij} = \boldsymbol{r}_i - \boldsymbol{r}_j$ and $\boldsymbol{v}_{ij} = \boldsymbol{v}_i - \boldsymbol{v}_j$ where $\boldsymbol{r}_i$ and $\boldsymbol{v}_i$ are, respectively, the position and velocity vectors of particle i, h is the SPH smoothing length, and c_i and c_j are the sound speeds in cells i and j. Quantities $\alpha = 1$ and $\beta = 2$ are standard values of the coefficients of artificial viscosity, known to result in acceptable dissipative behavior across a wide variety of test problems.

In protoplanetary disks, Mach numbers $n_m = v/c$, where v is the bulk velocity of the flow, and c is the speed of sound, are high ($n_m = 10-20$). However, shocks are not as strong (i.e. the density jumps are not as pronounced) as when gas collapses or collides with other gas along nearly radial trajectories, like in cosmological structure formation. This is important because it means that the standard settings for the viscous coefficients may be higher than strictly necessary for correct evolution of the flow. Since artificial viscosity is essentially a nuisance with regard to improving the modeling of the hydrodynamical flow, any improvement which decreases the importance of unphysical side effects while retaining the required stability and dissipative effects of the code, is desirable.

Mayer et al. (2004) have therefore experimented with lowering the coefficients of artificial viscosity, finding that for e.g. $\alpha = 0.5$ and $\beta = 0$ or $\alpha = 0.5$ and $\beta = 1$, fragmentation is more vigorous. However, the simulations of binary protoplanetary disks in Mayer et al. (2005) were designed following a conservative approach, hence the standard values $\alpha = 1$ and $\beta = 2$ were employed. They do, however, include the Balsara (1995) modification of the computation of the velocity divergence from its usual form by multiplying the above equation by a correction factor

$$f_i = \frac{|(\nabla \cdot \mathbf{v}_i)|}{|(\nabla \cdot \mathbf{v}_i)| + |(\nabla \times \mathbf{v}_i)| + 0.0001 c_i / h_i}. \tag{8.11}$$

to reduce shear viscosity. This factor is near unity when the flow is strongly compressive, but near zero in shear flows. In the simulations of Nelson et al. (2000) and Nelson (2000), the typical reduction of viscosity due to this term is a factor of three or better.

Nelson's code starts from the same formulation of artificial viscosity shown above (except that surface density replaces volume density), modified by the same Balsara shear factor, but then also includes a second treatment to obtain a locally varying artificial viscosity. This treatment is due to Morris and Monaghan (1997), who implemented a time dependence to the coefficient $\bar{\alpha}$ that allows growth in regions where it is physically appropriate (strong compressions) and decay in quiescent regions where it is inappropriate. The decay takes place over distances of a few SPH smoothing lengths, after which the coefficient stabilizes to a constant, quiescent value. Nelson adopts a formulation including both the $\bar{\alpha}$ and β terms, where $\bar{\alpha}/\beta = 0.5$, but the magnitudes of the coefficients vary in time and space according to the Morris and Monaghan (1997) formulation. Thus, except in strongly compressing regions (shocks) where it is required to stabilize the flow, artificial viscous dissipation is minimized.

In GASOLINE, only the Balsara correction term is added. Therefore, in general the GASOLINE simulations presented in Mayer et al. (2005) should be more diffusive, which should go in the direction of suppressing fragmentation if all other aspects of the modeling are the same. Figure 8.2 supports the claim of Mayer et al. (2004) on the effect of artificial viscosity by showing that even the simple omission of the Balsara term to reduce spurious shear viscosity can suppress fragmentation in an otherwise fragmenting disk. However, we note that the effect of artificial viscosity is more subtle than this, and that tests such as those of Fig. 8.2 probably address only the impact of viscosity on the transition between mild overdensities and those strong enough to begin collapsing. Once they start collapsing, the prevailing effect of artificial viscosity could actually enhance the growth and survival of the clumps by taking kinetic energy out of the system. The issue of how viscosity can affect the collapse of the clumps is addressed in Pickett and Durisen (2007) (however, note that the formulation of the latter viscosity is different from the standard Monaghan viscosity+Balsara switch used in GASOLINE by Mayer et al. (2004)).

Artificial viscosity, be it explicitly inserted or implicit to the code, is one way in which actual physical processes occurring on the sub-grid scale, such as shock

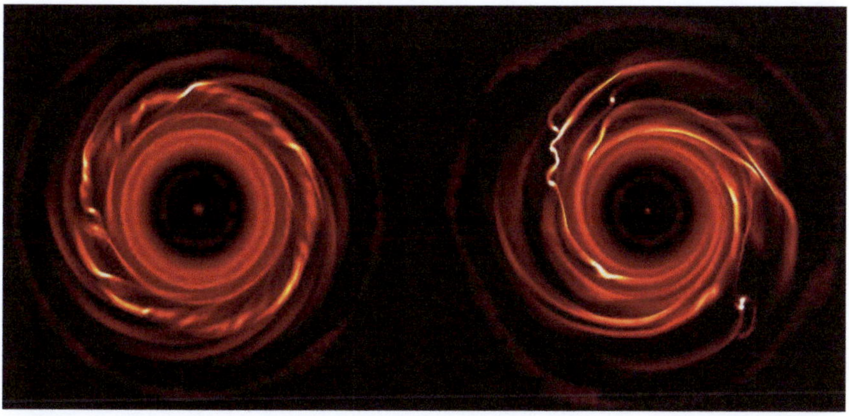

Fig. 8.2 Color coded density maps of two isothermal runs evolved with 1 million SPH particles. The disk starts with a minimum Toomre parameter approaching $Q \sim 1$. The snapshots after about eight orbital times at 10 AU, i.e. about 240 years, are shown. On the *left*, a run without the Balsara correction term is shown. On the *right*, the same disk is evolved with the Balsara term on

front heating and dissipation, can be included in the calculation. A tensor formulation artificial viscosity is included in the Boss and Myhill (1992) code, employing the diagonal terms Q_r^r, Q_θ^θ, and Q_ϕ^ϕ, but is generally not used in the disk instability models, except to discern the extent to which artificial viscosity can suppress fragmentation (Boss 2006). Numerical stability is maintained instead by using as small a fraction of the Courant time step as is needed, sometimes as low as 1% of the Courant value. Boss (2006) showed that his code reproduced an analytical (Burgers) shock wave solution nearly as well without any artificial viscosity as when a standard amount ($C_Q = 1$, see Boss and Myhill 1992) of artificial viscosity was employed, implying that the intrinsic numerical viscosity of his code was sufficient to stabilize such a shock front. Tests on the full set of hydrodynamic equations have not been undertaken because of the lack of a similarly analytical shock wave solution in three dimensional spherical coordinates.

8.2.5 Internal Energy Equation

Both GASOLINE and Nelson's code employ the following energy equation (called "asymmetric"), advocated by Evrard (1990) and Benz (1990), which conserves energy exactly in each pairwise exchange and is dependent only on the local particle pressure,

$$\frac{du_i}{dt} = \frac{P_i}{\rho_i^2} \sum_{j=1}^{n} m_j \mathbf{v}_{ij} \cdot \nabla_i W_{ij}, \tag{8.12}$$

where u_i is the internal energy of particle i, which is equal to $1/(\gamma - 1)P_i/\rho_i$ for an ideal gas (γ is the adiabatic index). In this formulation entropy is closely conserved making it similar to alternative entropy integration approaches, such as that proposed by Springel and Hernquist (2002). The equation above includes only the part related to adiabatic heating and cooling due to mechanical compression and expansion, while the full energy equation has a term due to artificial viscosity and one due to cooling to be discussed below.

The adiabatic index γ is different in Nelson (2000) and Mayer et al. (2005) because Nelson (2000) performs only two dimensional simulations. In both cases, the assumed value of γ is for a gas at temperatures less than 1,000 K in which rotational transitions are active, but not the vibrational ones. Because of the differences in dimensionality, this assumption yields $\gamma = 1.4$ in 3D for the pure hydrogen gas (i.e. average molecular weight of 2.0) assumed by Mayer et al. (2004, 2005) and $\gamma \sim 1.53$ in 2D, for the solar composition gas (average molecular weight 2.31) used by Nelson (see also Nelson 2006 for a discussion of issues that affect the ratio of specific heats in 2D calculations). A value of $\gamma = 1.42$ was incorporated into the vertical structure models used in Nelson (2000) in order to remain consistent, since the assumptions implicit in that calculation required a 3D treatment. Other published works adopt $\gamma = 5/3$ (Rice et al. 2003a, 2005; Cai et al. 2006a,b). As explained in general in Lodato and Rice (2004, 2005), and shown by Mayer et al. (2005) for the case of binary protoplanetary disks, the value of γ can have an impact on whether fragmentation occurs or not in a self-gravitating disk. In particular, a higher γ favors stability versus fragmentation by delivering stronger compressional heating in spiral shocks Lodato and Rice (2004, 2005). A recent paper (Boley et al. 2007b) explains how fixing a value of the adiabatic index without relating the choice to the local temperature and density of the gas is a poor approximation to the actual thermodynamics of molecular hydrogen (the main constituent of protoplanetary disks). Future work will need to asses that.

A term dependent on artificial viscosity in the form of

$$\sum_{j=1}^{n} m_j \frac{1}{2} \Pi_{ij} \mathbf{v}_{ij} \cdot \nabla_i W_{ij} \tag{8.13}$$

is added to Eq. (8.7). This term allows the modeling of irreversible heating occurred in shocks and the related changes in the entropy of the fluid.

If no radiative cooling is included, the resulting model is sometimes dubbed "adiabatic" to distinguish it from the "isentropic" cases in which no irreversible heating is included (Durisen et al. 2007). However, in the simulations of binary protoplanetary disks of Mayer et al. (2005) and Nelson (2000), a cooling term is always present. The cooling term is described in the following section.

Boss and Myhill (1992) code an equation for the specific internal energy with explicit time differencing, in the same manner as the other hydrodynamic equations are solved. This energy equation includes the effects of compressional heating and cooling and of radiative transfer, in either the diffusion or Eddington approximations.

In the latter case, a separate equation for the mean intensity must be solved. In the diffusion approximation, the energy equation is solved by Boss' code (Boss 2001) in the for

$$\frac{\partial(\rho E)}{\partial t} + \nabla \cdot (\rho E \mathbf{v}) = -p\nabla \cdot \mathbf{v} + \nabla \cdot \left[\frac{4}{3\kappa\rho} \nabla(\sigma T^4) \right], \tag{8.14}$$

where $E(\rho, T)$ is the specific internal energy, $\kappa(\rho, T)$ is the Rosseland mean opacity of the gas and dust, T is the gas and dust temperature, and $\sigma = 5.67\times10^{-5}$ cgs is the Stefan–Boltzmann constant. The diffusion approximation energy equation has been used in all Boss' disk instability models with radiative transfer to-date. However, the Eddington approximation energy equation was used to derive the initial quasi-steady state thermal profiles (Boss 1996), used for the initial conditions in Boss' disk instability models. In the Eddington approximation code, the energy equation is

$$\frac{\partial(\rho E)}{\partial t} + \nabla \cdot (\rho E \mathbf{v}) = -p\nabla \cdot \mathbf{v} + L, \tag{8.15}$$

where L is the rate of the change of internal energy due to radiative transfer. The formulation of L depends on the optical depth τ as

$$L = 4\pi\kappa\rho(J - B), \quad \tau < \tau_c, \tag{8.16}$$

$$L = \frac{4\pi}{3} \nabla \cdot \left(\frac{1}{\kappa\rho} \nabla J \right), \quad \tau > \tau_c, \tag{8.17}$$

where τ_c is a critical value for the optical depth ($\tau_c \sim 1$), and $B = \sigma T^4/\pi$ is the Planck function. The mean intensity J is determined by the equation

$$\frac{1}{3}\frac{1}{\kappa\rho} \nabla \cdot \left(\frac{1}{\kappa\rho} \nabla J \right) - J = -B. \tag{8.18}$$

The computational burden associated with the iterative solution of the mean intensity equation in the Eddington approximation has so far precluded its use in disk instability models with the high spatial resolution needed to follow the evolution over many orbital periods. Because of the high optical depths at the midplane (up to $\tau \sim 10^4$) of these disks, however, the diffusion approximation is valid near the critical disk midplane, and radiative transfer in the diffusion approximation imposes little added computational burden.

8.2.6 Cooling in the Simulations

Nelson et al. (2000) and Nelson (2000) require that the disk be in instantaneous vertical entropy equilibrium and instantaneous vertical thermal balance in order to determine its structure. This implicitly assumes that the disk will be convectively unstable vertically over a short timescale and quickly restores thermal balance. Convection is expected given the high optical depths of massive gravitationally unstable disks (Ruden and Pollack 1991). Vigorous vertical currents with features resembling convective instabilities have indeed been observed in massive protoplanetary disks modeled with both SPH and grid codes (Boss 2003, 2004a; Mayer et al. 2007), though Boley et al. (2006) found convective-like motions early in their grid code disk simulations but not at later times when the disk became strongly gravitationally unstable. Under these assumptions, the gas is locally (and instantaneously) adiabatic as a function of z. In an adiabatic medium, the gas pressure and density are related by $p = K\rho^{\gamma}$ (K is the adiabatic constant) and the heat capacity of the gas, C_V, is a constant (by extension, so is the ratio of specific heats, γ, see above). In fact, this will not be the case in general because, in various temperature regimes, molecular hydrogen will have active rotational or vibrational modes. It may dissociate into atomic form or it may become ionized.

From the known (ρ, T) structure, Nelson et al. (2000) derive the temperature of the disk photosphere by a numerical integration of the optical depth, τ, from $z = \infty$ to the altitude at which the optical depth becomes $\tau = 2/3$

$$\tau = 2/3 = \int_{\infty}^{z_{\rm phot}} \rho(z)\kappa(\rho, T)dz. \tag{8.19}$$

In optically thin regions, for which $\tau < 2/3$ at the midplane, these authors assumed that the photosphere temperature is equal to that of the midplane. The photosphere temperature was then tabulated as a function of the three input variables radius, surface density, and specific internal energy. At each time the photosphere temperature was determined for each particle from such a table and used to cool the particle as a blackbody at that temperature. The cooling of any particular particle proceeds as

$$\frac{du_i}{dt} = \frac{-2\sigma T_{\mathit{eff}}^4}{\Sigma_i} \tag{8.20}$$

where Σ_i is the specific surface density of particle i and T_{eff} is its photospheric temperature. In the optically thin regime ($\tau < 2/3$), the above expression was multiplied by τ. The factor of two accounts for the two surfaces of the disk. On every particle, the condition that the temperature (both midplane and photosphere) never falls below the 3 K cosmic background temperature was enforced. Rosseland mean opacities from the tables of Pollack et al. (1985) were used, and opacities for packets of matter above the grain destruction temperature were taken from Alexander and Ferguson (1994).

In parts of the disk where the calculated midplane temperature was greater than dust vaporization temperature, the opacity was temporarily reduced to ~5% of its tabulated values over the entire column above and below that point in the disk. This accounts for the fact that dust formation, after once being vaporized, may occur at rates slower than the timescales for vertical transport through the column. In other regions of the disk it was assumed that the opacity remains unaffected.

Cooling is treated very differently in Mayer et al. (2005). It is independent of distance from the midplane (so there is no dependence on z, as if there was constant thermal equilibrium vertically) but there is an explicit dependence on the distance from the center. The cooling term is proportional to the local orbital time, $P_{\rm orb} = 2\pi/\Omega$, where Ω is the angular velocity, via the following equation

$$\Lambda = dU/dt = U/A\Omega^{-1} \tag{8.21}$$

The disk orbital time is a natural characteristic timescale for spiral modes developing in a rotating disk. Cooling is switched off inside 5 AU in order to maintain temperatures high enough to be comparable to those in protosolar nebula models (e.g. Boss 1998), and in regions reaching a density above 10^{-10} g/cm^3 to account for the local high optical depth. Indeed according to the simulations of Boss (2002) with flux-limited diffusion, the temperature of the gas evolves nearly adiabatically above such densities. In practice, in these regions the gas simply obeys Eq. (8.7) with the artificial viscosity term (9). Mayer et al. (2005) considered cooling times from 0.3 to 1.5 the local orbital time. The jury is still out on whether these cooling times are credible or excessively short. Calculations by Boss (2002, 2004a) and Mayer et al. (2007), which use different approximate treatments of radiative transfer, do find cooling times of this magnitude through a combination of radiative losses and convection, but other works employing different codes and different radiation schemes, such as those of Cai et al. (2006a) and Boley et al. (2006, 2007a,b), have encountered longer cooling times and have not found fragmentation.

The cooling times naturally arising in Nelson (2000) as a result of his radiative scheme were about 25 times longer than the typical orbital time in the region (5–10 AU). Hence they were much longer than those assumed in Mayer et al. (2005). As we shall see, the different cooling time has profound implications on the final outcome of the simulations. Finally, we note that Rafikov (2005, 2007) uses an analytical model including convection, and finds that cooling times are longer than the orbital time, and thus fragmentation should not happen at a distance less than 50 AU from the center of a massive protoplanetary disk. We caution, however, that these analytical models are one-dimensional by nature while gravitational instability is inherently a three-dimensional problem (Mayer and Gawryszczak 2008).

In a recent work, one of the authors, L. Mayer, began performing simulations of binary protoplanetary disks using the new flux-limited diffusion scheme for radiative transfer adopted in Mayer et al. (2007). The latter uses the flux-limiter of Bodenheimer et al. (1991) to model the transition between optically thick and optically thin regions of the disk. The disk then radiates as a blackbody at the edge, with the radiative efficiency being modulated by a parameter that defines how large is the emitting surface area, or in other words, the part of the disk that qualifies as edge. In Section 8.5 we briefly describe some preliminary results of these new calculations.

Boss (2001) noted that in his diffusion approximation models, the radiative flux term is set equal to zero in regions where the optical depth τ drops below 10, so that the diffusion approximation does not affect the solution in low optical depth regions where it is not valid. This assumption is intended to err on the conservative side of limiting radiative cooling. A test model (Boss 2001) that varied this assumption by using a critical $\tau_c = 1$ instead of 10, led to essentially the same results as with $\tau_c = 10$, implying insensitivity of the results to this assumption. Another test model (Boss 2001) used a more detailed flux-limiter to ensure that the radiative energy flux did not exceed the speed of light (specifically, that the magnitude of the net flux vector $\boldsymbol{H}$ did not exceed the mean intensity J), and also yielded essentially the same results as the model with the standard assumptions. In low optical depth regions such as the disk envelope, the gas and dust temperature is assumed to be 50 K.

In the Boss (2006) models, the disk was assumed to be embedded in an infalling envelope of gas and dust that formed a thermal bath with a temperature of 50 K. Thus the effective surface boundary condition on the disk was 50 K in the radiative diffusion calculation. Boss (2004a) found that convective-like motions occurred in the models with a vigor sufficient to transport the heat produced at the midplane by clump formation to the surface of the disk, where it is effectively assumed to be radiated away into the protostellar envelope. This code also employed a full thermodynamical description of the gas, including detailed equations of state for the gas pressure, the specific internal energy, and the dust grain and atomic opacities in the Rosseland mean approximation.

In all of the Boss code models to date, the dissociation of molecular hydrogen into atomic hydrogen is calculated using a dissociation constant, and the transition between para- and ortho-hydrogen is included in the specific internal energy as follows. Linear interpolation from 100 K to 200 K is used to represent the variations between a specific internal energy per gram of hydrogen, i.e., $3R_gT/2\mu_g$ (where R_g is the gas constant $= 8.314 \times 10^7$ cgs, and μ_g is the mean molecular weight for the gas) for temperatures less than 100 K, and $5R_gT/2\mu$ for temperatures above 200 K. No discontinuity is present in this internal energy equation of state, which has been used in all of Boss' disk instability models, though the first adiabatic index Γ_1 is discontinuous at 100 K and 200 K.[1]

Rafikov (2007) has noted that while vigorous convection is possible in disks, the disk photosphere will limit the disk's radiative losses and so may control the outcome of a disk instability. Numerical experiments designed to further test the radiative transfer treatment employed in the Boss models (beyond the tests described above) have been completed (Boss 2007, 2008). Understanding the extent to which the surface of a fully 3D disk, with optical depths that vary in all directions and with corrugations that may shield the disk's surface from the central protostar and other regions of the disk, requires a fully numerical treatment and is not amenable to a simple analytical approach.

[1] The first adiabatic index Γ_1 for a simple perfect gas is the same as γ and is equal to $1 + R_g/(\mu_g C_V)$.

8.3 Initial and Boundary Conditions

Here we describe the initial and boundary conditions for the models, focusing on the different choices made by different workers. These choices include (1) density and temperature profiles of the disks, (2) disk masses and Toomre parameters, (3) spatial resolution, (4) the boundary conditions, and (5) the orbital configuration in the binary experiments.

One important difference between the initial conditions used in Mayer et al. (2005) and those employed in the other two works is that only in the former paper the individual disks are grown in isolation before the binary configuration is initialized. The disk mass is grown until it reaches the desired value while keeping the temperature of the particles constant over time. The Toomre parameter is prevented from falling below 2 by setting the temperature of the disk sufficiently high at the start. This way the initial conditions used for the binary experiments are those of a gravitationally stable disk. The Toomre parameter is then lowered to a value in the range 1.4–2 by resetting the temperature of the particles before placing the star+disk system on the binary orbit (the temperature is set to 65 K, hence the Toomre parameter will depend on the mass of the disk, see below in the next two sections). As the disk evolves in isolation, it expands slightly, losing the sharp outer edges. The inner hole also fills up partially, but most of the particles remain on very similar orbits. Once the star + disk system is placed on an orbit around a companion disk, the system is further evolved using the full energy equation plus a cooling term dependent on the orbital time, therefore including adiabatic compression and expansion, irreversible shock heating and radiative cooling. Nelson (2000) does not grow the disk but starts with a treatment that also solves the full energy equation but treats the cooling term differently (see above).

In both Mayer et al. (2005) and Nelson (2000) matter is set up on initially circular orbits assuming rotational equilibrium in the disk. The central star is modeled by a single massive, softened particle. Radial velocities are set to zero. Gravitational and pressure forces are balanced by centrifugal forces including the small contribution of the disk mass.The magnitudes of the pressure and self-gravitational forces are small compared to the stellar term, therefore the disk is nearly Keplerian in character. No explicit initial perturbations are assumed beyond computational roundoff error in either Mayer et al. (2005) or Nelson (2000). Due to the discrete representation of the fluid variables, this perturbation translates to a noise level of order $\sim 10^{-3} - 10^{-2}$ for the SPH calculations. The relatively large amplitude of the noise originates from the fact that the hydrodynamic quantities are smoothed using a comparatively small number of neighbors (see Herant and Woosley 1994). An increase in the number of particles does not necessarily decrease the noise unless the smoothing extends over a larger number of neighbors. This perturbation provides the initial seed that can be amplified by gravitational instability. The initial disk model in Boss (2006) is rotating with a near-Keplerian angular velocity chosen to maximize the stability of the initial equilibrium state, with zero translational motions. The envelope above the disk, however, is assumed to be falling down with free-fall velocities but with the same angular velocity profile as the rotating disk, the same assumptions as are used in the single star and disk models (e.g., Boss 2005). Random cell-to-cell noise at the level of 10% is introduced to the disk density to seed the cloud with non-axisymmetry at a controlled level.

8.3.1 Density and Temperature Profiles

Mayer et al. (2005) grow the disks slowly over time until the desired mass is reached. Their initial disk models extend from 4 to 20 AU and have a surface density profile $\Sigma(r) \sim r^{-1.5}$ with an exponential cut-off at both the inner and outer edge. The initial vertical density structure of the disks is imposed by assuming hydrostatic equilibrium for an assumed temperature profile $T(r)$.

Nelson (2000) adopts a power law for the initial disk surface density profile of the form

$$\Sigma(r) = \Sigma_0 \left[1 + \left(\frac{r}{r_c} \right)^2 \right]^{-p/2}, \tag{8.22}$$

where p is 3/2, Σ_0 is the central surface density of the disk, which is determined once the total disk mass is assigned, and r_c is the core radius. The Nelson (2000) disk extends from 0.3 to 15 AU and the core radius used for the power laws has the value of $r_c = 1$ AU. The initial temperature profile of this disk is given by

$$T(r) = T_0 \left[1 + \left(\frac{r}{r_c} \right)^2 \right]^{-q/2}, \tag{8.23}$$

where q is 1/2, and T_0 is the central temperature, which is determined once the minimum desired Toomre parameter (and hence the minimum temperature) is determined. The stars in Nelson (2000) have Plummer softening of 0.2 AU each, while a softening length of 2 AU was used with the spline kernel softening in Mayer et al. (2005).

The shape of the initial temperature profile in Mayer et al. (2005) is similar to that used by Boss (1998, 2001) and is shown in Fig. 8.3 together with the profiles of several runs with either binary or isolated disks after a few orbits of evolution. As shown in this figure, the temperature depends only on radius, so there is no difference between midplane and an atmosphere. Between 5 and 10 AU the temperature varies as $\sim r^{-1/2}$, which resembles the slope obtained if viscous accretion onto the central star is the key driver of disk evolution (Boss 1993). Between 4 and 5 AU, the temperature profile rises more steeply in agreement with the 2D radiative transfer calculations of Boss (1996). Beyond 10 AU, however, the temperature smoothly flattens out and reaches a constant minimum (an exponential cut-off is used).

In Mayer et al. (2005) the minimum temperature is fixed to 65 K. It is implicitly assumed that the disk temperature is related to the temperature of the embedding molecular cloud core from which the disk would be accreting material (Boss 1996). Note that, at least for the protosolar nebula, 50 K is probably a conservative upper

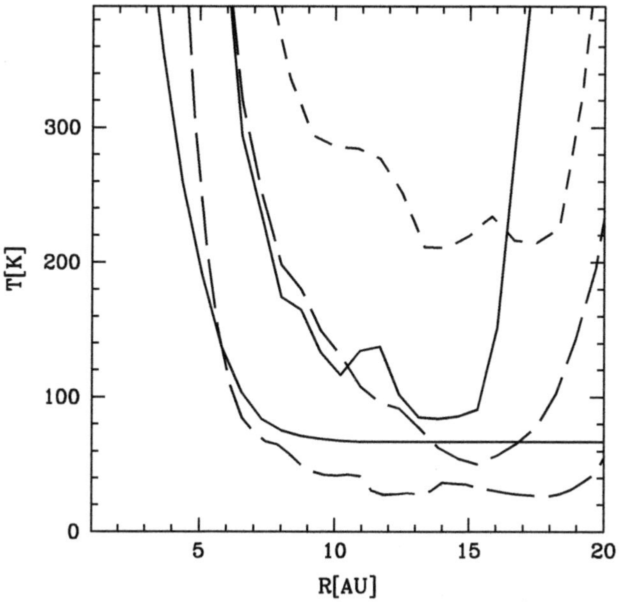

Fig. 8.3 Azimuthally averaged mid-plane temperature profiles at the time of maximum amplitude of the overdensities (at between 120 years and 2,000 years of evolution depending on the model) in some of the runs described in Mayer et al. (2005). The initial temperature profile is shown by the thick solid line. We show the results for a run with two massive disks ($M = 0.1 M_{\odot}$, *thick short-dashed line*, the disks do not fragment) at a separation of about 60 AU, a run with just one of these two disks in isolation (*thin long-dashed line*, disk fragments), a run with the same massive disks at a larger separation of 116 AU (*thin solid line*, disks fragment) and a run with two light disks ($M = 0.01 M_{\odot}$, the disks do not fragment) at a separation of 60 AU (*thick long-dashed line*). The runs adopt cooling times in the range $0.5 - 1.5$ the local orbital time. A smaller separation and a larger disk mass both favor stronger spiral shocks and hence a larger temperature increase during the evolution

limit for the characteristic temperature at $r > 10$ AU based on the chemical composition of comets in the Solar System (temperatures as low as 20 K are suggested in the recent study by Kawakita et al. 2001). Temperatures in the outer part of the disk between 30 and 70 K are found also for several T Tauri disks by modeling their spectral energy distribution assuming a mixture of gas and dust and including radiative transfer (D'Alessio et al. 2001).

In the Boss (2006), the initial disk is an approximate semi-analytic equilibrium model (Boss 1993) with a temperature profile derived from the Eddington approximation radiative transfer models of Boss (1996). The outer disk temperatures were taken to be 40, 50, 60, 70, or 80 K, in order to test the effects of starting with disks that were either marginally gravitationally unstable (40, 50 K) or gravitationally stable (60, 70, 80 K). The disk temperature is not allowed to fall below its initial value, an approximation that errs on the side of suppressing fragmentation.

Figure 8.4 shows the initial radial temperature profile for the models with an outer disk temperature of 80 K, as was assumed in the Boss (2006) and shown in Fig. 8.7.

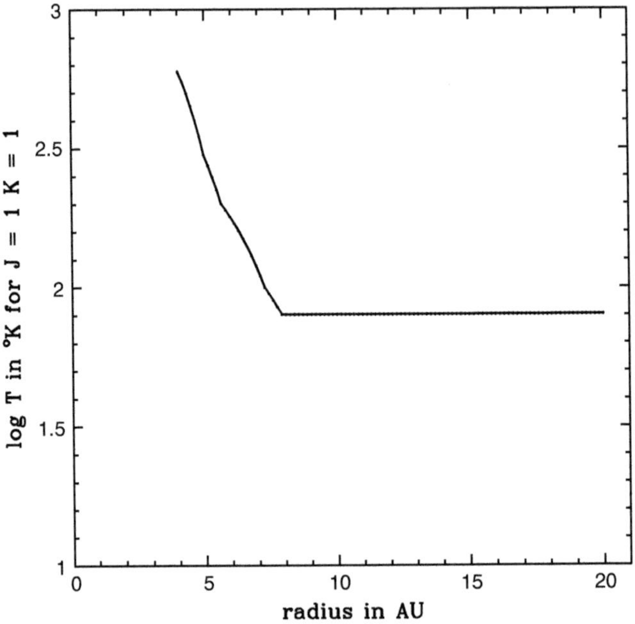

Fig. 8.4 Initial radial temperature profile for the midplane of the Boss (2006) models with outer disk temperatures of 80 K. Each *dot* corresponds to a radial grid point (100 in all) distributed between 4 AU and 20 AU. This initial profile was used for the model shown in Fig. 8.7

The temperature rises strongly toward the protostar at 0 AU because of the heating associated with mass accretion onto the disk from the protostellar envelope and onto the central protostar from the disk (Boss (1993, 1996). The temperature distributions in Boss (1993, 1996) are steady state solutions for axisymmetric (2D) protoplanetary disks with varied disk and stellar masses, opacities, and other parameters.

8.3.2 *Disk Masses and Toomre Parameters*

Both Nelson (2000) and Mayer et al. (2005) consider a system composed of two disks with their central stars. Boss (2006) considers different disk-star systems with stellar companions that are introduced later during the disk's evolution. In Boss (2006) models, the back-reaction of a star-disk system on its binary companion (which is assumed to be a point masses) is not calculated. The two disks in Nelson (2000) have equal masses of $0.05M_{\odot}$ whereas Mayer et al. (2005) consider a range of disk masses encompassing models from as light as the lowest expected values of the minimum mass solar nebula ($0.012M_{\odot}$) to as massive as the heaviest among T Tauri disks ($0.1M_{\odot}$). A few simulations with disks having unequal masses were also performed by Mayer et al. (2005). However, in the majority of the runs, the disks have the same mass. In the Boss (2006) models, the disk mass is $0.091M_{\odot}$

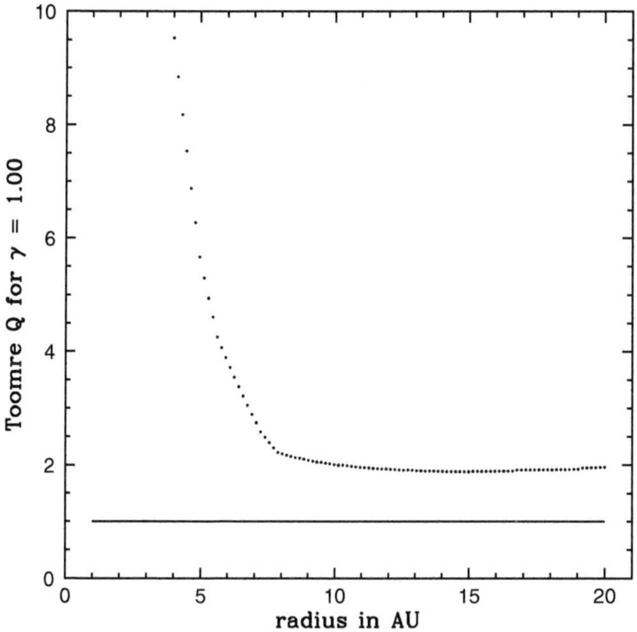

Fig. 8.5 Initial Toomre Q profile for the midplane of the Boss (2006) models with outer disk temperatures of 80 K, used for the model shown in Fig. 8.7

Boss (2006) considers systems with $1M_{\odot}$ equal-mass stars. The mass of the central star is usually $1M_{\odot}$ in Mayer et al. (2005) and $0.5M_{\odot}$ in Nelson (2000). The minimum value of the Toomre parameter varies from 1.3 to 1.9 in Boss (2006) models as the outer disk temperatures in those models are varied from 40 to 80 K. Figure 8.5 shows the radial profile of the Toomre parameter for the Boss (2006) models with an outer disk temperature of 80 K, the same as for the model shown in Figs. 8.4 and 8.7. The disk is very stable to gravitational perturbations in its inner regions, because of the high inner disk temperatures (Fig. 8.4). However, Q drops to a minimum value of 1.9 in the outer region.

The minimum Toomre parameter is ~1.5 in Nelson (2000), achieved just inside the outer edge of the disk at 10–12 AU, whereas it is approximately 1.4 or higher in Mayer et al. (2005) at the disk edge where the temperature also falls to its minimum. The details of the individual models can be found in Mayer et al. (2005). The initial surface densities and Toomre profiles differ between the two works and give rise to a different susceptibility to various channels by which non-axisymmetric models can grow. In Nelson (2000), Q is nearly flat over the largest portion of the disk, with a steep rise at small radii and a shallow increase towards the outer edge of the disk. In Mayer et al. (2005), disks are constructed in such a way that they begin with a steep inner rise of Q, which decreases outwards and reaches its minimum at the disk edge (Mayer et al. 2004). However, as the disk grows in mass, the Q profile changes; its minimum shifts inwards near 15 AU so that the overall profile becomes quite similar to that of Nelson (2000), by the beginning of the simulations. We refer to Mayer et al. (2004) and Nelson et al. (2000) for details on the Q profiles.

8.3.3 Numerical Resolution

Nelson (2000) employs 60,000 particles per disk in his 2D simulations, while Mayer et al. (2005) uses 200,000 particles per disk. Due to the differences in dimensionality, the mass resolution along the disk midplane in both models is similar. So in this respect, the two simulations are quite comparable. The gravitational softening is fixed at $\sim$0.06 AU in Mayer et al. (2005), but it can become smaller than that in Nelson (2000) in overdense regions forming during the simulation.

Boss (2006) used a grid with either $100 \times 22 \times 256 = 0.56 \times 10^6$, or $100 \times 22 \times 512 = 1.1 \times 10^6$ grid points, distributed over the top hemisphere of the calculational grid, and either 32 or 48 terms in the spherical harmonic expansion for the gravitational potential.

8.3.4 Boundary Conditions

Mayer et al. (2005) adopt no boundary conditions at all in their disks. The central star is free to move in response to deviations of the gravitational potential of the disk from the initial equilibrium and gas particles can get as close as resolution allows. Although such a choice is not ideal from the point of view of computational efficiency, since particles nearest to the center have the shortest timesteps, this should reduce fluctuations in the inner density and pressure profiles due to the sudden removal of particles. Nelson (2000) instead implements an inner boundary condition by treating the central star as a sink particle, namely a particle that absorbs the mass and momentum of particles falling below some threshold radius. He uses an accretion radius of 0.2 AU as a compromise between the numerical requirement that the integration timestep not be so small that long period evolution cannot be followed, and the desire to model as large a radial extent of the disk as possible. The gravitational softening of the central star in Mayer et al. (2005) is 2 AU, which for a spline kernel softening means that effectively, the force resolution is 4 AU. In both Mayer et al. (2005) and Nelson (2000) the initial location of the innermost ring of particles lies slightly outside the inner accretion radius or gravitational softening of the star. There is no outer boundary condition in either of these works.

In the Boss (2006) models, the inner boundary at 4 AU is allowed to remove mass and angular momentum from the grid and deposit it onto the central protostar. The outer boundary at 20 AU is fixed in space, and attempts to capture gas which reaches it while suppressing its tendency to bounce back inward. As a result of the strong tidal forces by the binary companions in Boss (2006), disk gas which attempts to flow outward, becomes artificially trapped in the outermost shell of cells. Thus any clumps observed on the or close to the outer boundary of the Boss (2006) are artifacts of the outer boundary conditions and should be disregarded. In most cases, these clumps contain only a small fraction of the disk's mass, and so their influence on the evolution of the rest of the disk should be minor.

8.3.5 Orbital Parameters

Both Mayer et al. (2005) and Nelson (2000) consider coplanar disks corotating with their orbital motion as expected from fragmentation of a cloud core (Bate 2000). If core formation and fragmentation is a highly dynamical process, as recent simulations of gravoturbulent molecular cloud collapse suggest, so that several cores interact strongly during collapse, more complicated orbital configurations will arise. These will need to be explored in the future. Fast close encounters between disks will also occur in the latter scenario, as Lodato et al. (2007) have investigated. In Nelson (2000) the semi-major axis of the relative orbit is set to an initial value of 50 AU and the orbit has an eccentricity of 0.3. Mayer et al. (2005) consider a more circular orbit ($e = 0.14$) and two different values of semimajor axes, 58 and 116 AU. The calculation starts with the companion disk being at the apoastron of the orbit.

Boss (2006) considers only a single disk in his models, with the binary companion being a point mass protostar. The orbit of the binary companion was chosen to have a semimajor axis of either 50 or 100 AU [comparable to those chosen by Mayer et al. (2005)], and eccentricities of either 0.25 or 0.5. The calculations start off with the binary companion at either apoastron or periastron in its orbit. The models assume that the mass of the binary companion is the same as that of the central protostar, (i.e., $1 M_\odot$).

Note that in the Boss (2006) models, the binary semimajor axis is defined to be equal to the radial separation between the two protostars, so that in a model with a semimajor axis of 50 AU and an eccentricity of 0.5, the closest approach between the two protostars is 25 AU, just beyond the edge of the 20 AU-radius of the disk that is being perturbed by the binary companion. If the binary companion had a similar size disk, these disks would collide, but the binary companion is assumed to be a diskless point mass in all of the Boss (2006) models.

8.4 Gravitational Instability in Binary Systems

8.4.1 Does Binarity Help or Suppress Disk Fragmentation?

The three studies seemingly reached very different conclusions regarding the role of binary companions in disk instabilities. Nelson (2000) found no aid to fragmentation, Mayer et al. (2005) found fragmentation in a few cases but also found an indication that binarity might reduce the susceptibility to fragment, while

Boss (2006) found fragmentation to be enhanced by binarity. In the following we analyze in more detail these differences, trying to understand their causes.

As we have seen in the previous sections, there are many differences in the codes and setup of the numerical experiments. Mayer et al. (2005) performed a larger number of experiments, thus exploring a larger parameter-space in terms of initial conditions. Nelson (2000) and Boss (2006) had more realistic treatments of radiation transfer. Boss (2006) ran the experiments with the highest number of resolution elements. We note, however, that while they begin with a smaller number of resolution elements, the SPH simulations by Nelson (2000) and Mayer et al. (2005) are adaptive, hence the resolution increases with time, especially in the most overdense regions. In addition, simulations in Nelson et al. (2000) are two-dimensional, which implies a higher effective resolution in the midplane compared to 3D calculations for equal number of particles.

Disk thermodynamics is crucial for the outcome of gravitational instability. Fragmentation will occur only if cooling times are comparable to the orbital time. Therefore, leaving alone all the other differences, the simple fact that Nelson (2000) had cooling times in a large fraction of his disks that were longer than ten times the orbital times can explain why fragmentation did not occur in his models. With such long cooling times, disks will not fragment in isolation, no matter how strong the spiral structure appearing in the disk is, and irrespective of whether this structure is spontaneous or is tidally triggered by a nearby companion. Nevertheless, fragmentation is not determined by the longest cooling times in the disk, but by the shortest. Nelson's disks did exhibit short cooling times at larger radii, but in spite of this fact, did not fragment.

Rapid cooling of the disk midplane by convective-like motions in 3D disks has been shown to occur with several different codes (Boss 2004a, 2005; Mayer et al. 2007) and can lead to disk fragmentation (Boss 2004a, 2005), provided that the heat transported upward by these motions to the disk's surface can be radiated away to the protostellar envelope, a condition disputed by Rafikov (2007). The vertical structure model of Nelson (2000) assumed efficient vertical energy transport via convection, but did not produce fragmentation. The thermal boundary conditions on the disk surface then become of critical importance, and these boundary conditions are the subject of current research. Until the issue of disk thermal boundary conditions can be further clarified, it is useful to ask whether for relatively short cooling times, comparable to or less than the orbital period, binarity promotes or suppresses fragmentation. The latter question is what both Mayer et al. (2005) and Boss (2006) tried to answer.

Mayer et al. (2005) explored a range of cooling times, from less than 0.5 to 1.5 times the orbital period. They found that the effect of binarity changed with the disk mass; except for the shortest cooling time ($0.3T_{\mathrm{orb}}$), massive disks ($M_d = 0.1\ M_{\odot}$) that fragment in isolation, do not fragment when in a binary with a separation of $\sim$60 AU whereas disks with masses of $0.05 - 0.08\ M_{\odot}$ that do not fragment in isolation, do fragment in such binaries provided that the cooling time is somewhat shorter than the disk orbital time. When the separation grows from 60 AU to 116 AU, the behavior of disks becomes almost indistinguishable from that seen in isolation and fragmentation becomes possible in disks with $0.1\ M_{\odot}$. Finally, very

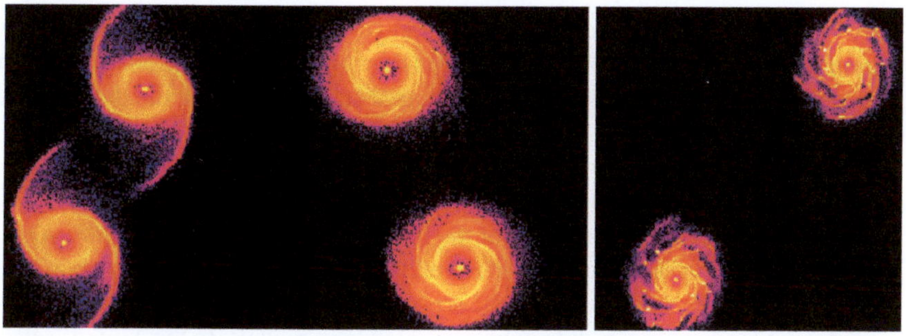

Fig. 8.6 Face-on density maps for two simulations of interacting $M = 0.1 M_{\odot}$ protoplanetary disks in binaries with $t_{\rm cool} = 0.5 P_{\rm rot}$. The binary in the top panel has a nearly circular orbit with an initial separation of 60 AU. A snapshot of this system is shown on left after 150 years (the time of the first pericentric passage) and on right after 450 years. Large tidally induced spiral arms are visible at 150 years. The panel in the bottom shows a snapshot of the simulation of a system with an initial orbital separation twice larger than the system in the top. The time of the snapshot is 160 years after the start of the simulation. In this case, fragmentation into permanent clumps occurs after a few disk orbital times. Figures adapted from Mayer et al. (2005).

light disks, with masses of $\sim$0.01 $M_{\odot}$ do not fragment irrespective of whether they are in isolation or in a binary system. Simulations from this work are presented in Fig. 8.6, which shows how larger separations are more favorable to fragmentation in the case of massive disks. Mayer et al. (2005) interpreted the different behavior of disks having different masses as the product of different net cooling times at different mass scales. In more massive disks, the spiral arms grow stronger as they are better amplified by self-gravity. As a result, shocks are more oblique and disk material acquires higher eccentricities, resulting in overall higher Mach numbers and stronger heating. For a given cooling time, the "net cooling", namely the ratio between cooling and heating, is higher for lighter disks. This explains the higher susceptibility towards fragmentation for disks with intermediate masses of $\sim$0.05 – 0.08 $M_{\odot}$. Then, as the mass decreases further, a third regime is reached. When $M_d \sim 0.01\ M_{\odot}$, the self-gravity is so low that spiral arms are too weak in the first place and cannot form significant overdensities, no matter how strong the perturbation of the companion, and even if the cooling time is comparable to or shorter than the disk orbital time. Figure 8.3 shows the temperature evolution of the disk in some of the runs performed in Mayer et al. (2005). It shows that temperature increase in the outer disk, which opposes fragmentation, is larger in more massive disks and at smaller binary separations, supporting the interpretation of the authors concerning why fragmentation can be suppressed. The results of Mayer et al. (2005) are not in conflict with those of Nelson (2000) for runs that have similar orbital parameters and comparable disk masses (i.e. disk/star systems with mass-ratios of 0.1 and separations of 50–60 AU). It is true that in some of the latter runs disks fragment in Mayer et al. (2005) while they never fragment in Nelson (2000), but this discrepancy is seen only for the shortest cooling times (0.3 – 0.5 $T_{\rm orb}$) used in Mayer et al. (2005), these are more than an order of magnitude shorter than the typical cooling times of Nelson (2000).

Boss (2006) used his standard radiative transfer approach (i.e., diffusion approximation radiation transport, Rosseland mean dust opacities, and detailed equations of state for the gas pressure and specific internal energy)to handle the disk thermodynamics. One model from Boss (2006) was particularly similar to that of Nelson (2000). In that model, the binary companion has a semimajor axis of ~50 AU, and an eccentricity of 0.25. Similar to Nelson (2000), the initial value of Toomre's Q in that model was ~2 throughout most of the inner disk implying that the disk in both models were initially stable. While the total mass of the system was twice as high in Boss (2006) as in Nelson (2000), the ratio of the disk-mass to the protostar mass was the same (and approximately equal to 0.1) in both models. In the case of Nelson (2000), the disk formed strong spiral arms but never fragmented. The disk heated up as a result of viscous dissipation and also partially because of the spiral shocks, finally reaching a steady state characterized by a Toomre Q in the range of 4 to 5 (i.e., quite stable to the growth of gravitational perturbations, Fig. 8.8). By comparison, in the Boss (2006) model (see Fig. 8.7) the disk also formed strong

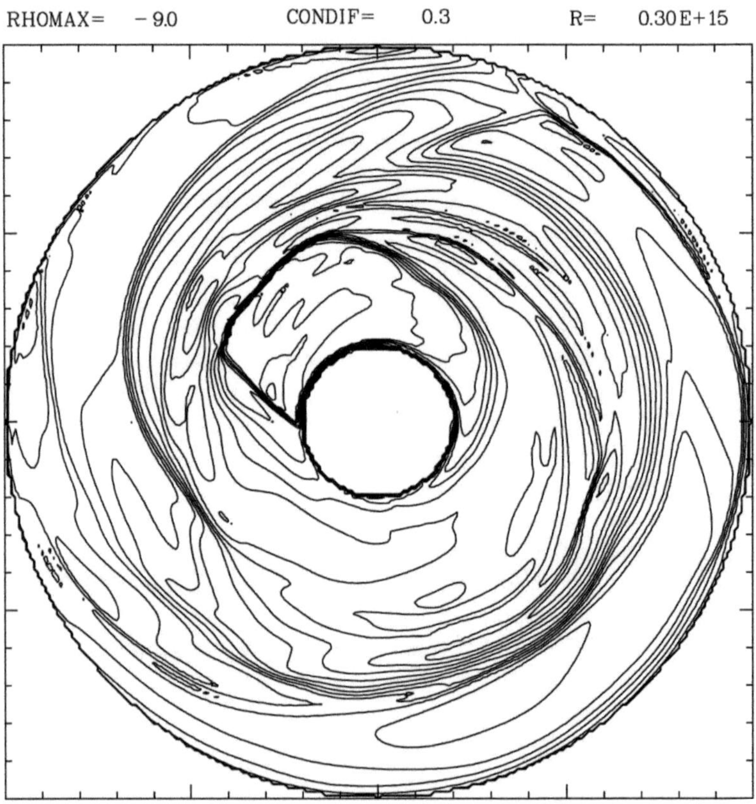

Fig. 8.7 Midplane density contours after 241 year of evolution of a $0.091 M_{\odot}$ disk in an orbit around a member of a binary $1 M_{\odot}$ protostar system, showing the formation of a self-gravitating clump of mass $4.7 M_{\text{Jup}}$ at 10 o'clock (Boss 2006)

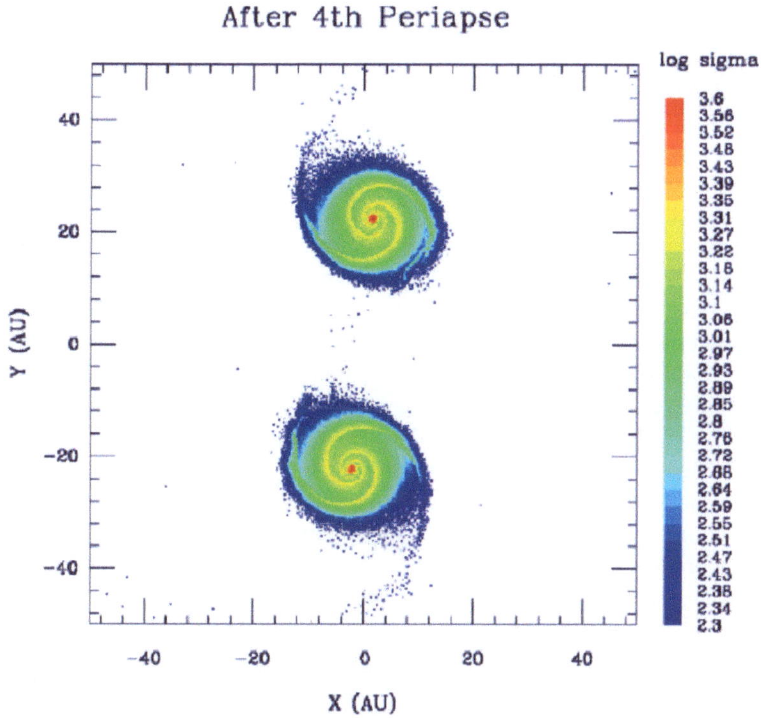

Fig. 8.8 Particle distribution of the binary system after periapse passage. Mass surface density units are in log (g cm^{-2}). The trajectory of each component is counterclockwise, and periapse occurs when the stars (at each disk center) reach the $y = 0$ axis and are 35 AU apart. The tidal torques have caused the appearance of two-armed spiral structures in the disks along with a significant mass redistribution, but no fragmentation. The figure has been adopted from Nelson (2000)

spiral arms, but a self-gravitating clumps was able to form as well, with a mass of 4.7 M_{Jup}. When Boss (2006) ran an identical model, except with the binary companion in a more eccentric orbit ($e = 0.5$), the clump that formed at a similar time to the one in Fig. 8.5 was not massive enough (only 0.68 M_{Jup}) to be self-gravitating, though later in the evolution, self-gravitating clumps did form. These two models show that in Boss (2006), the ability of a binary companion to induce disk fragmentation depends strongly on the orbital parameters.

It is important to note that while Nelson (2000) was able to follow the evolution of the disk through several periastron passages and monitor the resultant disk heating, Boss (2006) only followed a single periastron, largely because of the pile-up of disk-mass at the outer edge of the disk, an obvious numerical artifact that greatly reduced the value of carrying the models any further in time. Thus, it is uncertain what would happen to the clumps that formed after the first periastron passage in the Boss (2006) models, if subsequent periastron passages were calculated as well.

The models of Boss (2006) included binary companions having eccentric orbits with semimajor axes of 50 AU or 100 AU. Mayer et al. (2005), on the other hand, considered 13 models with stellar companions at 58 AU, and four models with semimajor axes of 116 AU. The models with $a = 116$ AU fragmented similar to the 100 AU models in Boss (2006). For the models with $a = 58$ AU, Mayer et al. (2005) found that whether or not fragmentation occurred depended on the disk-masses and the assumed cooling times, as described previously. The models with $a = 58$ AU, disk-masses of 0.1 $M_{\odot}$, and protostars with masses of 1 $M_{\odot}$, can be compared directly with some of Boss's models having essentially the same parameters. While such models never fragmented in Mayer et al. (2005), or only produced transient clumps that disappeared in a few disk orbital times, fragmentation always occurred in Boss (2006). A subtle issue in this comparison is that all fragments obtained by Boss (2006) were in fact transient fragments, because the finite-difference code used in Boss (2006) simulations is unable to provide the enhanced local spatial resolution that is needed to allow self-gravitating clumps to survive and orbit indefinitely. On the other end, Mayer et al. (2005) did not run the same model at higher resolution as they had done for isolated disk models in previous works (e.g. Mayer et al. 2004), and therefore one cannot exclude that their clumps would survive longer or fragmentation would be aided in the first place with an increased number of particles and proportionally smaller softening length. The azimuthal resolution in Boss (2006) is indeed higher than the hydrodynamical resolution and even more than the gravitational force resolution adopted by Mayer et al. (2005). In Mayer et al. (2005), mild overdensities build up along the spiral arms and after the periastron of the relative orbit even in the runs that do not fragment (Fig. 8.6) but they are immediately dissolved. It is possible that based on the preliminary results of the aforementioned code comparison between AMR and SPH codes (Mayer and Gawryszczak 2008), with higher resolution, these overdensities become more nonlinear and collapse. Nevertheless, the difference remains that a companion on a tighter orbit suppresses fragmentation according to Mayer et al. (2005), whereas it promotes fragmentation according to Boss (2006).

The intense heating can be considered as the reason for the suppression of fragmentation in Mayer et al. (2005). Such heating is apparently not present in Boss (2006). Whether the SPH artificial viscosity is biasing the results too much with the stronger shocks present in binary systems or whether the disks in Boss (2006) cool too fast, is at the moment unclear. It is especially noteworthy that Nelson (2000) studied a 50 AU binary system with SPH, at lower resolution than in Mayer et al. (2005). This system should therefore be even more strongly affected by the presence of heating from artificial viscosity. Instead, Nelson (2000) finds that the disks are in fact too cold compared to the observed L1551 IRS5 system implying that still more heating is required than artificial viscosity provides, making his disks even less likely to fragment.

Another difference in the setup that might explain the above-mentioned discrepancies is the fact that in Boss (2006), the companion is simply a protostar, whereas in Mayer et al. (2005) the companion is a protostar with a disk. In tight binary systems, the presence of the other disk might have an effect; near periastron of the orbit, the two disks almost touch each other, possibly enhancing tidal and compressional heating on one another. Since the orbits in Mayer et al. (2005) are nearly circular, in the binaries with 58 AU separations, the two disks are almost always in the latter situation. This would also mean that the behavior of real tight binary systems, which will normally have two disks orbiting each other, should be closer to what found by Mayer et al. (2005) and Nelson (2000).

8.4.2 Disk Evolution: Internal vs. External

Mayer et al. (2005) saw very different behavior when they compared the a disk evolving in isolation and with a companion. Since gravitational instability is regulated by thermal pressure and mass density, remarkable differences should be seen in both quantities when comparing isolated and binary disks. Indeed, we have already seen the effect of temperature (Fig. 8.3). Here we discuss the effect on mass density on the surface density profile of a disk. In particular, the following question arises; how much of the restructuring of the disk-mass distribution is due to the disk's self-gravity and how much is due to tidal torques induced by the companion?

Figure 8.9 shows the evolution of the disk surface density profile in one of the binary disk simulations of Mayer et al. (2005) evolved both with and without

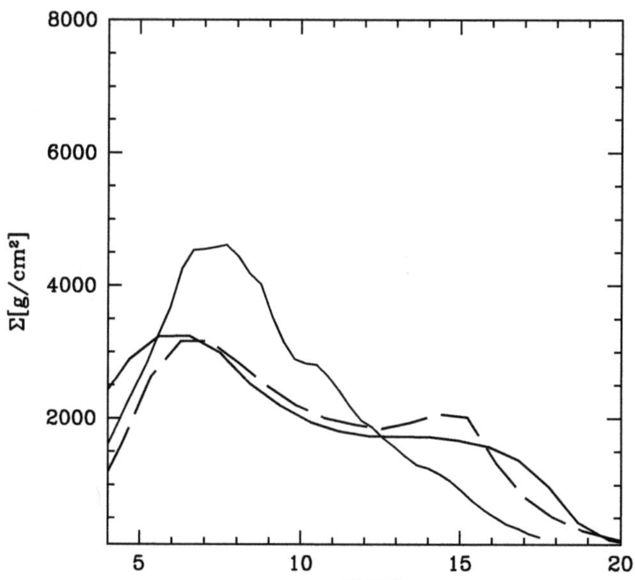

Fig. 8.9 *Thick solid line*: The azimuthally averaged surface density profile of a disk with a mass of $0.1\,M_{\odot}$ at $t = 0$. *Dashed line*: the evolved state of the same disk after being run without self-gravity for two orbits around an equally massive companion. The *thin solid line* shows the disk's surface density with self-gravity. A cooling time equivalent to 0.3 times the orbital time was adopted in both runs. Figure from Mayer et al. (2005)

self-gravity. The tidal interaction modifies the disk structure in both cases, but such changes are moderate compared to those occurring when self-gravity is introduced. When self-gravity is included, the density peak developing as a result of inflow driven by internal gravitational torques is almost a factor of 2–3 higher than the maximum density in similar disk models evolved in isolation (Mayer et al. 2004). This statement applies to all of the runs in Mayer et al. (2005). The density in the tidally perturbed disks grows not only in the center but even out to several AUs, which explains why models with masses lower than 0.1 $M_{\odot}$ become more prone to fragmentation in the region around 5–10 AU when perturbed by a binary companion. Evidently in these lighter disks, the heating from shocks is not enough to compensate such a large density increase. Since disks are truncated within 15 AU, when clumps form, they do so within such radius, typically between 8 AU and 12 AU. The locations where the clumps form correspond to the location of the density maximum and are slightly closer to the star compared to those of clumps in the isolated disks studied by Mayer et al. (2004). In fact, in isolation, gravitationally unstable disks typically develop a density maximum between 12 and 15 AU, and that is where Q drops below 1 and fragmentation occurs (Mayer et al. 2004). The conclusion is that in all the simulations of Mayer et al. (2005), the restructuring of the disk results from a combination of tidal torques and self-gravity of the disks.

The mass transport present even without self-gravity is the result of tidal torques induced by the gravitational interaction with the companion. These tidal torques produce a two-armed spiral mode in the otherwise passive disk. The disk becomes truncated to a smaller radius and more mass piles up in the inner few AU as the arms redistribute angular momentum. We recall that the disks in Mayer et al. (2005) have been slowly grown in mass in isolation before being evolved with a companion. While the disk evolves in isolation, the inner hole, present in the initial conditions, is gradually filled and therefore the rapid accumulation of mass seen in the binary case is not an artificial result of the inner boundary condition. The mass inflow produces compressional heating, raising the temperature of the disk inside 10 AU. Exchange of mass between the two disks occurs but their mass varies by only $\sim$10%.

8.4.3 Temperatures in Binary Self-Gravitating Disks and Effects on Dust Grains

Both Nelson (2000) and Mayer et al. (2005) found significant heating along spiral shocks in binary systems. Mayer et al. (2005) found that the temperatures can be a factor of 2–3 higher relative to the same disk in isolation for disks in the mass range of $\sim$0.05–0.1 solar-masses. This has important implications for the destruction of dust grains, hence on the formation of planetesimals and thus of those Earth-sized rocky cores that are a necessary step to form giant planets in the core-accretion model. The consequence of the high temperatures in the GI active outer region of the disk is the vaporization of ice grains, which constitute as much as 30–40% of the dust content in the disk.

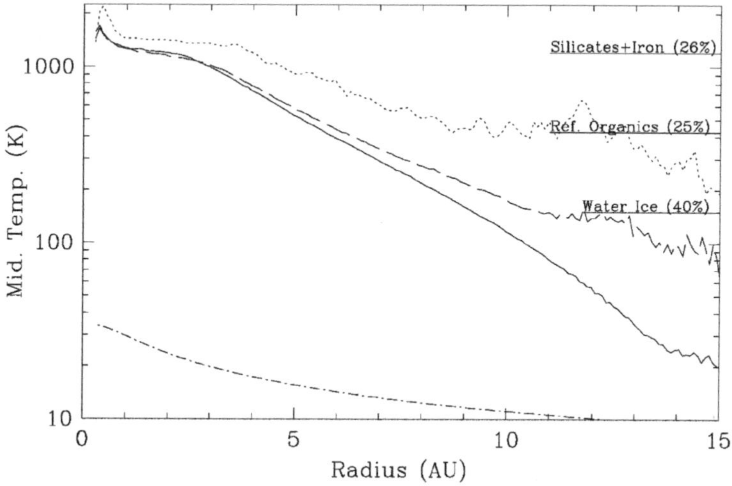

Fig. 8.10 Temperature profiles of the disks in Nelson (2000) shown before (*solid line*) and after (*dashed line*) the 4th periapse. The initial profile is shown with a *dashed–dotted line*. The *dotted line* shows the maximum temperature reached inside the spiral arms at that radius. At the right are vaporization temperatures of the major grain species in the solar nebula and their fraction of the total grain mas, as discussed in Pickett et al. (1994)

The results of Nelson (2000) are shown in Fig. 8.10. The actual surface density of solids might be reduced by up to 40% as a result of vaporization compared to isolated disks. The direct consequence would be that core accretion will be less efficient in binary systems compared to isolated disks. Strong heating is instead absent in light disks ($M_d = 0.01 M_\odot$) whose temperatures increase by less than 50%. Therefore core accretion should be favored in such light disks, with masses comparable to the minimum mass solar nebula, because a larger relative dust content would be maintained. Those, however, are also the disks in which the growth of rocky cores of a few Earth masses (necessary to trigger runaway gas accretion) would take longer owing to their low surface densities. Vaporization would have other effects as well that are not included in the simulations carried out so far and might be relevant for the outcome of gravitational instability. For instance, vaporization will lower the opacity. This leads to even stronger shocks (and heating) but, at the same time, may reduce the cooling time. Which effect wins in terms of fragmentation is unclear. In addition, vaporization might lower the local molecular weight, which goes in the direction of favoring fragmentation (Mayer et al. 2007). Future simulations will have to investigate all these effects simultaneously.

A caveat in the above discussion is that the surface density of massive disks ($M_d > 0.05 M_\odot$) at distances of 10–15 AU from the center is 50% higher than it would be without a companion by the end of the simulation. Assuming a uniform gas-to-dust ratio in the disk, the increase in surface density could compensate for the vaporization of dust grains, making massive disks not less favorable then light disks for giant planet formation via core accretion (see above). Spiral arms might also gather solids as a result of pressure gradients (Haghighipour and Boss 2003a,b; Rice et al. 2004, 2006; Haghighipour 2005) leading to an enhanced gas-to-dust ratio inside them, another effect that could favor core accretion. Given the wide parameter space involved here (including both disk initial conditions and composition changes), only more realistic calculations incorporating directly both vaporization and dust particle dynamics within the gaseous disk will be able to settle this issue.

8.4.4 Effects of Artificial Viscosity

Boss (2006) did not employ artificial viscosity in his standard models, but did include the artificial viscosity in a subset of models designed to determine to what extent the use of artificial viscosity in either a finite-difference code (e.g., Pickett et al. 2000) or an SPH code (e.g., Nelson 2000) might affect the disk instability process. Boss (2006) ran four models with varying amounts of artificial viscosity and found that only when the artificial viscosity was set to a value of 10 times higher than the standard value ($C_Q = 1$) did the disk become so hot as to appreciably stifle fragmentation.

One possible source of the different outcomes between the results of Boss (2006), Nelson (2000) and Mayer et al. (2005), is the amount and effect of artificial viscosity assumed in these models. An artificial viscosity equivalent to an effective viscosity ($\alpha_{\rm eff}$) with $\alpha_{\rm eff} = 0.002$ to 0.005 was included in Nelson (2000) to correctly model shock heating in his SPH code. The same applies to the Mayer et al. (2005) models [the value of the corresponding $\alpha_{\rm eff}$ is comparable to that in Nelson (2000)]. In the Boss (2006) models, however, artificial viscosity did not have an explicit term, but the level of implicit numerical viscosity appears to be equivalent to $\alpha \sim 10^{-4}$ (Boss 2004b), approximately 20 to 50 times lower than that in Nelson (2000). Given the experiments of Boss (2006) with artificial viscosity, the use of this level of artificial viscosity in Nelson (2000) is consistent with the absence of fragmentation and the difference in his net cooling times with that in Boss (2006). Relatively short cooling times are obtained in models without artificial viscosity [e.g., $\sim$1 to 2 orbital periods (Boss 2004a)] compared to the effective cooling time reported in Nelson (2000) [approximately 5–15 orbital periods for orbital distances from 10 AU to 5 AU, respectively]. Of course this leaves us with an open question; how viscous will be disks in reality, and therefore which of the three works produces the answer closest to reality.

A hint on how realistic a chosen amount of artificial viscosity is might be given by the amount of heating that it produces. The use of artificial viscosity by Nelson (2000) was motivated by a good reason, namely to model shock dissipation in the disks, and produced substantial heating. Nevertheless, an analysis of the flux densities derived from his simulations fell nearly an order of magnitude short of that required to reproduce the observations of the L1551 IRS5 binary system, on which his initial conditions were based.

It is of some interest that Boss and Yorke (1993, 1996) were able to match spectral energy distributions of the eponymous T Tauri system T Tau, with axisymmetric disk models similar to those that form the basis for Boss (2006) models, without using the artificial viscosity that Nelson (2000) found necessary to model L1551 IRS5. A part of this difference can perhaps be explained by the fact that the system modeled by Boss and Yorke was T Tau, a system at a much later evolutionary stage than L1551 IRS5, with correspondingly different energy output. We look to future observations using the ALMA telescope with great interest, because of the likelihood for observing younger, and much more deeply embedded objects, of greater relevance to the earliest stages of disk evolution where gravitational instabilities are more probable.

8.4.5 Initial Conditions in the Context of Star Formation

Are the initial conditions adopted in the existing simulations of binary, self-gravitating protoplanetary disks realistic? In reality, the two disks will be communicating since their beginning, undergoing mass transfer, and growing out of gas flowing from the periphery of the molecular cloud core. This is quite different from the setup assumed in the simulations discussed so far. Tidal perturbations and mass transfer might be too sudden in the computations described in this chapter, where as in reality they will be achieved gradually. However, if star formation occurs in gravoturbulent clouds, such as those modeled by Bate et al. (2002), rather than in isolated cores, disks will not have time to slowly adjust to such an extremely dynamic environment by the time they become gravitationally unstable. In the gravoturbulent model, the collapse of the individual cores would occur on a timescale much smaller than the average collapse timescale of the larger star-forming region. A short collapse time of cores is also suggested by observations. Among the latter, we recall the large column densities of prestellar cores and the fact that molecular abundances in such cores are consistent with a rapid collapse (Aikawa 2004). The resulting systems would have undergone several tidal interactions with bound or unbound companions since their birth. Turbulent molecular clouds have velocity dispersions of order 2–5 km s^{-1}, which means the typical crossing time of a region 10^{-2} pc in size will be $\sim 10^3$ years. The characteristic timescale of encounters between cores in such a turbulent cloud has to be of the same order of magnitude, i.e. once again comparable with the binary orbital time.

Despite the above, Mayer et al. (2005) studied the case of two self-gravitating disks reaching gradually the conditions present at the beginning of the simulations by starting with very light, nearly non-self gravitating disks and growing the disk slowly over the course of a few binary orbital periods (Fig. 8.11). This way the disk profile had time to adjust. The spiral arms, tidally induced on the third orbit, were slightly weaker than those in the standard run, and transient localized overdensities were apparent which were not present before. However, no gravitationally bound clumps occurred in the disk and the outer disk temperature after one orbit ($\sim$300 K) became comparable to that in the original run.

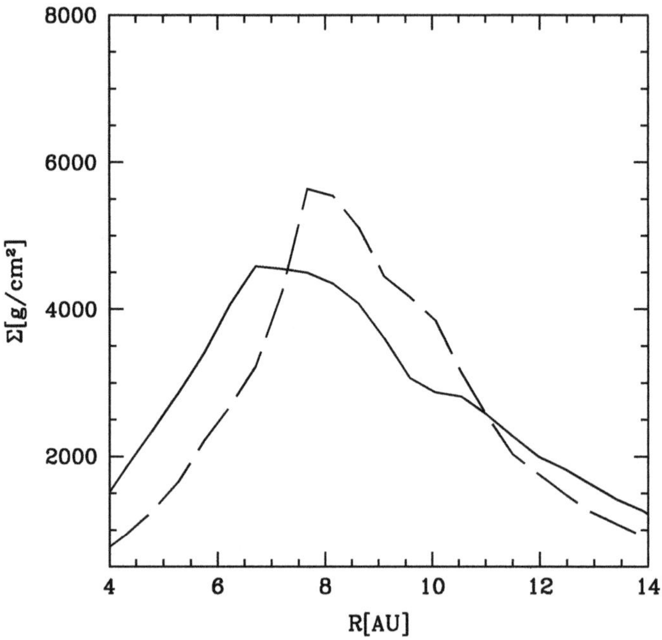

Fig. 8.11 Azimuthally averaged surface density profile after two binary orbits for a run employing two massive disks with $M_d = 0.1 M_{\odot}$. The disks were evolved with a cooling time equivalent to 0.3 times the local orbital time with self-gravity (*solid line*) and with self-gravity switched off on the first binary orbit (*dashed line*)

The final surface density profiles of the disks in the two runs were also quite similar. Mass redistribution due to gravitational torques led to a profile which could not be described by a single power-law, had a remarkable density peak close to 7–8 AU, and was steeper than r^{-2} outside this radius (Figs. 8.9 and 8.11). The surface density profiles were steeper than those produced by gravitational instabilities when there is no companion. The mass inflow towards the center was greater. One would be tempted to conclude that the viscous evolution of the disk, where the "viscosity" is due to gravitational instability, is faster in binary systems. This might lead to a faster dissipation of the disk and a faster growth of the star since gas flows outside in. However, while we believe that our results are qualitatively correct, we caution that they might be still somewhat uncertain on the quantitative side because the large, initially massless hole in the middle of the disks might affect the mass transfer (gas particles in the inner annulus do not feel any pressure gradient and will tend to flow inward if they are slightly perturbed from the initial centrifugal equilibrium).

In the simulations of Mayer et al. (2005), the restructuring of the disk results from a combination of tidal torques and intrinsic self-gravity (see Section 8.4.5). Since in the early stages, protoplanetary disks should be massive enough to be self-gravitating (Yorke and Bodenheimer 1999), it seems that this profound restructuring driven by the two simultaneous effects will likely occur in binary systems, and will occur early. Inflow driven by gravitational torques in the early stages of protostellar disk evolution is now seen in many different calculations for the case of low-mass stars as well as massive stars (Krumholz et al. 2007; Voroboyov and Basu 2007).

Mayer has recently performed new SPH simulations with a density-dependent polytropic equation of state (Bate 2001) that use a variable mass resolution technique to reach down to achieve a spatial resolution of ~0.1 AU in a rotating (non-turbulent) collapsing molecular core of a fraction of a parsec in size. These simulations show substantial evolution of binary disks and mass inflow towards the central star just as a result of self-gravity. At the beginning of the simulations a rapidly rotating core collapses, becomes bar unstable and fragments into two clumps (see also Bate and Burkert 1997) that later become a pair of pre-stellar cores surrounded by a fairly large, tidally truncated disk (about 30 or 80 AU in size), as shown in Fig. 8.12. The two disk + core systems have unequal masses; while each system starts out with more than 2/3 of the mass being in the disk and the rest in a dense central clump, the precursor of the star, less than 0.1 AU in size, after about two binary orbits (corresponding to ~2,000 years and to the time at which we stopped the

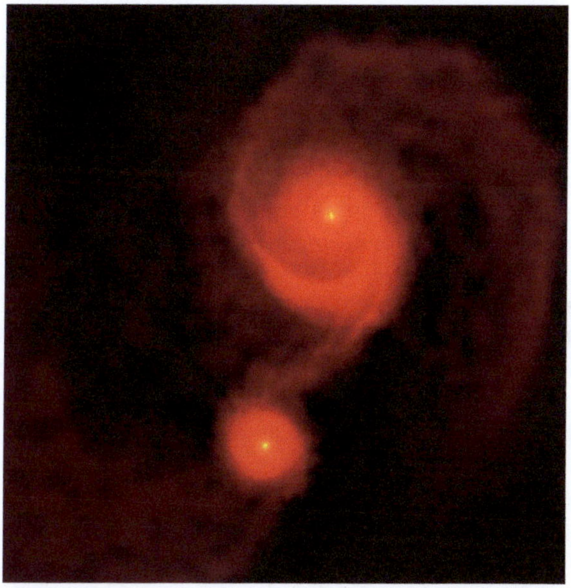

Fig. 8.12 Color coded density map of a binary protostar+disk system resulting from the collapse of a rotating molecular cloud core with initial density profile proportional to r^{-1}. The *box* is 200 AU on a side, and the system is shown about 5×10^4 years after the collapse has been initiated and also a couple of binary orbits after the two disks have formed from the fragmentation of a bar-unstable, rapidly rotating protostellar core. The mass of the larger system is $\sim 0.5\, M_\odot$ whereas the smaller system has a mass of only $\sim 0.15\, M_\odot$. The total mass of the molecular core was $1 M_\odot$ and its size was 10,000 AU at $t = 0$. The simulation employed 500,000 particles in total, but because the mass resolution in the inner 500 AU was higher than in the surrounding volume, as many as 4/5 of the particles were used only in this inner region

simulation) more than half of the disk mass has accreted onto the central clump. At this stage disks are slightly lighter than the central clump. The gas in the protostellar disks loses angular momentum from a combination of spiral instabilities in the disks and tidal torques from the companion, and the spiral arms are also strengthened by the tidal perturbation of the companion. The physical driver of accretion is just gravity in these simulations since no other mechanisms to remove angular momentum are present except gravitational torques (both intrinsic and tidal). The only caveat is that artificial viscosity, while required to model physical dissipation in shocks correctly, might also enhance angular momentum transport. The accretion rate from the disk onto the central protostar at the end of the simulation is nearly $5 \times 10^{-5} M_{\odot}/\text{year}$, or about ten times higher than the accretion rate from the core onto the disk for the lightest disk + core system (this system is also the one that suffers the strongest tidal perturbation among the two systems), and about a factor of two lower for the other, more massive system.

8.5 Conclusions

The feasibility of disk instability as a formation mechanism for giant planets is subject of active debate. The results for the specific case of binary systems must thus be considered and interpreted in light of such more general debate. It is recognized that the ultimate outcome of gravitational instability, whether it produces fragmentation and bound condensations or only spiral arms and transient overdensities, depends on the balance between heating and cooling and on how accurately gravity is resolved at scales both large and small (Durisen et al. 2007; Boss 2007). It is then crucial to understand how reliable is the modeling of thermodynamics and gravity employed by different codes.

The artificial viscosity used in SPH codes (Mayer et al. 2005; Nelson 2000) acts as a source of heating in disks. There are indications that artificial viscosity generally tends to suppress fragmentation (Mayer et al. 2004). However, when artificial viscosity is such that simulations can match fluxes observed from protostellar systems, the resulting high level of disk heating can prevent fragmentation (Nelson 2000). Using a gravitational softening length that is smaller than the SPH smoothing length throughout most of the disk evolution, as in Mayer et al. (2005), favors fragmentation, as does the numerical noise associated with the particle realization of the fluid in SPH (for example, numerical noise can introduce spurious density fluctuations that then grow as a result of gravitational instability). Sharp disk edges promote fragmentation, while low resolution (mass resolution in SPH, as set by the number of particles, or grid size in grid codes) seems to suppress fragmentation (Mayer et al. 2004; Boss 2000), although Nelson (2006) show cases where the opposite happens. The dependence on resolution and different numerical techniques to model the gravitational and hydrodynamical forces is being investigated systematically in an on-going code comparison that involves both SPH and adaptive mesh refinement (AMR) codes. Preliminary results confirm an increasing susceptibility towards fragmentation with increasing resolution for isothermal disks with $Q \sim 1$ (Mayer and Gawryszczak 2008). Finally, and perhaps most importantly, short cooling times promote fragmentation (Mayer et al. 2005), whereas long cooling times prevent it altogether (Nelson 2000). Similarly, Boley et al. (2007b) propose a suite of tests for the radiative physics schemes employed by different codes that should ensure that the right amount of energy is transported from the midplane to the edge of the disk and then radiated away at the disk boundary.

In summary, the three works discussed in this chapter yielded the following results;

- Mayer et al. (2005) found that binary companions with semimajor axes of 58 AU prevented disk fragmentation, unless the disks had moderate masses (0.05–0.08 $M_{\odot}$) and cooled even more rapidly than convective cooling would permit (Boss 2002, 2003; Mayer et al. 2007). The same conclusion, with similar but not identical initial conditions, is reached by Nelson (2000). Instead, Boss (2006) found fragmentation and transient clump formation for semimajor axis of 50 AU, although even closer encounters (i.e., higher eccentricities) tended to work against the formation of self-gravitating clumps.
- Both Boss (2006) and Mayer et al. (2005) found fragmentation or transient clump formation to occur in binary systems with semimajor axis of 100 AU, Nelson (2000) did not run calculations with such large orbital separations.

There are some differences in the design of the experiments in the three works that need to be taken into account in order to compare them appropriately. We note that most of the disks used by Mayer et al. (2005) were marginally unstable by construction ($Q < 2$), and therefore the fragmentation seen for larger semimajor axes might simply reflect the fact that as the separation increases the results tend to converge to those for disks with no binary companion. This in other words means that the fragmentation seen in such cases could have nothing to do with the tidal perturbation of a companion. Conversely, Boss (2006) uses mostly disks that would be stable in isolation ($Q \sim 2$ or larger), and hence the only logical outcome of his calculations in the presence of a companion is either that the disk remains stable or that fragmentation is enhanced (i.e. no experiment was constructed to see whether fragmentation could be suppressed). This different logic behind the design of the simulations in the two works complicates the comparison and calls for future attempts by these and other authors to perform and compare exactly the same experiment. Finally, the disks used in Nelson (2000) were not able to fragment in isolation while many of those used by Mayer et al. (2005) were fragmenting for the range of cooling times adopted by these authors.

A major source of the differences obtained by Nelson (2000) and Mayer et al. (2005) relative to Boss (2006) could be the use of artificial viscosity in the first two works based on SPH and its neglect by Boss (2006) since, as we discussed previously, artificial viscosity has a direct effect on fragmentation (see Section 8.2.4). The much longer cooling times in Nelson (2000) are also expected to prevent fragmentation. Clearly, when artificial viscosity is used to heat a disk, and this heat is unable to escape on an orbital time scale, the chances for clump formation by disk instability are severely reduced. As mentioned above, this is a general issue for the disk instability model, irrespective of the presence of a binary companion. Some of the heating associated with artificial viscosity will indeed arise in nature as a result of turbulence and other unresolved aspects of hydrodynamical flows. It remains for future work to determine what the proper amount of viscous heating should be for using in simulations of realistic protoplanetary disks, and to determine the proper boundary conditions at the surfaces of protoplanetary disk that will allow the disk to radiate away energy at the correct rate (Boley et al. 2007a,b).

The sensitivity of fragmentation to the cooling time is readily shown in a series of new simulations of binary disks with the algorithm for radiative transfer described in Mayer et al. (2006). Intermediate mass disks 0.05 $M_{\odot}$ that were fragmenting in binaries for separations of 58 AU and sufficiently short cooling times (Mayer et al. 2005) develop strong spiral arms but no clumps when flux-limited diffusion plus atmospheric cooling via blackbody radiation is used (see Fig. 8.13). This is not surprising since this latest radiation physics model yields cooling times of order or slightly longer than the orbital time (apparently via convection) while in Mayer et al. (2005), these disks were fragmenting only for cooling times half or less than half the local orbital time. Indeed the average outer disk temperatures (outside 5 AU) in this new simulation is larger than 100 K while it was 60–70 K in the corresponding run of Mayer et al. (2005). These new results agree superficially with Nelson (2000), although whether the agreement holds at a deeper level of detail is unclear at the moment.

Moreover, we recall that for intermediate semimajor axes, orbital eccentricities might also be important. This might be one of the reasons, in addition to heating by artificial viscosity, behind the discrepant results at small orbital separations. Indeed,

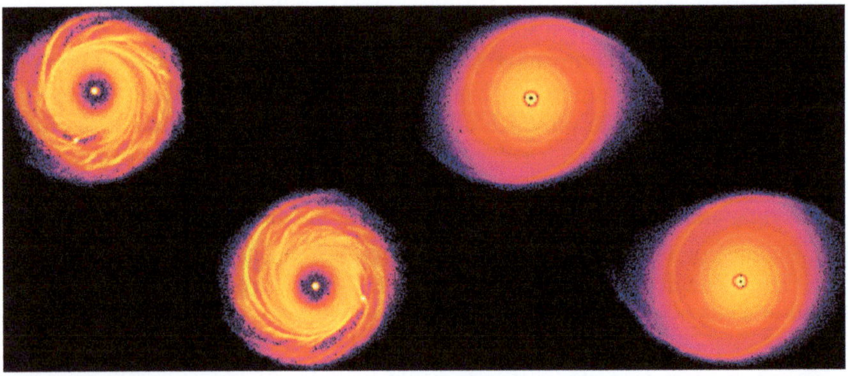

Fig. 8.13 Color coded density maps of two runs employing two disks with masses 0.05 $M_{\odot}$ moving on a binary orbit with average separation of 60 AU. The results are shown after 1.5 binary orbits. On the *left* a run in which the cooling time is fixed to 0.3 orbital time is shown. On the *right* a newer run is shown in which flux-limited diffusion is employed and the disk cools at the surface as a blackbody resulting in a cooling time slightly longer than the orbital time. Clump formation has occurred in the run with the short cooling time while in the other run, the two disks have achieved higher temperatures, lower densities, and much weaker spiral structures

while Mayer et al. (2005) adopt nearly circular orbits, Boss (2006) chooses eccentric orbits. Hence in Boss (2006), for a given semimajor axis, the disks will spend a larger fraction of the orbital time far away from each other. In such simulations, the tidal perturbation will be more impulsive rather than continuous. In other fields of astrophysics which deal with similar problems, such as in the study of galaxy interactions, it is well known that impulsive or continuous tidal heating give rise to quite different responses in a self-gravitating system, to the point of determining a completely different structural evolution Mayer et al. (2001). What is seen in particular in the case of galaxies is that impulsive encounters can generate "cold" features such as bars that then survive for many orbital times, but the same features are erased as the object increases too much kinetic energy and/or thermal energy owing to a continuous tidal perturbation such as that associated with circular or nearly circular orbits. Similarly, one could speculate that, in encounters between disks on eccentric orbits, transient overdensities might have a better chance to survive as the same tidal force that triggered their formation fades away later towards the apocenter of the orbit. Again, simulations exploring a larger parameter space are needed to assess if eccentricity is such an important parameter and might help to partially reconcile the disagreement between Boss (2006) and Mayer et al. (2005). Disks perturbed by fast-flybys of other stars or brown dwarfs also suffer significant tidal heating and do not fragment unless the cooling time is very short (Lodato et al. 2007).

The results of Mayer et al. (2005) and Boss (2006) suggest that the formation of gas giant planets around binary stars with semimajor axes of 100 AU or larger may be possible by the disk instability mechanism. Note that these are also the systems for which core-accretion is not disfavored either since spiral shocks do not heat the gas to temperatures high enough to vaporize major dust grain species (Mayer et al. 2005). For smaller semimajor axes, the situation is much more complex. For the latter, Mayer et al. (2005) conclude that, while disk instability is unlikely, core-accretion might take place once the disk is light enough, $M_d \sim 0.01\ M_\odot$ (e.g. as a result of accretion onto the star) that only weak spiral shocks arise so that the temperature is low enough for ice grains to survive. Nelson (2000) claimed that both disk instability and core-accretion would be unlikely in such systems, drawing the same conclusion that Mayer et al. (2005) would have reached had they not considered binary systems composed of light, weakly self-gravitating disks (see Section 8.4.3 for possible caveats). Finally, Boss (2006) finds that these systems would fragment, although he cannot follow the clumps for a long enough time to show that they are long lasting. Despite the fact that disagreements exist between the different works, it is clear for now that the tightest binary systems might become an ideal testbed for theories of planet formation. It is thus important for observers to refine their estimates of the semimajor axes of binary systems containing gas giant planets, in order to learn if these systems could have been formed by disk instability. Post-formational orbital evolution of multiple systems (e.g., decay of an unstable triple system) might be another means to explain the observed binary systems with gas giant companions.

The observational fact that binary stars with separations small enough for mutual tidal interactions to be important, are orbited by gas giant planets means that somehow these planets can indeed form even in these systems. Given the problems that core accretion encounters as well in binary systems (Thébault et al. 2004), disk instability would seem to remain a possible means for forming gas giants in binary systems.

Acknowledgments We thank Richard H. Durisen and Nader Haghighipour for their improvements to the manuscript. APB's work is supported in part by the NASA Planetary Geology and Geophysics Program under grants NNG05GH30G and NNX07AP46G, and is contributed in part to the NASA Astrobiology Institute under grant NCC2-1056. Parts of this work were carried out under the auspices of the National Nuclear Security Administration of the U.S. Department of Energy at Los Alamos National Laboratory under Contract No. DE-AC52-06NA25396, for which this is publication LA-UR-07-7370.

References

Aikawa, Y., 2004, Proceedings of the Conference "Star Formation at High Angular Resolution", International Astronomical Union Symposium 221, XXV General Assembly of the IAU, Sydney, Australia, 22–25 July 2003. Eds: M. Burton and R. Jayawardhana, T. Bourke, 67
Alexander, D. R. & Ferguson, J. W. 1994, ApJ, 437, 879
Balsara, D. S. 1995, J. Comput. Phys., 121, 357
Bate, M. R. 2000, MNRAS, 314, 33
Bate, M. R. 2001, in "From Darkness to Light: Origin and Evolution of Young Stellar Clusters", ASP Conference Proceedings, eds. T. Montmerle and P. André (San Francisco, Publications of the Astronomical Society of the Pacific) 243, 233
Bate, M. R., Bonnell, I. A. & Bromm, V. 2002, MNRAS, 332, L65
Bate, M. R. & Burkert, A. 1997, MNRAS, 228, 1060
Benz, W. 1990, in The Numerical Modeling of Stellar Pulsation, ed. J. R. Buchler (Dordrecht: Kluwer), 269
Binney, Ê. J. & Tremaine, Ê. S. 1987, *Galactic dynamics*, (Princeton, Princeton University Press)
Boley, A. C. et al. 2006, ApJ, 651, 517
Boley, A. C. & Durisen, R. H. 2006, ApJ, 641, 534
Boley, A. C., Hartquist, T. W., Durisen, R. H. & Scott, M. 2007a, ApJ, 656, L89
Boley, A. C., Durisen, R. H., Nordlund, Å. & Lord, J. 2007b, ApJ, 665, 1254
Boss, A. P. 1993, ApJ, 417, 351
Boss, A. P. 1996, ApJ, 469, 906
Boss, A. P. 1997, Science, 276, 1836
Boss, A. P. 1998, ApJ, 503, 923
Boss, A. P. 2000, ApJ, 545, L61
Boss, A. P. 2001, ApJ, 563, 367
Boss, A. P. 2002, ApJ, 576, 462
Boss, A. P. 2003, ApJ, 599, 577
Boss, A. P. 2004a, ApJ, 610, 456
Boss, A. P. 2004b, ApJ, 616, 1265
Boss, A. P. 2005, ApJ, 629, 535
Boss, A. P. 2006, ApJ, 641, 1148
Boss, A. P. 2007, ApJ, 661, L73
Boss, A. P. 2008, ApJ, 677, 607
Boss, A. P. & Myhill, E. A. 1992, ApJS, 83, 311
Boss, A. P. & Yorke, H. W. 1993, ApJ, 411, L99
Boss, A. P. & Yorke, H. W. 1996, ApJ, 469, 366

Cai, K. et al. 2006a, ApJ, 636, L149
Cai, K. et al. 2006b, ApJ, 624, L173 (Erratum)
Cameron, A. G. W. 1978, Moon Planets, 18, 5
Chauvin, G. et al. 2006, A&A, 456, 1165
D'Alessio, P., Calvet, N. & Hartmann, L. 2001, ApJ, 553, 321
Duquennoy, A. & Mayor, M. 1991, A&A, 248, 485
Durisen, R. H. et al. 2007, in Protostars & Planets V, eds. B. Reipurth, D. Jewitt & K. Keil (Tucson: Univ. Arizona Press), 607
Eggenberger, A., Udry, S. & Mayor, M. 2004, A&A, 417, 353
Evrard, A.E. 1990, ApJ, 363, 349
Gammie, C. F. 2001, ApJ, 553, 174
Gingold, R. A. & Monaghan, J. J. 1977, MNRAS, 181, 375
Haghighipour, N. & Boss, A. P. 2003a, ApJ, 583, 996
Haghighipour, N. & Boss, A. P. 2003b, ApJ, 598, 1301
Haghighipour, N. 2005, MNRAS, 362, 1015
Herant, M. & Woosley, S. 1994, ApJ, 425, 814
Hernquist, L. & Katz, N. 1989, ApJS, 70, 419
Johnson, B. & Gammie, C. F. 2003, ApJ, 597, 131
Kawakita, H., et al. 2001, Science, 294, 1089
Krumholz, M.R., Klein, R.I. & McKee, C.F., 2007, ApJ, 656, 959
Kuiper, G. P. 1951, Proc. Nat. Acad. Sci., 37, 1
Laughlin, G., Korchagin, V. & Adams, F. C. 1997, ApJ, 477, 410
Laughlin, G., Korchagin, V. & Adams, F. C. 1998, ApJ, 504, 945
Lodato, G. & Rice, W. K. M. 2004, MNRAS, 351, 630
Lodato, G. & Rice, W. K. M. 2005, MNRAS, 358, 1489
Lodato, G., Meru, F. Clarke, C. & Rice, W. K. M. 2007, MNRAS, 374, 590
Lucy, L. B. 1977, AJ, 82, 1013
Mayer, L., Colpi, M., Governato, F., Moore, B., Quinn, T., Stadel, J. & Lake, G., 2001, ApJ, 559, 754
Mayer, L., Quinn, T., Wadsley, J. & Stadel, J. 2002, Science, 298, 1756
Mayer, L., Quinn, T., Wadsley, J. & Stadel, J. 2004, ApJ, 609, 1045
Mayer, L., Wadsley, J., Quinn, T. & Stadel, J. 2005, MNRAS, 363, 641
Mayer, L., Lufkin, G., Quinn, T. & Wadsley, J., 2007, ApJ, 661, L77
Mayer, L. & Gawryszczak, A., 2008, Proceedings of the Conference "Extreme Solar Systems", Santorini, Greece, June 25–29, 243. Eds: F. Rasio and D. Fischer, 243
Mejía, A. C., Durisen, R. H., Pickett, M. K. & Cai, K. 2005, ApJ, 619, 1098
Monaghan, J. J. 1992, ARAA, 30, 543
Morris, J. P. & Monaghan, J. J. 1997, J. Comp. Phys., 136, 41
Nelson, A. F. 2000, ApJ, 537, L65
Nelson, A. F. 2006, MNRAS, 373, 1039
Nelson, A. F., Benz, W., Adams, F. C. & Arnett, W. D. 1998, ApJ, 502, 342
Nelson, A. F., Benz, W. & Ruzmaikina, T. V. 2000, ApJ, 529, 357
Papaloizou, J. C. & Savonije, G. J. 1991, MNRAS, 248, 353
Patience, J. et al. 2002, ApJ, 581, 654
Pickett, B. K., Cassen, P., Durisen, R. H. & Link, R. 1998, ApJ, 504, 468
Pickett, B. K. et al. 2000, ApJ, 529, 1034
Pickett, B. K. et al. 2003, ApJ, 590, 1060
Pickett, M.K. & Durisen, R. H, 2007, ApJL, 654, L155
Pollack, J. B., McKay, C. & Christofferson, B. M., 1985, Icarus, 64, 471
Pollack, J.B., Hollenbach, D., Beckwith, S., Simonelli, D.P., Roush, T. & Fong, W. 1994, ApJ, 421, 615
Quinn, T., Katz, N., Stadel, J. & Lake, G. 1997, eprint arXiv:astro-ph/9710043
Rafikov, R. R. 2005, ApJ, 621, L69
Rafikov, R. R. 2007, ApJ, 662, 642

Raghavan, D. et al. 2006, ApJ, 646, 523
Rice, W. K. M., Armitage, P. J., Bate, M. R. & Bonnell, I. A. 2003a, MNRAS, 339, 1025
Rice, W. K. M. et al. 2003b, MNRAS, 346, L36
Rice, W. K. M., Lodato, G., Pringle, J. E., Armitage, P. J. & Bonnell, I. A. 2004, MNRAS, 355, 543
Rice, W. K. M., Lodato, G. & Armitage, P. J. 2005, MNRAS, 364, L56
Rice, W. K. M., Lodato, G., Pringle, J. E., Armitage, P. J. & Bonnell, I. A. 2006, MNRAS, 372, L9
Ruden, S. P. & Pollack, J. B. 1991, ApJ, 375, 740
Springel, V. & Hernquist, L. 2002, MNRAS, 333, 649
Stadel, J. G. 2001, Thesis (PhD), University of Washington
Thébault, P. et al. 2004, A&A, 427, 1097
Tomley, L., Cassen, P. & Steiman-Cameron, T. 1991, ApJ, 382, 530
Tomley, L., Steiman-Cameron, T. Y. & Cassen, P. 1994, ApJ, 422, 850
Toomre, A. 1964, ApJ, 139, 1217
Tremain, S. 2003, AJ, 125, 894
Udry, S. Eggenberger, A., Beuzit, J.L., Lagrange, A.M., Mayor, M. & Chauvin, G., 2004, in "The Environment and Evolution of Double and Multiple Stars", Proceedings of IAU Colloquium 191, held 3–7 February, 2002 in Merida, Yucatan, Mexico. Eds: C. Allen and C. Scarfe. Revista Mexicana de Astronomi
Vorobyov, E.I. & Basu, S. 2007, MNRAS, 381, 1009
Wadsley, J., Stadel, J. & Quinn, T.R. 2004, New Astronomy, 9, 137
Yorke, H. W. & Bodenheimer, P. 1999, ApJ, 525, 330

Chapter 9
N-Body Integrators for Planets in Binary Star Systems

John E. Chambers

9.1 Introduction

The discovery of planets orbiting in binary star systems represents an exciting new field of astrophysics. The stability of planetary orbits in binary systems can only be addressed analytically in special cases, so most researchers have studied stability using long-term N-body integrations of test particles, examining binary systems with a range of masses and orbits (e.g. Wiegert and Holman 1997; Haghighipour and Wiegert 1999; Haghighipour 2006). This has led to a good understanding of the likely regions of stability and instability in binary systems. Integrators can also been used to study the more complex problem of several finite-mass planets orbiting in a binary system, where interactions between the planets are significant. However, at the time of writing, this problem has been explored in less detail than the test-particle case, and we still lack a general theory for the stability of these systems.

N-body integrators have found a second application modelling the formation of planetary systems in and around binary stars (e.g. Quintana et al. 2002; Quintana and Lissauer 2006; Haghighipour and Raymond 2007). Typically, these studies have paralleled those of planet formation around single stars, examining a particular stage of growth such as the formation of planetesimals, oligarchic growth, or late-stage accretion of terrestrial planets. The results of these studies are discussed extensively in the chapters in this volume by Quintana et al. and Haghighipour et al.

Most conventional integrator algorithms can be applied to binary star systems with little or no modification. Runge–Kutta, Bulirsch–Stoer and Everhart's RADAU integrators fall into this category for example (Press et al. 1992; Stoer and Bulirsch 1980; Everhart 1985). These algorithms contain no built-in information about the system of differential equations they are solving, so they can be applied to binary systems and single-star systems equally well.

Over the last decade and a half, symplectic integrator algorithms have become increasingly popular and are widely used to study the dynamics of planetary and

J.E. Chambers (✉)
Carnegie Institution for Science, 5241 Broad Branch Road, NW, Washington DC 20015
e-mail: chambers@dtm.ciw.edu

N. Haghighipour (ed.), *Planets in Binary Star Systems*, Astrophysics and Space Science Library 366, DOI 10.1007/978-90-481-8687-7_9,

satellite orbits. These algorithms have two advantages over conventional integrators. First, symplectic integrators typically have good long-term energy conversation properties. While energy is not conserved in most problems, the energy error typically makes high frequency oscillations about zero, while exhibiting no long-term trend beyond that generated by computer round-off error. Secondly, in problems involving a dominant, primary mass, such as the Sun in the solar system, the motion of other objects about the central body can be "built in" (Wisdom and Holman 1991). A relatively small amount of computation is required to calculate the accelerations due to the central body. For roughly circular orbits, a large stepsize (up to about 10% of the shortest orbital period) can be used. In this case, the stepsize only needs to be small enough to resolve the perturbations between the smaller bodies and to avoid instability generated by overlapping resonances involving the stepsize (Wisdom and Holman 1992). In the case of highly eccentric orbits, the stepsize must also be small enough to adequately resolve periapse passages (Rauch and Holman 1999), an issue we will return to in Section 9.5.

The generally advantageous properties of symplectic integrators mean that they have become the tool of choice for many researchers at present, and they will be the focus of this chapter. The rest of the chapter is organized as follows. Section 9.2 contains a review of the original symplectic mapping developed by (Wisdom and Holman 1991). Section 9.3 shows how this algorithm has been modified for use specifically in binary-star systems. Section 9.4 shows how symplectic integrators can be improved by developing symplectic correctors, and describes a new corrector for binary-star algorithms. Section 9.5 discusses problems that can arise when planets come close to one or both binary stars, and what might be done to overcome these problems. Finally Section 9.6 contains a summary.

9.2 Mixed-Variable Symplectic Integrators

The most widely used symplectic integrators applied to planetary systems are "mixed-variable symplectic" (MVS) mappings, so called because they separate a problem into two parts, each of which is solved using a different set of variables (typically Cartesian coordinates and orbital elements). These algorithms were first introduced by Wisdom and Holman (1991) and described independently by Kinoshita et al. (1991).

To understand how these integrators work, it is easiest to start by considering Hamilton's equations for a system of N bodies:

$$\frac{dx_i}{dt} = \frac{\partial H}{\partial p_i}$$
$$\frac{dp_i}{dt} = -\frac{\partial H}{\partial x_i} \tag{9.1}$$

where $\mathbf{r} = (x, y, z)$ and $\mathbf{p} = (p_x, p_y, p_z)$ are the coordinates and momenta of the bodies respectively, and H is the Hamiltonian of the system. Using Hamilton's equations, the evolution of any quantity q can be expressed as

$$\begin{aligned}\frac{dq}{dt} &= \sum_{i=1}^{3N}\left(\frac{\partial q}{\partial x_i}\frac{dx_i}{dt} + \frac{\partial q}{\partial p_i}\frac{dp_i}{dt}\right) \\ &= \sum_{i=1}^{3N}\left(\frac{\partial q}{\partial x_i}\frac{\partial H}{\partial p_i} - \frac{\partial q}{\partial p_i}\frac{\partial H}{\partial x_i}\right) \\ &= \{q, H\} \\ &= Fq \end{aligned} \tag{9.2}$$

where $\{,\}$ are Poisson brackets, and F is an operator that depends on the Hamiltonian. The evolution of q can be found by solving Eq. (9.2), which gives

$$q(\tau) = e^{\tau F} q(0) = \left(1 + \tau F + \frac{\tau^2 F^2}{2} + \cdots\right) q(0) \tag{9.3}$$

Symplectic integrators usually divide the Hamiltonian into several parts each of which can be solved efficiently in the absence of the others. Most algorithms divide H into parts that can be solved analytically although this is not strictly necessary. If we separate the Hamiltonian so that $H = H_\mathrm{A} + H_\mathrm{B}$, with operators A and B corresponding to the Hamiltonians H_A and H_B, then

$$q(\tau) = e^{\tau(\mathrm{A}+\mathrm{B})} q(0) \tag{9.4}$$

where

$$\begin{aligned} e^{\tau(\mathrm{A}+\mathrm{B})} &= 1 + \tau(\mathrm{A}+\mathrm{B}) + \frac{\tau^2(\mathrm{A}+\mathrm{B})^2}{2} + \cdots \\ &= 1 + \tau(\mathrm{A}+\mathrm{B}) + \frac{\tau^2(\mathrm{A}^2 + \mathrm{AB} + \mathrm{BA} + \mathrm{B}^2)}{2} + \cdots \end{aligned} \tag{9.5}$$

In general, the operators A and B will not commute so that $\mathrm{AB} \neq \mathrm{BA}$.

A symplectic integrator is generated by concatenating several terms of the form $\exp(a_k \tau \mathrm{A})$ and $\exp(b_k \tau \mathrm{B})$, where $k = 0, \ldots, N$, and a_k and b_k are constant coefficients. The goal is to make the resulting expression equal to Eq. (9.5) up to some order in the stepsize τ. This is most easily accomplished using the Baker–Campbell–Hausdorff (BCH) formula which expresses the product of two exponential operators as a single new exponential operator:

$$e^{\mathrm{A}} e^{\mathrm{B}} = \exp\left\{\mathrm{A} + \mathrm{B} + \frac{1}{2}[\mathrm{A}, \mathrm{B}] + \frac{1}{12}[\mathrm{A}, \mathrm{A}, \mathrm{B}] + \frac{1}{12}[\mathrm{B}, \mathrm{B}, \mathrm{A}] + \cdots\right\} \tag{9.6}$$

where $[\mathrm{A}, \mathrm{B}] = \mathrm{AB} - \mathrm{BA}$ and $[\mathrm{A}, \mathrm{B}, \mathrm{C}] = [\mathrm{A}, [\mathrm{B}, \mathrm{C}]]$ etc. (Yoshida 1990).

The most commonly used algorithm is the second-order leapfrog integrator, so named because the operators A and B are applied alternately with different substep sizes:

$$\exp\left(\frac{\tau}{2}\mathrm{A}\right)\exp(\tau\mathrm{B})\exp\left(\frac{\tau}{2}\mathrm{A}\right) = \exp\left\{\tau(\mathrm{A}+\mathrm{B}) + \frac{\tau^3}{12}[\mathrm{B},\mathrm{B},\mathrm{A}] - \frac{\tau^3}{24}[\mathrm{A},\mathrm{A},\mathrm{B}] + O(\tau^5)\right\} \quad (9.7)$$

This differs from the true evolution [Eq. (9.5)] by terms proportional to τ^3 and higher. Over the course of a long integration, the number of steps will be inversely proportional to τ. The total error will be $O(\tau^2)$, meaning that leapfrog is a 2nd-order integrator.

Each timestep using the leapfrog algorithm consists of advancing the system corresponding to H_{A} for a time $\tau/2$, then advancing H_{B} for τ, and finally advancing H_{A} for $\tau/2$. The integrator consists of three substeps. However, the first and last of these both involve A, so the last substep of one timestep can be combined with the first substep of the following step. Over the course of a long integration, each timestep is effectively composed of only two substeps.

Wisdom and Holman (1991) split the Hamiltonian into a part H_{Kep} containing terms corresponding to the Keplerian motion of each planet about the central star, and a second part H_{Int} containing direct and indirect perturbation terms due to interactions between the planets. This is accomplished using Jacobi coordinates, where the position of the innermost planet is measured with respect to the central star, and the positions of the remaining planets are measured with respect to the centre of mass of the central star and planets with lower indices. Evolution under the Keplerian part of the Hamiltonian can be calculated using Gauss's f and g functions (Danby 1988):

$$\begin{aligned}\mathbf{r}(t) &= f\mathbf{r}(0) + g\mathbf{v}(0)\\ \mathbf{v}(t) &= \dot{f}\mathbf{r}(0) + \dot{g}\mathbf{v}(0)\end{aligned} \quad (9.8)$$

where f and g are functions of the initial and final coordinates and velocities. These functions provide an efficient way to determine the new position on a Keplerian orbit, given the initial coordinates and a time interval, without having to calculate all of the orbital elements. If Jacobi coordinates are used, H_{Int} is a function of the coordinates only, so this part of the problem can also be solved analytically.

When applied to planetary systems, $H_{\mathrm{Int}} \sim \epsilon H_{\mathrm{Kep}}$ in the Wisdom–Holman mapping, where ϵ is the planetary to stellar mass–ratio, which is typically small. The error over a long integration is therefore $O(\epsilon\tau^2)$, and the small value of ϵ ensures that the Wisdom–Holman mapping performs well even though it is only a second-order integrator.

Jacobi coordinates are the natural choice for systems like the planets in the solar system, where the planetary orbits do not cross and there is a dominant central mass. In systems with a single dominant mass, Jacobi coordinates typically lead to smaller errors than other canonical coordinate systems, such as barycentric coordinates, for a given stepsize. Barycentric coordinates are an especially poor choice for the solar system since the inner planets are less massive than the outer ones. As a result, the guiding centre for the motion of the inner planets is much closer to the Sun itself than the barycentre of the Solar System. In principle, Jacobi coordinates can also be used in binary and multiple star systems, again provided that orbits do not approach one another, and all the orbits have a "hierarchical" or nested arrangement. However, ϵ will not be small in cases where two or more bodies have comparable masses, and this may lead to large errors in some cases.

9.3 Binary-Star Algorithms

9.3.1 Modifying the Wisdom–Holman Mapping

The integrators described in the previous section can be applied to any hierarchical system of bodies. This means they can be used to calculate the orbital evolution of planets in a binary star system provided that the radial ordering of the planets doesn't change. The Wisdom–Holman mapping can also be adapted to systems that contain multiple hierarchies, such as two binary systems in orbit about each other, by using a generalized version of Jacobi coordinates (Beust 2003).

Whenever planets come close to one another, the condition $H_{\mathrm{Int}} \ll H_{\mathrm{Kep}}$ is violated and the Wisdom–Holman mapping performs poorly. In situations where planets have eccentric orbits that cross those of their neighbours, or where close encounters are possible, new algorithms must be developed. Duncan et al. (1998) and Chambers (1999) have described two ways to do this. Here we will describe the latter approach since this leads directly to a new class of symplectic integrators that can be applied to orbits in binary systems. For reasons that will become clear, the new method requires a new set of coordinates. Ideally, these should include three spatial coordinates for the centre of mass of the system (as Jacobi coordinates do), and treat all the planets equivalently, that is, make no assumptions about their radial ordering (in contrast to Jacobi coordinates).

Canonical heliocentric coordinates (also called democratic heliocentric coordinates) meet both of these requirements (Duncan et al. 1998). Here, the position of planet i with respect to the barycentre $\mathbf{r}_i = (x_i, y_i, z_i)$ is replaced by its position measured with respect to the central star, and the stellar coordinates $\mathbf{r}_0$ are replaced with those of the centre of mass of the system. The new coordinates $\mathbf{R}_i$ and $\mathbf{R}_0$ are

$$\begin{aligned} \mathbf{R}_0 &= \frac{m_0\mathbf{r}_0 + \sum_{j=1}^{N} m_j\mathbf{r}_j}{m_{\mathrm{tot}}} \\ \mathbf{R}_i &= \mathbf{r}_i - \mathbf{r}_0 \end{aligned} \tag{9.9}$$

where subscript 0 refers to the star, m_i is the mass of planet i, and m_{tot} is the total mass of the system.

The canonically conjugate momenta (which correspond to barycentric velocities) are

$$\mathbf{P}_0 = \mathbf{p}_0 + \sum_{j=1}^{N} \mathbf{p}_j$$

$$\mathbf{P}_i = \mathbf{p}_i - \frac{m_i}{m_{\text{tot}}} \left(\mathbf{p}_0 + \sum_{j=1}^{N} \mathbf{p}_j \right) \tag{9.10}$$

where $\mathbf{p}_i = (p_{xi}, p_{yi}, p_{zi})$ etc.

Using these coordinates, the Hamiltonian for a system of N planets orbiting a single star can be split into three parts:

$$H = H_{\text{Kep}} + H_{\text{Int}} + H_{\text{Jump}} \tag{9.11}$$

where

$$H_{\text{Kep}} = \sum_{i=1}^{N} \left(\frac{P_i^2}{2m_i} - \frac{Gm_0 m_i}{R_i} \right)$$

$$H_{\text{Int}} = -\sum_{i=1}^{N} \sum_{j>i} \frac{Gm_i m_j}{R_{ij}}$$

$$H_{\text{Jump}} = \frac{1}{2m_0} \left(\sum_{i=1}^{N} \mathbf{P}_i \right)^2 \tag{9.12}$$

where $R_i = |\mathbf{R}_i|$ and $R_{ij} = |\mathbf{R}_j - \mathbf{R}_i|$. Note that we have dropped a term $P_0^2/2m_{\text{tot}}$ which simply acts to move the centre of mass at a constant velocity.

Several second-order symplectic integrators can be constructed using canonical heliocentric coordinates, for example:

$$\exp\left(\frac{\tau I}{2}\right) \exp\left(\frac{\tau J}{2}\right) \exp(\tau K) \exp\left(\frac{\tau J}{2}\right) \exp\left(\frac{\tau I}{2}\right) \tag{9.13}$$

where I, J and K are operators associated with H_{Int}, H_{Jump} and H_{Kep} respectively. Other second-order algorithms are similar except that the operators are permuted, making sure that the arrangement is symmetrical in each case.

Advancing the system under H_{Int} is straightforward since this part of the Hamiltonian is a function of the coordinates only. As a result, the positions of all planets remain constant while the velocities change due to perturbations from the other planets. Advancing under H_{Jump} is trivial since this is a function of the momenta only. In this case, each planet's velocity remains constant but its spatial coordinates jump by a small amount. This jump is the same for all planets, so it becomes relatively more important for objects close to the central star, a point we will return to in Section 9.5. Advancing H_{Kep} is best done using Gauss's f and g functions as before, noting that $\mathbf{P}_i$ and $\mathbf{R}_i$ are canonically conjugate.

The integrator described by Eq. (9.13) performs quite well for planetary systems in which the planets do not undergo close encounters. Typical energy errors are intermediate between those using Jacobi coordinates and barycentric coordinates for hierarchical systems, although Jacobi coordinates lose their advantage if the planets have crossing orbits. Despite consisting of five substeps rather than three, the integrator described above involves slightly less computational effort than the Wisdom–Holman mapping since advancing under H_{Jump} is trivial, while H_{Int} only contains direct terms whereas indirect terms are also present when using Jacobi coordinates.

As before, problems arise when a pair of planets has a close encounter because H_{Int} can become comparable in size to H_{Kep}. Chambers (1999) showed that this difficulty can be overcome using a hybrid algorithm. Here, each term in H_{Int} is split between H_{Int} and H_{Kep} so that the former always remains much smaller than the latter. Under this new arrangement, the Hamiltonian is divided as follows:

$$
\begin{aligned}
H_{\mathrm{Large}} &= \sum_{i=1}^{N}\left(\frac{P_i^2}{2m_i} - \frac{Gm_0m_i}{R_i}\right) - \sum_{i=1}^{N}\sum_{j>i}\frac{Gm_im_j}{R_{ij}}[1-\Gamma(R_{ij})] \\
H_{\mathrm{Small}} &= -\sum_{i=1}^{N}\sum_{j>i}\frac{Gm_im_j}{R_{ij}}\Gamma(R_{ij}) \\
H_{\mathrm{Jump}} &= \frac{1}{2m_0}\left(\sum_{i=1}^{N}\mathbf{P}_i\right)^2 \qquad (9.14)
\end{aligned}
$$

where Γ is a partition function.

A second-order hybrid integrator has the same form as (9.13), that is:

$$
\exp\left(\frac{\tau S}{2}\right)\exp\left(\frac{\tau J}{2}\right)\exp(\tau L)\exp\left(\frac{\tau J}{2}\right)\exp\left(\frac{\tau S}{2}\right) \qquad (9.15)
$$

where L and S are operators associated with H_{Large} and H_{Small} respectively.

The partition function is chosen so that $\Gamma(R) = 1$ when R is large and $\Gamma(R) \to 0$ as $R \to 0$. With this choice of Γ, it is always the case that $H_{\mathrm{Large}} \gg H_{\mathrm{Small}}$, and the resulting integrator remains accurate during close encounters between planets. However, H_{Large} is no longer analytically soluble during a close encounter since it now includes a three-body problem. These three-body terms can be calculated using a conventional N-body algorithm such as Bulirsch–Stoer. Provided this is done to an accuracy level close to machine precision, the user shouldn't be able to tell in practice whether the solution was derived analytically or numerically.

The reason for avoiding Jacobi coordinates becomes apparent when studying the terms that are partitioned between H_{Large} and H_{Small}. If Jacobi coordinates are used, R_{ij} is a function of all the Jacobi coordinates with indices between i and j. This means that in many cases, the orbits of several planets will have to be integrated numerically using Bulirsch–Stoer. When using canonical heliocentric coordinates, however, R_{ij} is a function of $\mathbf{R}_i$ and $\mathbf{R}_j$ only.

Using Bulirsch–Stoer (or any other strategy) to evolve the system through a close encounter will slow down an integration. However, only the pair of planets involved in the encounter need to be integrated in this way. All the other planets are advanced analytically under H_{Large} using Gauss's functions. When there are no encounters, the algorithm becomes identical to Eq. (9.13) and there is no loss of speed.

The algorithm given by Eq. (9.15) works well for most systems of planets orbiting a single star, and has also been used extensively to study the formation of planets from a disk of smaller bodies, since these bodies undergo many close encounters (e.g. Chambers 2001; Thebault et al. 2002; Raymond et al. 2004; Fogg and Nelson 2005; Ford and Chiang 2007; Mandell et al. 2007). However, the algorithm performs poorly when applied to planets orbiting in binary systems. The problem arises because H_{Jump} now contains a momentum contribution from one of the binary stars. The stellar momentum is typically large and this leads to large changes in position for all the planets via H_{Jump}. It is no longer true that $H_{\mathrm{Jump}} \ll H_{\mathrm{Kep}}$, and the error per step becomes large unless an unacceptably small timestep is chosen.

As Chambers et al. (2002) have shown, the solution to this problem is to devise new coordinate systems for binary systems such that all large terms can be incorporated into a single part of the Hamiltonian. Stable planetary orbits typically fall into one of two classes: (i) those that are tightly bound to one member of a binary, or (ii) those that orbit both stars at a distance that is considerably larger than the semimajor axis of the binary orbit. Each configuration will require a different set of coordinates and we will consider the two cases separately.

9.3.2 Wide Binary Case

The Hamiltonian for a system containing N planets orbiting one member of a binary star is

$$H = \frac{p_A^2}{2m_A} + \frac{p_B^2}{2m_B} + \sum_{i=1}^{N} \frac{p_i^2}{2m_i} - \frac{Gm_A m_B}{r_{AB}} - Gm_A \sum_{i=1}^{N} \frac{m_i}{r_{iA}} - Gm_B \sum_{i=1}^{N} \frac{m_i}{r_{iB}} - G \sum_{i=1}^{N} \sum_{j>i} \frac{m_i m_j}{r_{ij}}, \tag{9.16}$$

where the planets orbit star A while star B is a distant companion.

Making use of the hierarchical arrangement of the binary system, we define a new set of coordinates, called wide-binary coordinates, as follows:

$$\begin{aligned}
\mathbf{R}_A &= \left(\frac{m_A\mathbf{r}_A + m_B\mathbf{r}_B + \sum_j m_j\mathbf{r}_j}{m_{\rm tot}}\right),\\
\mathbf{R}_i &= \mathbf{r}_i - \mathbf{r}_A,\\
\mathbf{R}_B &= \mathbf{r}_B - \left(\frac{m_A\mathbf{r}_A + \sum_j m_j\mathbf{r}_j}{m_A + \sum_j m_j}\right),
\end{aligned}\tag{9.17}$$

where $m_{tot} = m_A + m_B + \sum_j m_j$ is the total mass of the system, and each of the summations run from 1 to N. Using these coordinates, the position of each planet is measured with respect to star A, while the position of star B is measured with respect to the center of mass of all the other objects.

The conjugate momenta $\mathbf{P}$ are

$$\begin{aligned}
\mathbf{P}_A &= \mathbf{p}_A + \mathbf{p}_B + \sum_{i=1}^{N}\mathbf{p}_j,\\
\mathbf{P}_i &= \mathbf{p}_i - m_i\left(\frac{\mathbf{p}_A + \sum_j \mathbf{p}_j}{m_A + \sum_j m_j}\right),\\
\mathbf{P}_B &= \mathbf{p}_B - m_B\left(\frac{\mathbf{p}_A + \mathbf{p}_B + \sum_j \mathbf{p}_j}{m_{\rm tot}}\right),
\end{aligned}\tag{9.18}$$

where the summations run from 1 to N.

In terms of the new coordinates, the Hamiltonian can be written as

$$H = H_{\rm Kep} + H_{\rm Int} + H_{\rm Jump},\tag{9.19}$$

where

$$\begin{aligned}
H_{\rm Kep} &= \left(\frac{P_B^2}{2\mu_{\rm bin}} - \frac{Gm_{\rm tot}\mu_{\rm bin}}{R_B}\right) + \sum_{i=1}^{N}\left(\frac{P_i^2}{2m_i} - \frac{Gm_Am_i}{R_i}\right),\\
H_{\rm Int} &= -\sum_{i=1}^{N}\sum_{j>i}\frac{Gm_im_j}{R_{ij}} + Gm_Bm_A\left(\frac{1}{R_B} - \frac{1}{|\mathbf{R}_B + \mathbf{S}|}\right)\\
&\quad + Gm_B\sum_{i=1}^{N}m_i\left(\frac{1}{R_B} - \frac{1}{|\mathbf{R}_B - \mathbf{R}_i + \mathbf{S}|}\right),\\
H_{\rm Jump} &= \frac{1}{2m_A}\left(\sum_{i=1}^{N}\mathbf{P}_i\right)^2
\end{aligned}\tag{9.20}$$

where $\mu_{\rm bin} = (m_A + \sum m_i)m_B/m_{\rm tot}$ is the reduced mass of the binary system (including the mass of the planets), and

$$\mathbf{S} = \frac{\sum_{i=1}^{N} m_i \mathbf{R}_i}{m_A + \sum_{i=1}^{N} m_i} \tag{9.21}$$

The terms in $H_{\rm Kep}$ consist of those due to the Keplerian motion of the binary (adding the masses of the planets to star A) and those due to the Keplerian motion of the planets about star A. The terms in $H_{\rm Int}$ represent the interactions between planets and also the tidal perturbations on the planets due to star B. Finally, $H_{\rm Jump}$ contains indirect perturbation terms.

In the absence of close encounters, $H_{\rm Int} \ll H_{\rm Kep}$ and $H_{\rm Jump} \ll H_{\rm Kep}$. Each part of the Hamiltonian can be advanced efficiently using analytic solutions. For example, the x component of the acceleration of planet k under $H_{\rm Int}$ is given by

$$\begin{aligned}
\frac{dV_{x,k}}{dt} &= -\frac{1}{m_k}\frac{\partial H_{\rm Int}}{\partial X_k} \\
&= -\left(\frac{Gm_A m_B}{m_A + \sum_i m_i}\right)\frac{X_B + S_x}{|\mathbf{R}_B + \mathbf{S}|^3} \\
&\quad -\left(\frac{Gm_B}{m_A + \sum_i m_i}\right)\sum_{i=1}^{N} m_i \frac{X_B - X_i + S_x}{|\mathbf{R}_B - \mathbf{R}_i + \mathbf{S}|^3} \\
&\quad + Gm_B \frac{X_B - X_k + S_x}{|\mathbf{R}_B - \mathbf{R}_k + \mathbf{S}|^3} - \sum_{i \neq k} \frac{Gm_i}{R_{ik}^3}(X_k - X_i)
\end{aligned} \tag{9.22}$$

where X is the x component of $\mathbf{R}$, and $\mathbf{V}$ is the velocity. Note that the acceleration on planet k does not involve any terms proportional to $1/m_k$, so test particles can be integrated in exactly the same way as massive planets.

The x component of the acceleration on star B is given by

$$\frac{dV_{x,B}}{dt} = Gm_A\left[\frac{X_B}{R_B^3} - \frac{X_B + S_x}{|\mathbf{R}_B + \mathbf{S}|^3}\right] + G\sum_{i=1}^{N} m_i \left[\frac{X_B}{R_B^3} - \frac{(X_B - X_i + S_x)}{|\mathbf{R}_B - \mathbf{R}_i + \mathbf{S}|^3}\right] \tag{9.23}$$

Close encounters between planets can be dealt with in the same way as for systems with a single star by partitioning the planet interaction terms between $H_{\rm Kep}$ and $H_{\rm Int}$ as in Eq. (9.14).

One step of the new wide-binary algorithm consists of five substeps:

- Advance $H_{\rm Int}$ for $\tau/2$, where τ is the timestep.
- Advance $H_{\rm Jump}$ for $\tau/2$.

- Advance H_{Kep} for τ.
- Advance H_{Jump} for $\tau/2$.
- Advance H_{Int} for $\tau/2$.

The first and last substeps can be combined into a single substep except at the beginning of the integration or whenever output is required.

The wide-binary integrator is a second-order algorithm since the three pieces of the Hamiltonian are applied in a symmetric order (see Yoshida 1990). Each timestep has an error $O(\epsilon\tau^3)$, where ϵ is the ratio of the planetary mass to the stellar mass so that $\epsilon \ll 1$.

Figure 9.1 compares the accuracy of the wide-binary algorithm with the hybrid symplectic integrator of Eq. (9.15) when integrating the four giant planets of the Solar System in the presence of a binary companion. The giant planets orbit the Sun, while a second Solar-mass star orbits the combined system moving on an orbit with a semimajor axis of 160 AU, an eccentricity of 0.25, and an inclination of 0. The figure shows the energy error as a function of time for a 100,000-year integration using a stepsize of 50 days. The upper panel shows the performance of the hybrid integrator, while the lower panel shows the wide-binary algorithm. The hybrid algorithm performs poorly since it treats the binary companion as the equivalent of an additional planet, so that H_{Jump} is no longer small compared to H_{Kep}. The wide-binary algorithm, which treats the binary companion as a special body, has an energy error about three orders of magnitude lower, as a result.

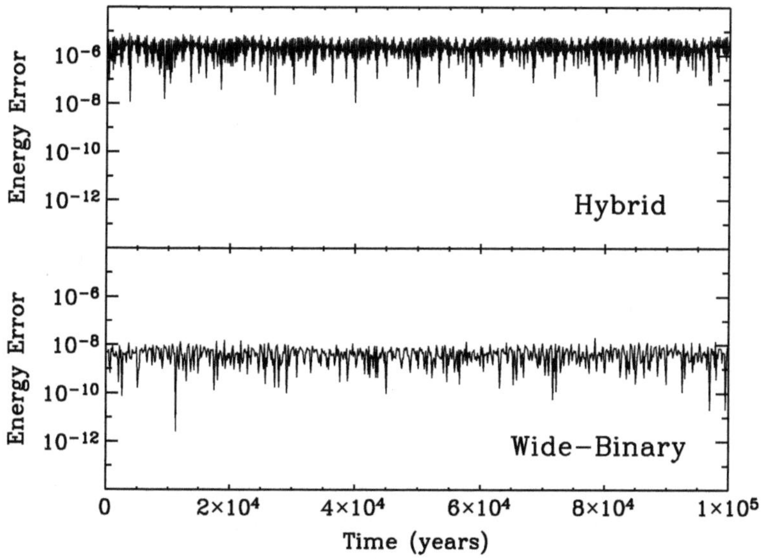

Fig. 9.1 A comparison between the energy errors (accuracies) of hybrid and wide-binary integrators. The system consists of the four giant planets of the solar system orbiting the Sun, and a second Sun-like star in an orbit with a semimajor axis of 160 AU and eccentricity of 0.25. As shown here, the hybrid integrator has a poor performance compared to the wide-binary algorithm

9.3.3 Close Binary Case

The Hamiltonian for a system of N planets orbiting a close binary has the same form as Eq. (9.16) except that now it is understood that the planets orbit both members of the binary. The hierarchical nature of the system suggests we switch to the following close-binary coordinates:

$$\begin{aligned} \mathbf{R}_A &= \frac{m_A \mathbf{r}_A + \sum_j m_j \mathbf{r}_j + m_B \mathbf{r}_B}{m_{\text{tot}}}, \\ \mathbf{R}_i &= \mathbf{r}_i - (\nu_A \mathbf{r}_A + \nu_B \mathbf{r}_B), \\ \mathbf{R}_B &= \mathbf{r}_B - \mathbf{r}_A, \end{aligned} \tag{9.24}$$

where m_{tot} is the total mass of all the bodies, the summations run from 1 to N, and

$$\begin{aligned} \nu_A &= m_A / m_{\text{bin}} \\ \nu_B &= m_B / m_{\text{bin}} \end{aligned} \tag{9.25}$$

where $m_{\text{bin}} = m_A + m_B$ is the mass of the binary,

Using these coordinates, the position of each planet is measured with respect to the center of mass of the two stars, while $\mathbf{R}_B$ is the relative coordinates of the binary companion.

The conjugate momenta $\mathbf{P}$ are

$$\begin{aligned} \mathbf{P}_A &= \mathbf{p}_A + \mathbf{p}_B + \sum_{j=1}^{N} \mathbf{p}_j, \\ \mathbf{P}_i &= \mathbf{p}_i - m_i \left(\frac{\mathbf{p}_A + \mathbf{p}_B + \sum_j \mathbf{p}_j}{m_{\text{tot}}} \right), \\ \mathbf{P}_B &= \mathbf{p}_B - \nu_B (\mathbf{p}_A + \mathbf{p}_B), \end{aligned} \tag{9.26}$$

where summation indices run from 1 to N.

The new Hamiltonian is

$$H = H_{\text{Kep}} + H_{\text{Int}} + H_{\text{Jump}}, \tag{9.27}$$

where

$$\begin{aligned} H_{\text{Kep}} &= \left[\frac{P_B^2}{2\mu_{\text{bin}}} - \frac{G m_{\text{bin}} \mu_{\text{bin}}}{R_B} \right] + \sum_{i=1}^{N} \left[\frac{P_i^2}{2m_i} - \frac{G m_{\text{bin}} m_i}{R_i} \right], \\ sH_{\text{Int}} &= G m_{\text{bin}} \sum_{i=1}^{N} m_i \left[\frac{1}{R_i} - \frac{\nu_A}{|\mathbf{R}_i + \nu_B \mathbf{R}_B|} - \frac{\nu_B}{|\mathbf{R}_i - \nu_A \mathbf{R}_B|} \right] - \sum_{i=1}^{N} \sum_{j>i} \frac{G m_i m_j}{R_{ij}}, \\ H_{\text{Jump}} &= \frac{1}{2m_{\text{bin}}} \left(\sum_{i=1}^{N} \mathbf{P}_i \right)^2 \end{aligned} \tag{9.28}$$

where $\mu_{\text{bin}} = m_A m_B / (m_A + m_B)$ is the reduced mass of the binary.

Using the close-binary coordinates, terms in H_{Kep} correspond to the Keplerian motion of the two binary stars about their common centre of mass, and of the planets about this centre of mass. In addition, H_{Int} contains terms due to interactions between the planets, and perturbations on the planetary orbits caused by higher order moments of the binary potential. As before, H_{Jump} contains indirect correction terms.

In the absence of close encounters, H_{Int} and H_{Jump} are small compared to H_{Kep}, and each part of the Hamiltonian can be advanced analytically. The x component of the acceleration on planet k, when advancing H_{Int}, is given by

$$\begin{aligned}\frac{dV_{x,k}}{dt} &= \frac{Gm_{\mathrm{bin}}X_k}{R_k^3} - \sum_{i\neq k}\frac{Gm_i}{R_{ik}^3}(X_k - X_i)\\ &\quad -G\left[m_A\frac{(X_k+\nu_B X_B)}{|\mathbf{R}_k+\nu_B\mathbf{R}_B|^3} + m_B\frac{(X_k-\nu_A X_B)}{|\mathbf{R}_k-\nu_A\mathbf{R}_B|^3}\right] \qquad (9.29)\end{aligned}$$

while the corresponding acceleration on star B is given by

$$\frac{dV_{x,B}}{dt} = G\sum_i m_i\left[\frac{(X_i-\nu_A X_B)}{|\mathbf{R}_i-\nu_A\mathbf{R}_B|^3} - \frac{(X_i+\nu_B X_B)}{|\mathbf{R}_i+\nu_B\mathbf{R}_B|^3}\right] \qquad (9.30)$$

where $\mathbf{P}_B = \mu_{\mathrm{bin}}\mathbf{V}_B$.

Close encounters between planets can be included in the same way as before by dividing the planet interaction terms between H_{Kep} and H_{Int}.

One could devise a second-order scheme using close-binary coordinates that is analogous to the second-order wide-binary integrator described above. This scheme would contain five terms arranged symmetrically. However, this scheme would have to use a small stepsize in order to accurately integrate the orbit of the binary star. It is more efficient to assign a separate small stepsize τ/N_{bin} to the binary star, and choose a larger global sizestep τ to integrate the planets. (This is analogous to the individual-timestep procedure described by Saha and Tremaine 1994.) We can do this by splitting H_{Kep} into a part $H_{B\mathrm{Kep}}$ that involves terms in $\mathbf{R}_B$ and $\mathbf{P}_B$ and a part $H_{P\mathrm{Kep}}$ that does not. In a similar manner, we split H_{Int} into two new parts $H_{B\mathrm{Int}}$ and $H_{P\mathrm{Int}}$, where

$$\begin{aligned}H_{B\mathrm{Kep}} &= \frac{P_B^2}{2\mu_{\mathrm{bin}}} - \frac{Gm_{\mathrm{bin}}\mu_{\mathrm{bin}}}{R_B},\\ H_{P\mathrm{Kep}} &= \sum_{i=1}^{N}\left[\frac{P_i^2}{2m_i} - \frac{Gm_{\mathrm{bin}}m_i}{R_i}\right],\\ H_{B\mathrm{Int}} &= Gm_{\mathrm{bin}}\sum_{i=1}^{N}m_i\left[\frac{1}{R_i} - \frac{\nu_A}{|\mathbf{R}_i+\nu_B\mathbf{R}_B|} - \frac{\nu_B}{|\mathbf{R}_i-\nu_A\mathbf{R}_B|}\right],\\ H_{P\mathrm{Int}} &= -\sum_{i=1}^{N}\sum_{j>i}\frac{Gm_im_j}{R_{ij}}. \qquad (9.31)\end{aligned}$$

An efficient second-order close-binary scheme has the following form:

- Advance $H_{P\mathrm{Int}}$ for $\tau/2$, where τ is the timestep.
- Repeat the following N_{bin} times:
  - Advance $H_{B\mathrm{Int}}$ for $\tau/(2N_{\mathrm{bin}})$.
  - Advance $H_{B\mathrm{Kep}}$ for $\tau/(2N_{\mathrm{bin}})$.
- Advance H_{Jump} for $\tau/2$.
- Advance $H_{P\mathrm{Kep}}$ for τ.
- Advance H_{Jump} for $\tau/2$.
- Repeat the following N_{bin} times:
  - Advance $H_{B\mathrm{Kep}}$ for $\tau/(2N_{\mathrm{bin}})$.
  - Advance $H_{B\mathrm{Int}}$ for $\tau/(2N_{\mathrm{bin}})$.
- Advance $H_{P\mathrm{Int}}$ for $\tau/2$.

Chambers et al. (2002) suggested making N_{bin} smaller than the global timestep by a factor equal to the ratio of the binary orbital period to the period of the innermost planet. One could also use individual timesteps for the binary companion and the planets in the wide-binary algorithm described earlier. However, the amount of computer time saved would be modest since most of the effort is required to calculate the direct perturbations between the planets and this would not change using individual timesteps.

9.4 Symplectic Correctors

Wisdom et al. (1996) showed that the performance of the Wisdom–Holman mapping can be improved at little extra cost by applying "symplectic correctors". If exp(M) represents a single step of an integrator, given by Eq. (9.7) for example, the addition of a corrector modifies the step to become

$$e^{\mathrm{C}} e^{\mathrm{M}} e^{-\mathrm{C}} \tag{9.32}$$

where C is chosen in order to remove the leading order error terms. Over the course of multiple timesteps, the corrector and inverse corrector terms cancel out, so these only need to be applied at the beginning and end of an integration and when output is required. Thus, the addition of a corrector involves little extra computational expense over a long integration.

Following Yoshida (1990), one uncorrected step of the Wisdom–Holman mapping can be expressed as

$$\begin{aligned} e^{\mathrm{M}} &= \exp\left(\frac{\tau \mathrm{I}}{2}\right) \exp(\tau \mathrm{K}) \exp\left(\frac{\tau \mathrm{I}}{2}\right) \\ &= \exp\left\{\tau(\mathrm{K}+\mathrm{I}) + \frac{\tau^3}{12}[\mathrm{K},\mathrm{K},\mathrm{I}] - \frac{\tau^5}{720}[\mathrm{K},\mathrm{K},\mathrm{K},\mathrm{K},\mathrm{I}] + O(\epsilon\tau^7) + O(\epsilon^2\tau^3)\right\} \end{aligned} \tag{9.33}$$

where K and I are operators associated with H_{Kep} and H_{Int} respectively, $\epsilon \sim \mathrm{I}/\mathrm{K}$ is the planet to star mass-ratio, and we have explicitly listed only commutator terms where I occurs once.

We wish to devise a corrector that eliminates terms in [K, K, I] and [K, K, K, K, I]. In this way we can reduce the error per step from the usual $O(\epsilon\tau^3)$ to $O(\epsilon^2\tau^3)$. Using the identity

$$e^{\mathrm{C}}e^{\mathrm{M}}e^{-\mathrm{C}} = \exp\left\{\mathrm{M} + [\mathrm{C},\mathrm{M}] + \frac{1}{2}[\mathrm{C},\mathrm{C},\mathrm{M}] + \frac{1}{6}[\mathrm{C},\mathrm{C},\mathrm{C},\mathrm{M}] + \cdots\right\} \tag{9.34}$$

we find that a corrector of the form

$$e^{\mathrm{C}} = \exp\left\{a\tau^2[\mathrm{K},\mathrm{I}] + b\tau^4[\mathrm{K},\mathrm{K},\mathrm{K},\mathrm{I}] + O(\epsilon\tau^6) + O(\epsilon^2\tau^3)\right\} \tag{9.35}$$

where a and b are constants, results in a corrected step

$$\begin{aligned} e^{\mathrm{C}}e^{\mathrm{M}}e^{-\mathrm{C}} = \exp\Bigg\{&(\mathrm{K}+\mathrm{I})\tau + \left(\frac{1}{12} - a\right)\tau^3[\mathrm{K},\mathrm{K},\mathrm{I}] \\ &- \left(\frac{1}{720} + b\right)\tau^5[\mathrm{K},\mathrm{K},\mathrm{K},\mathrm{K},\mathrm{I}] + O(\epsilon\tau^7) + O(\epsilon^2\tau^3)\Bigg\} \end{aligned} \tag{9.36}$$

so that the [K, K, I] and [K, K, K, K, I] terms can be eliminated if $a = 1/12$ and $b = -1/720$. To be useful in practice, we need to be able to express the corrector as the product of terms involving exp(K) and exp(I) separately. Following Wisdom et al. (1996), correctors of the form of the Eq. (9.35) can be developed by noting that

$$\begin{aligned} \mathrm{Y}(i,k) &= e^{k\tau\mathrm{K}}e^{i\tau\mathrm{I}}e^{-2k\tau\mathrm{K}}e^{-i\tau\mathrm{I}}e^{k\tau\mathrm{K}} \\ &= \exp\left\{2ik\tau^2[\mathrm{K},\mathrm{I}] + \frac{ik^3\tau^4}{3}[\mathrm{K},\mathrm{K},\mathrm{K},\mathrm{I}] + \cdots\right\} \end{aligned} \tag{9.37}$$

where we retain only commutators in which I appears once.

Combining two such expressions, we obtain a suitable series of operators for the corrector (9.35):

$$\begin{aligned} \mathrm{Y}(i_1,k_1)\cdot\mathrm{Y}(i_2,k_2) = \exp\Bigg\{&2(i_1k_1 + i_2k_2)\tau^2[\mathrm{K},\mathrm{I}] \\ &+ \frac{1}{3}(i_1k_1^3 + i_2k_2^3)\tau^4[\mathrm{K},\mathrm{K},\mathrm{K},\mathrm{I}] + \cdots\Bigg\} \end{aligned} \tag{9.38}$$

where to match Eq. (9.35), we require that

$$i_1 k_1 + i_2 k_2 = \frac{1}{24}$$

$$i_1 k_1^3 + i_2 k_2^3 = -\frac{1}{240} \tag{9.39}$$

There are many possible solutions to Eqs. (9.39), for example

$$i_1 = -\frac{\sqrt{10}}{72} \qquad k_1 = \frac{3\sqrt{10}}{10}$$

$$i_2 = \frac{\sqrt{10}}{24} \qquad k_2 = \frac{\sqrt{10}}{5} \tag{9.40}$$

which generates the symplectic corrector included in the *Mercury* integrator package (Chambers 1999). Wisdom et al. (1996) provide equivalent and higher order correctors for the alternative second-order mapping

$$e^{\mathrm{M}} = \exp\left(\frac{\tau \mathrm{K}}{2}\right) \exp(\tau \mathrm{I}) \exp\left(\frac{\tau \mathrm{K}}{2}\right). \tag{9.41}$$

Symplectic correctors can also be devised for the binary-star integrators described in Section 9.3. The problem is different in that the Hamiltonian consists of three parts rather than two, but this only complicates things slightly provided we want a corrector that eliminates only terms $O(\epsilon)$. We start with an expression for one step of the wide or close-binary algorithms:

$$\begin{aligned} e^{\mathrm{M}} &= \exp\left(\frac{\tau \mathrm{I}}{2}\right) \exp\left(\frac{\tau \mathrm{J}}{2}\right) \exp(\tau \mathrm{K}) \exp\left(\frac{\tau \mathrm{J}}{2}\right) \exp\left(\frac{\tau \mathrm{I}}{2}\right) \\ &= \exp\left\{ \tau(\mathrm{I}+\mathrm{J}+\mathrm{K}) + \frac{\tau^3}{12}[\mathrm{K},\mathrm{K},\mathrm{I}] + \frac{\tau^3}{12}[\mathrm{K},\mathrm{K},\mathrm{J}] - \frac{\tau^5}{720}[\mathrm{K},\mathrm{K},\mathrm{K},\mathrm{K},\mathrm{I}] \right. \\ &\quad \left. - \frac{\tau^5}{720}[\mathrm{K},\mathrm{K},\mathrm{K},\mathrm{K},\mathrm{J}] + O(\epsilon\tau^7) + O(\epsilon^2\tau^3) \right\} \end{aligned} \tag{9.42}$$

where I, J and K are operators associated with H_{Int}, H_{Jump} and H_{Kep} for the wide or close-binary integrator, and we list only commutators that contain I or J once.

We wish to eliminate the leading-order error terms involving [K, K, I] and [K, K, J], as well as [K, K, K, K, I] and [K, K, K, K, J]. This suggests we look for a corrector of the form

$$e^{\mathrm{C}} = \exp\left\{ h_1 \tau^2 [\mathrm{K},\mathrm{I}] + h_2 \tau^2 [\mathrm{K},\mathrm{J}] + h_3 \tau^4 [\mathrm{K},\mathrm{K},\mathrm{K},\mathrm{I}] \right.$$

$$+h_4\tau^4[\mathrm{K},\mathrm{K},\mathrm{K},\mathrm{J}] + O(\epsilon\tau^6) + O(\epsilon^2\tau^3)\Big\} \tag{9.43}$$

where h_1, h_2, h_3 and h_4 are constants, which gives a corrected step of the form

$$\begin{aligned}\exp\Big\{&(\mathrm{I}+\mathrm{J}+\mathrm{K})\tau + \left(\frac{1}{12}-h_1\right)\tau^3[\mathrm{K},\mathrm{K},\mathrm{I}] + \left(\frac{1}{12}-h_2\right)\tau^3[\mathrm{K},\mathrm{K},\mathrm{J}]\\ &-\left(\frac{1}{720}+h_3\right)\tau^5[\mathrm{K},\mathrm{K},\mathrm{K},\mathrm{K},\mathrm{I}] - \left(\frac{1}{720}+h_4\right)\tau^5[\mathrm{K},\mathrm{K},\mathrm{K},\mathrm{K},\mathrm{J}]\\ &+O(\epsilon\tau^7)+O(\epsilon^2\tau^3)\Big\}\,,\end{aligned} \tag{9.44}$$

so that terms in [K, K, I], [K, K, J], [K, K, K, K, I] and [K, K, K, K, J] can all be eliminated if we choose $h_1 = h_2 = 1/12$ and $h_3 = h_4 = -1/720$.

Following the same procedure used to obtain the corrector for the Wisdom–Holman mapping, we note that

$$\begin{aligned}e^{\mathrm{C}}e^{\mathrm{A}/2}e^{\mathrm{B}}e^{\mathrm{A}/2}e^{-\mathrm{C}} = \exp\Big\{&\mathrm{A}+\mathrm{B}+[\mathrm{C},\mathrm{A}]+[\mathrm{C},\mathrm{B}]+\frac{1}{2}[\mathrm{C},\mathrm{C},\mathrm{A}]+\frac{1}{2}[\mathrm{C},\mathrm{C},\mathrm{B}]\\ &+\frac{1}{6}[\mathrm{C},\mathrm{C},\mathrm{C},\mathrm{A}]+\frac{1}{6}[\mathrm{C},\mathrm{C},\mathrm{C},\mathrm{B}]+\cdots\Big\}\,,\end{aligned} \tag{9.45}$$

where we give only commutators that contain A or B once. Hence

$$\begin{aligned}Z(i,j,k) &= e^{k\tau\mathrm{K}}e^{j\tau\mathrm{J}/2}e^{i\tau\mathrm{I}}e^{j\tau\mathrm{J}/2}e^{-2k\tau\mathrm{K}}e^{-j\tau\mathrm{J}/2}e^{-i\tau\mathrm{I}}e^{-j\tau\mathrm{J}/2}e^{k\tau\mathrm{K}}\\ &= \exp\Big\{2ik\tau^2[\mathrm{K},\mathrm{I}]+2jk\tau^2[\mathrm{K},\mathrm{J}]+\frac{ik^3\tau^4}{3}[\mathrm{K},\mathrm{K},\mathrm{K},\mathrm{I}]\\ &\qquad+\frac{jk^3\tau^4}{3}[\mathrm{K},\mathrm{K},\mathrm{K},\mathrm{J}]+\cdots\Big\}\end{aligned} \tag{9.46}$$

retaining only commutators that contain I or J once. Combining two such expressions gives

$$\begin{aligned}Z(i_1,j_1,k_1)\cdot Z(i_2,j_2,k_2) = \exp\Big\{&2(i_1k_1+i_2k_2)\tau^2[\mathrm{K},\mathrm{I}]\\ &+2(j_1k_1+j_2k_2)\tau^2[K,J]\\ &+\frac{1}{3}(i_1k_1^3+i_2k_2^3)\tau^4[\mathrm{K},\mathrm{K},\mathrm{K},\mathrm{I}]\\ &+\frac{1}{3}(j_1k_1^3+j_2k_2^3)\tau^4[\mathrm{K},\mathrm{K},\mathrm{K},\mathrm{J}]+\cdots\Big\}\,.\end{aligned} \tag{9.47}$$

To provide a suitable corrector, we need to satisfy the following criteria

$$i_1k_1 + i_2k_2 = \frac{1}{24} \quad j_1k_1 + j_2k_2 = \frac{1}{24}$$
$$i_1k_1^3 + i_2k_2^3 = -\frac{1}{240} \quad j_1k_1^3 + j_2k_2^3 = -\frac{1}{240} \tag{9.48}$$

so clearly one possible solution is

$$i_1 = j_1 = -\frac{\sqrt{10}}{72}$$
$$k_1 = \frac{3\sqrt{10}}{10}$$
$$i_2 = j_2 = \frac{\sqrt{10}}{24}$$
$$k_2 = \frac{\sqrt{10}}{5} \tag{9.49}$$

Figure 9.2 shows the effect of including this corrector when rerunning the integration shown in Fig. 9.1. The upper panel shows the energy error versus time

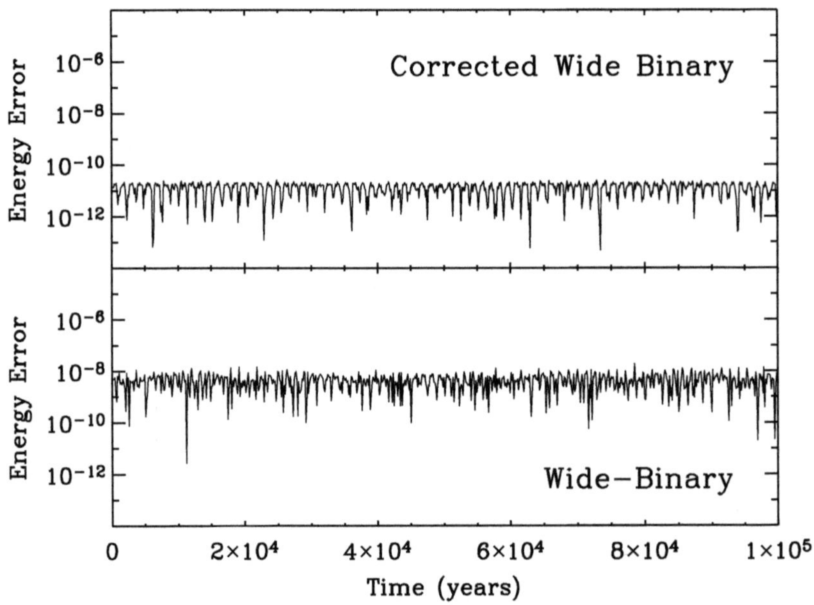

Fig. 9.2 A comparison between the energy errors (accuracies) of the wide-binary algorithm with and without the corrector. The integrated system is similar to that in Fig 9.1. As shown here, the corrector reduces the energy error by approximately three orders of magnitude

when integrating the four giant planets of the solar system with a binary companion using the corrector. The lower panel shows the case without a corrector– the same as in Fig. 9.1. The corrector reduces the energy error by roughly three orders of magnitude, comparable to the Jupiter/Sun mass ratio. However, because the corrector is only applied at the beginning of the integration and prior to each output, this improvement in accuracy is achieved at little computational cost. The corrector described above can also be used with the close- and wide-binary algorithms when close-encounter terms are partitioned along the lines of the hybrid integrator of (9.14), although the improvement in accuracy is generally not as high as seen in Fig. 9.2.

9.5 Stellar Encounters

The binary algorithms described in the previous sections are designed to work in particular circumstances. In the close-binary algorithm, the planets are assumed to orbit the centre of mass of a binary at a distance large compared to the binary separation. If a planet comes sufficiently close to the binary stars, this assumption will no longer be valid and the algorithm will break down. In the wide-binary algorithm, planets are assumed to orbit one member of a binary, while receiving small perturbations from the other star. If the distance between a planet and its central star ever becomes comparable to the distance to the other star, the wide-binary algorithm will also break down. The accuracy of each of the algorithms depends on the hierarchy of the system being preserved. For this reason the binary integrators are unable to follow the trajectories of planets moving on transfer orbits, where the centre of a planet's motion switches from one star to another or from one star to both stars.

The wide-binary algorithm suffers from a second limitation in that a planet cannot travel too close to its central star either. If this happens, the fixed stepsize of the integrator will be too large to properly follow the planet's periastron passage and accuracy will be lost. This is a well known limitation of symplectic integrators in general, including the Wisdom–Holman mapping (Rauch and Holman, 1999). Accuracy can be restored by regularizing the motion, so that the stepsize effectively depends on a planet's distance from the star (e.g. Preto and Tremaine, 1999), but regularization becomes highly inefficient for problems involving more than a few bodies.

Levison and Duncan (2000) have suggested an alternative solution which involves a new division of the Hamiltonian such that a planet's indirect perturbation terms are added to the Keplerian part of the Hamiltonian whenever these terms become large. This is analogous to the procedure described earlier for maintaining accuracy during a close encounter between two planets. We can see how this works by considering the Hamiltonian for the single-star case described by Eq. 9.12). In

Levison and Duncan's scheme, the terms in $H_{\rm Jump}$ are divided between $H_{\rm Jump}$ and $H_{\rm Kep}$ in such a way that the former always remains small compared to the latter:

$$H_{\rm Large} = \sum_{i=1}^{N} \left(\frac{P_i^2}{2m_i} - \frac{Gm_0m_i}{R_i} \right) + \frac{1}{2m_0} \left(\sum_{i=1}^{N} \mathbf{P}_i \right)^2 \Lambda(R_1, R_2 \ldots, R_N)$$

$$H_{\rm Int} = -\sum_{i=1}^{N} \sum_{j>i} \frac{Gm_im_j}{R_{ij}}$$

$$H_{\rm Small} = \frac{1}{2m_0} \left(\sum_{i=1}^{N} \mathbf{P}_i \right)^2 [1 - \Lambda(R_1, R_2 \ldots, R_N)] \tag{9.50}$$

Levison and Duncan (2000) advocate using a partition function of the form

$$\Lambda = 1 - \prod_{i=1}^{N} [1 - \lambda(R_i)] \tag{9.51}$$

where $\lambda(R)$ is chosen so that $\Lambda \to 1$ when any planet approaches the star and $\Lambda = 0$ when all the planets are far from the star. As with the hybrid integrator described by Eq. (9.14), $H_{\rm Large}$ has to be integrated numerically whenever $\Lambda \neq 0$. In addition, for this choice of Λ, $H_{\rm Small}$ must also be integrated numerically. An obvious shortcoming of this procedure is that an integration will proceed slowly whenever any planet passes close to the star since all the planets have to be integrated numerically in this case. However, if close periastron passages are relatively rare, this shortcoming is not severe.

The scheme of Levison and Duncan (2000) can be applied to the wide-binary algorithm to improve the accuracy whenever a planet passes close to the central star. A similar procedure could be developed to cope with planets that stray far from the central star in a wide binary, or come close to the stars in a close binary. In either case, the planet's orbit is likely to be unstable, so this state of affairs will be short-lived, compensating for the fact that all objects will have to be integrated numerically during this stage of the evolution.

However, this approach to dealing with stellar encounters suffers from two other drawbacks that limit its usefulness. The fact that $H_{\rm Small}$ must be integrated numerically in addition to $H_{\rm Large}$ can be overcome by choosing a partition function $\lambda(V)$ that depends on the planets' velocities rather than their positions. Since a planet's velocity will typically become large whenever it approaches a star, velocity can be used instead of position to identify a close encounter with a star.

A more serious problem arises when one considers low-mass planets or test particles. When advancing the system under $H_{\rm Small}$, the rate of change of the x component of velocity of planet k is given by Hamilton's equations:

$$\frac{dV_{x,k}}{dt} = -\frac{1}{m_k}\frac{\partial H_{\text{Small}}}{\partial X_k}$$
$$= \frac{1}{m_k}\frac{1}{2m_0}\left(\sum_{i=1}^{N}\mathbf{P}_i\right)^2\left(\frac{X_k}{R_k}\right)\frac{\lambda'(R_k)}{1-\lambda(R_k)}\prod_{i=1}^{N}\left[1-\lambda(R_i)\right]. \quad (9.52)$$

Note that the righthand side of this expression is proportional to $1/m_k$, so the rate of the change of the planet's velocity will become large if the planet's mass is small, and will become infinite for massless test particles. This means the scheme cannot be used to integrate test particles, and the accuracy will be severely degraded when integrating low-mass planets.

Unfortunately, choosing a partition function that depends on velocity rather than position does not overcome this problem. A better way to tackle this issue is to use a set of coordinates that doesn't give rise to momentum cross terms like those in H_{Jump} or H_{Small}, so that indirect terms are a function of the coordinates only. Barycentric coordinates have this property, but as noted earlier, symplectic integrators that use barycentric coordinates tend to perform poorly. Jacobi coordinates will not work since they do not have the necessary properties for treating close encounters between planets.

Chambers (2003) described a new set of coordinates with the right properties to apply Levison and Duncan's scheme for integrating close encounters with a star. These coordinates, dubbed "Yosemite coordinates" by the author (after the location of the 2000 Division on Dynamical Astronomy meeting where they were developed), are given by

$$\mathbf{R}_0 = \left(\frac{m_0\mathbf{r}_0 + \sum_j m_j\mathbf{r}_j}{m_{\text{tot}}}\right)$$
$$\mathbf{R}_i = (\mathbf{r}_i - \mathbf{r}_0) + \frac{\beta}{m_0}\sum_j m_j(\mathbf{r}_j - \mathbf{r}_0) \quad (9.53)$$

for the case of planets orbiting a single star, where the subscript 0 refers to the star, and

$$\beta = \frac{1-\sqrt{1+\mu}}{\mu\sqrt{1+\mu}} \sim -\frac{1}{2} \quad (9.54)$$

where

$$\mu = \frac{\sum_{j=1}^{N} m_j}{m_0}. \quad (9.55)$$

The canonically conjugate momenta in this case are

$$\mathbf{P}_0 = \mathbf{p}_0 + \sum_{j=1}^{N} \mathbf{p}_j$$

$$\mathbf{P}_i = \mathbf{p}_i - \frac{m_i}{m_{\text{tot}}} \left[\frac{\mathbf{p}_0 + (1+\beta+\beta\mu)\sum_j \mathbf{p}_j}{1+\beta\mu} \right]. \tag{9.56}$$

Using these coordinates, the Hamiltonian for a system of planets orbiting a single star is

$$H = H_{\text{Kep}} + H_{\text{Int}}, \tag{9.57}$$

where

$$H_{\text{Kep}} = \sum_{i=1}^{N} \left(\frac{P_i^2}{2m_i} - \frac{Gm_0 m_i}{R_i} \right)$$

$$H_{\text{Int}} = -\sum_{i=1}^{N} \sum_{j>i} \frac{Gm_i m_j}{R_{ij}} + \sum_{i=1}^{N} Gm_0 m_i \left(\frac{1}{R_i} - \frac{1}{|\mathbf{R}_i - \mathbf{S}|} \right), \tag{9.58}$$

and

$$\mathbf{S} = \frac{\beta \sum_j m_j \mathbf{R}_j}{m_0(1+\beta\mu)}. \tag{9.59}$$

The second set of terms in H_{Int} represent indirect perturbations on each planet. These terms can be divided between H_{Int} and H_{Kep}, in an analogous manner to the scheme of Levison and Duncan (2000), as follows:

$$H_{\text{Large}} = \sum_{i=1}^{N} \left(\frac{P_i^2}{2m_i} - \frac{Gm_0 m_i}{R_i} \right) + \sum_{i=1}^{N} Gm_0 m_i \left(\frac{1}{R_i} - \frac{1}{|\mathbf{R}_i - \mathbf{S}|} \right) \Lambda(R_1, R_2, \ldots, R_N)$$

$$H_{\text{Small}} = -\sum_{i=1}^{N} \sum_{j>i} \frac{Gm_i m_j}{R_{ij}} + \sum_{i=1}^{N} Gm_0 m_i \left(\frac{1}{R_i} - \frac{1}{|\mathbf{R}_i - \mathbf{S}|} \right) \left[1 - \Lambda(R_1, R_2, \ldots, R_N) \right], \tag{9.60}$$

where Λ is chosen so that $\Lambda \to 1$ whenever any planet approaches the star and $\Lambda = 0$ when all planets are far from the star.

The advantage of using Yosemite coordinates now becomes clear: each of the partitioned indirect terms in Eq. (9.60) is proportional to m_i. As a result, low-mass planets and test particles can be integrated to the same accuracy as massive planets without the problems encountered when using Eq. (9.50). The same scheme can be extended to binary systems by using new coordinate systems analogous to Yosemite coordinates with the binary hierarchy built in, as in the wide-binary and close-binary coordinate systems.

9.6 Conclusions

Conventional integration algorithms such as Runge–Kutta and Bulirsch–Stoer can be applied to planetary systems orbiting binary stars with little or no modification. However, these algorithms tend to be slow and exhibit long-term growth in energy errors. For these reasons, symplectic integration algorithms have become the tool of choice for many researchers. In this chapter, it was described how the symplectic map of Wisdom and Holman (1991) can be adapted for use with planets in binary systems, including modifications to handle close approaches between planets with each other and with the stars themselves. A more detailed description of the derivation and testing of these new algorithms is given by Chambers et al. (2002). In addition, it was shown that the performance of these algorithms can be improved at little extra cost by using symplectic correctors. However, there is still scope for future improvements. In particular, the current method for following close encounters between a planet and a star is slow and cumbersome, and there remains no easy way to handle planets whose orbital motion switches from one star to the other. Efficient solutions to these problems remain to be developed.

The adaptations described in this chapter mean that symplectic algorithms can now be applied to a wide variety of problems involving planets in binary-star systems. For example, Quintana et al. (2002, 2007) have modelled the formation of terrestrial planets in orbit around α Centauri and other binary systems using the wide-binary algorithm described in Section 9.3.2. These authors have shown that planet formation is similar to that around single stars provided that the binary separation exceeds a critical value, otherwise planetary accretion is strongly altered or curtailed. Similarly, Quintana and Lissauer (2006) have shown that planet formation can proceed in protoplanetary disks that orbit both members of a close binary, leading in some cases to planetary systems resembling the solar system. Turrini et al. (2004) and Thebault et al. (2004) have modelled the growth of the giant planet orbiting one member of the γ Cephei binary system using the new algorithms, and found that perturbations from the nearby companion would not prevent the in situ formation of the giant planet. Malmberg et al. (2007) have used the symplectic algorithms to examine the stability of planetary orbits in binary systems as a function of binary separation and binary lifetime, for transient binaries formed in stellar clusters, helping to constrain the circumstances under which planetary systems will survive. The coming years will surely provide additional exciting results based on the application of N-body integrators to planets in binary star systems.

References

H. Beust, Symplectic integration of hierarchical stellar systems, Astron. Astrophys. **400,** 1129–1144 (2003)

J.E. Chambers, A hybrid symplectic integrator that permits close encounters between massive bodies, Mon Not R. Astron. Soc. **304,** 793–799 (1999)

J.E. Chambers, Making more terrestrial planets, Icarus, **152,** 205–224 (2001)

J.E. Chambers, E.V. Quintana, M.J. Duncan and J.J. Lissauer, Symplectic integrator algorithms for modeling planetary accretion in binary stars systems, Astron. J. **123,** 2884–2894 (2002)

J.E. Chambers, Adventures in coordinate space, division on dynamical astronomy meeting #34, abstract 9.06, Bull. Am. Astron. Soc. **35,** 1043 (2003)

J.M.A. Danby, *Fundamentals of Celestial Mechanics*, (Richmond, VA., U.S.A.: Willmann-Bell, 1988)

M.J. Duncan, H.F. Levison and M.H. Lee, A multiple time step symplectic algorithm for integrating close encounters, Astron. J. **116,** 2067–2077 (1998)

E. Everhart, An efficient integrator that uses Gauss-Radau spacings. In *Dynamics of Comets: Their Origin and Evolution*, ed by A. Carusi and G.B. Valsecchi, Astrophysics and space Science Library, **115**, (Dordrecht: Reidel, 1985) pp 185

M.J. Fogg and R.P. Nelson, Oligarchic and giant impact growth of terrestrial planets in the presence of gas giant planet migration, Astron. Astrophys. **441,** 791–806 (2005)

E.B. Ford and E.I. Chiang, The formation of ice giants in a packed oligarchy: instability and aftermath, Ap. J. **661,** 602–615 (2007)

N. Haghighipour, Dynamical stability and habitability of the gamma Cephei binary planetary system, Ap. J. **644,** 543–550 (2006)

N. Haghighipour and S.N. Raymond, habitable planet formation in binary planetary systems, Ap. J. **666,** 436–446 (2007)

M.J. Holman and P.A. Wiegert, Long-term stability of planets in binary systems, Astron. J. **117,** 621–628 (1999)

H. Kinoshita, H. Yoshida and H. Nakai, Symplectic integrators and their application to dynamical astronomy, Cel. Mech. Dyn. Astron. **50,** 59–71 (1991)

H.F. Levison and M.J. Duncan, Symplectically integrating close encounters with the Sun, Astron. J. **120**, 2117–2123 (2000)

D. Malmberg, F. De Angeli, M.B. Davies, R.P. Church, D. Mackey and M.I. Wilkinson. Close encounters in young stellar clusters: implications for planetary systems in the solar neighbourhood, Mon. Not. R. Astron. Soc. **378,** 1207–1216 (2007).

A.M. Mandell, S.N. Raymond and S. Sigurdsson, Formation of Earth-like planets during and after giant planet migration, Ap. J. **660,** 823–844 (2007)

W.H. Press, S.A. Teukolsky, W.T. Vetterling and B.P. Flannery, *Numerical recipes in FORTRAN. The art of scientific computing*, (Cambridge, Cambridge University Press, 1992)

M. Preto and S. Tremaine, A class of symplectic integrators with adaptive time step for separable Hamiltonian systems, Astron. J. **118**, 2532–2541 (1999)

E.V. Quintana, J.J. Lissauer, J.E. Chambers and M.J. Duncan, Terrestrial planet formation in the alpha Centauri system, Ap. J. **576,** 982–96 (2002)

E.V. Quintana and J.J. Lissauer, Terrestrial planet formation surrounding close binary stars, Icarus, **185**, 1–20 (2006)

E.V. Quintana, F.C. Adams, J.J. Lissauer and J.E. Chambers, Terrestrial planet formation around individual stars within binary systems, Ap. J. **660,** 807–822 (2007)

K.P. Rauch and M. Holman, Dynamical chaos in the Wisdom – Holman integrator: origin and solutions, Astron. J. **117,** 1087–1102 (1999)

S.N. Raymond, T. Quinn and J.I. Lunine. Making other earths: dynamical simulations of terrestrial planet formation and water delivery, Icarus, **168,** 1–17 (2004)

P. Saha and S. Tremaine, Long-term planetary integration with individual time steps, Astron. J. **108,** 1962–1969 (1994)

J. Stoer and R. Bulirsch, *Introduction to Numerical Analysis* (Springer-Verlag, New York 1980)
P. Thébault, F. Marzari and H. Scholl. Terrestrial planet formation in exoplanetary systems with a giant planet on an external orbit, Astron. Astrophys. **384,** 594–602 (2002)
P. Thébault, F. Marzari, H. Scholl, D. Turrini and M. Barbieri, Planetary formation in the Gamma Cephei system, Astron. Astrophys. **427,** 1097–1104 (2004)
D. Turrini, M. Barbieri, F. Marzari and P. Tricarico, Planetary formation and orbital stability in the gamma Cephei system, Mem. S. A. It. Suppl. **5,** 127–130 (2004)
P.A. Wiegert and M.J. Holman, The stability of planets in the Alpha Centauri system, Astron. J. **113**, 1445–1450 (1997)
J. Wisdom and M. Holman, Symplectic maps for the N-body problem, Astron. J. **102**, 1528–1538 (1991)
J. Wisom and M. Holman, Symplectic maps for the N-body problem: stability analysis, Astron. J. **104**, 2022–2029 (1992)
J. Wisdom, M. Holman and J. Touma, in *Integration Algorithms and Classical Mechanics*, ed by J.E. Marsden, W.G. Patrick and W.F. Shadwick (Fields Inst. Commun. Providence, RI, USA., **10**, 1996) pp 217
H. Yoshida, Construction of higher order symplectic integrators, Phys. Lett. A. **150**, 262–268 (1990)

Chapter 10
Terrestrial Planet Formation in Binary Star Systems

Elisa V. Quintana and Jack J. Lissauer

10.1 Introduction

More than half of all main sequence stars, and an even larger fraction of pre-main sequence stars, reside in binary or multiple systems (Duquennoy and Mayor 1991; Mathieu et al. 2000). The presence of planet-forming material has been indirectly observed around one or both components of some young binaries, and (Mathieu et al. 2000) numerical simulations of the formation of binary stars suggest that disks form within these systems, as well (Bodenheimer et al. 2000). Terrestrial planets and the cores of giant planets are thought to form by an accretion process within a disk of dust and gas (Safronov 1969; Lissauer 1993), and therefore may be common in binary star systems. In this chapter, we present the results from numerical simulations of the final stages of terrestrial planet formation around one or both stars of a binary.

Our simulations were performed using two symplectic integration algorithms developed for terrestrial planet formation in dual-star systems (Chambers et al. 2002). These computations examine a large range of values for the stellar masses and binary orbital parameters (semimajor axes and eccentricities) to determine whether/where terrestrial planets can form and remain on stable orbits. Multiple realizations (from 3–30) were performed for each binary star configuration under study to account for the chaotic nature of these N-body systems. We statistically compare the resulting planetary systems formed to those that formed in additional simulations that we performed of the Sun–Jupiter–Saturn system which began with virtually identical initial disk conditions.

We show herein that terrestrial planets similar to those in the solar system can indeed form in a wide variety of binary star systems, and we have begun to delineate the range of binary star parameter-space which allow the formation of Earth-like

E.V. Quintana (✉)
SETI Institute, 515 N. Whisman Road, Mountain View, CA 94043, USA
e-mail: equintana@mail.arc.nasa.gov

J.J. Lissauer
Space Science and Astrobiology Division 245-3, NASA Ames Research Center, Moffett Field, CA 94035, USA
e-mail: Jack.Lissauer@nasa.gov

N. Haghighipour (ed.), *Planets in Binary Star Systems*, Astrophysics and Space Science Library 366, DOI 10.1007/978-90-481-8687-7_10,

planets. In Section 10.2 we describe our numerical model, including the initial conditions and physical assumptions of the circumstellar/circumbinary disk. Details of the algorithms that we helped to develop, and the corresponding performance tests, are presented in Chambers et al. (2002) and in the chapter in this volume by Chambers. We present a summary of the results of our simulations of planetary growth in S-type orbits around each star in the α Centauri system (Section 10.3), in other 'wide' binary star systems (Section 10.4), and in P-type orbits around both stars in close short-period binary star systems (Section 10.5). Note that the results from all of our simulations can be scaled for different star, disk, and individual planetesimal parameters with the formulae presented in Appendix C of Quintana et al. (2006). In Section 10.6, we summarize our results and provide a rough estimate of the fraction of solar-mass binary stars in the Galaxy that can potentially harbor Earth analogues.

10.2 Model and Initial Conditions

Our simulations are based on the conventional model of planet formation in which terrestrial planets form via pairwise accretion of rocky bodies from within a disk of gas and dust that has remained around a newly formed star (Safronov 1969; Lissauer 1993). For all of our simulations, we assume that rocky Moon- to Mars-sized planetesimals/embryos have already accreted from within this disk.

The growth stage from km-sized planetesimals to the Moon- to Mars-sized bodies via runaway and oligarchic growth, has been numerically simulated by Kortenkamp and Wetherill (2000) for the Sun–Jupiter–Saturn system, by Marzari and Scholl (2000) for the α Centauri AB binary star system, and by Thebault et al. (2004) for the γ Cephei binary/giant-planet system. In each case, the combined effects of the giant planets or stellar perturbations and gas-drag lead to periastron alignment of the planetesimal population, thereby reducing the relative and collisional velocities, and increasing the planetesimal accretion efficiency within $\sim$2.5 AU of the central star.

The initial conditions and physical assumptions of the rocky bodies in the circumstellar/circumbinary disk used in our binary star simulations are the same as those used in simulations of the final stages of terrestrial planet growth around the Sun (Chambers 2001). Our aim is to vary the binary star parameters – an enormous phase space – and use virtually the same disk properties in order to delineate the effects of various binary star systems on the terrestrial planet accretion. Although this formulation is neither complete nor definitive, this disk model reproduces our terrestrial planet system in simulations around the Sun (albeit with somewhat larger eccentricities), and can thus be used as a reference point, despite our approximations (e.g., the lack of very small bodies and the assumption of perfectly inelastic collisions).

Similar disk mass distributions is used for both the circumstellar and circumbinary simulations. In this disk model (originally presented in Chambers 2001), half of the disk mass is composed of 14 rocky Mars-sized 'embryos' (each with a mass of 0.0933 $M_\oplus$), and the remaining is distributed equally among 140 Moon-sized 'planetesimals' (each with a mass of 0.00933 $M_\oplus$). The total mass of the disk is $\sim$2.6 $M_\oplus$. The surface density profile of our disk model has the form $a^{-3/2}$ (where a is the semimajor axis), normalized to 8 g cm^{-2} at 1 AU, and follows from minimum-mass solar nebula models (Weidenschilling 1977). The bodies are distributed between 0.36 AU and 2.05 AU, and the radius of each body is calculated assuming a material density of 3 g cm^{-3}. Note that the larger embryos are widely spaced, in agreement with the oligarchic growth result of Kokubo and Ida (1998). The embryos and planetesimals begin with initial eccentricities $e \leq 0.01$ and inclinations $i \leq 0.5°$. Specific initial orbital elements were chosen at random from specified ranges; the same set of randomly selected values was used for all simulations.

The evolution of the accreting bodies, which are subject to gravitational forces from both stars, gravitational interactions with other bodies, and perfectly inelastic collisions, is followed for 200 Myr–1 Gyr, or until only one planet remains. Because these N-body systems are chaotic in nature, we performed multiple integrations (from 3–30) of each system with a very slight change in the initial conditions of one body in the disk. This tactic allows us to sample the range of possible outcomes for effectively equivalent initial conditions, and the result is a distribution of final planetary systems.

10.3 Planet Formation in the α Centauri AB Binary Star System

We first examined the late stage of planet formation in the α Centauri system. This system is comprised of a central binary consisting of the G2 star α Cen A (1.1 $M_\odot$) and the K1 star α Cen B (0.91 $M_\odot$) (Quintana et al. 2002). The stars of this system have an orbital semimajor axis of 23.4 AU and an eccentricity of 0.52. The M5 star α Cen C (Proxima Centauri) is thought to orbit this pair, but at a very large distance (12,000 AU). This star is neglected in our simulations. Observations at the Anglo-Australian telescope imply that no planet orbiting either star induces periodic velocity variations as large as 2 m/s (G. Marcy, personal communication, 2006). This upper bound, combined with dynamical stability calculations (Wiegert and Holman 1997) implies that any planet in an S-type orbit around either component of the α Cen AB binary must have a mass less than that of Saturn or orbit in a plane that is substantially inclined to the line of sight. Herein we present the results of simulations of the late stages of terrestrial planet growth around α Cen A and α Cen B for various initial inclinations of the circumstellar disk relative to the binary orbital plane.

In the majority of our simulations, the circumstellar disk is centered around α Cen A with α Cen B perturbing the system. The initial inclination of the midplane of the disk, i, was set to 0°, 15°, 30°, 45°, 60°, or 180° relative to the plane containing α Cen A and B. Although a stellar companion, present during the earlier stages of planet formation, would likely force the planetesimal disk into the plane of the binary orbit, many binary stars may originate as unstable triple star systems which could produce a binary system with an accretion disk at a high relative inclination. It is also possible that a companion may have been captured around a single star that possesses an accretion disk. The longitude of periastron of the stellar companion is set to either 90° or 180° for each run. We also performed a set of integrations with the disk centered around α Cen B, with α Cen A orbiting the system in the same plane ($i = 0°$) and direction. For comparison purposes, we performed a set of runs which follow the evolution of the disk around the Sun with neither giant planets nor a stellar companion perturbing the system.

Figure 10.1 shows the results from a simulation in which the disk is centered around α Cen A and coplanar to the binary orbital plane. Each panel shows the eccentricity of each body in the disk as a function of semimajor axis at the specified time, and the radius of each symbol is proportional to the radius of the body that it represents. Within 100 Myr of the integration, five terrestrial planets that are at least as massive as the planet Mercury have formed around α Cen A, with a single planetesimal remaining in a highly eccentric orbit. The planetesimal is ejected from the system soon thereafter (110 Myr), and the five terrestrial planets remain on stable orbits within 2 AU for the remainder of the 200 Myr simulation.

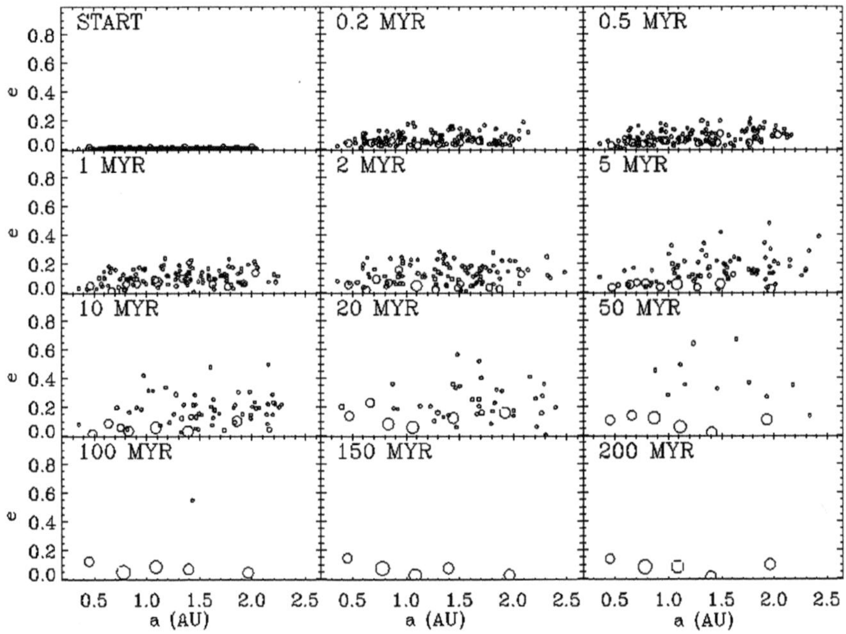

Fig. 10.1 The temporal evolution of planetary embryos and planetesimals in a circumstellar disk centered around α Cen A and coplanar with the binary orbital plane [simulation ACi0_3 in Quintana et al. (2002)]. The radius of each symbol is proportional to the radius of the body that it represents, and the eccentricities are displayed as a function of semimajor axis. By the end of the 200 Myr integration, five terrestrial planets have formed within 2 AU of α Cen A, accumulating approximately 89% of the initial disk mass

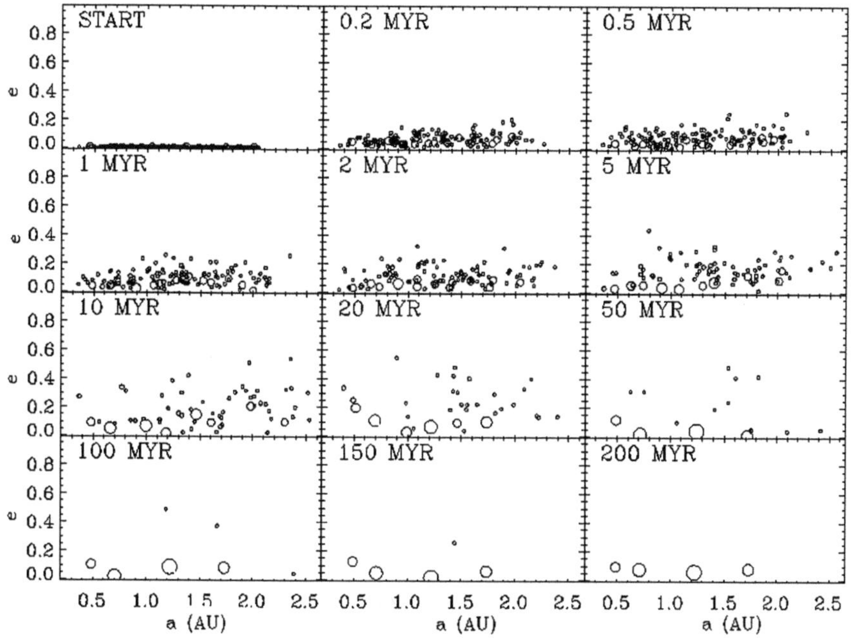

Fig. 10.2 The temporal evolution of virtually the same circumstellar disk around α Cen A as that shown in Fig. 10.1, but in this case a single planetesimal near 1 AU is shifted by a small amount (1 m along its orbit) prior to the integration [simulation ACi0_4 in Quintana et al. (2002)]. The stellar parameters and all other disk initial conditions are identical to the system shown in Fig. 10.1. The dynamics of the disk are generally the same in the earlier stages of the simulations shown here and in Fig. 10.1: the more massive embryos orbit with low eccentricities whereas the planetesimals are dynamically excited to much higher values. The stochastic nature of these N-body systems is evident in the final planetary system that formed, and demonstrates the sensitive dependence of the outcome on the initial conditions. Four terrestrial planets, comprised of approximately 88% of the initial disk mass, remain within 1.7 AU of α Cen A

The simulation shown in Fig. 10.2 begins with nearly identical initial conditions as the system in Fig. 10.1, with the exception of a very small (1 m) shift in the initial position of one planetesimal near 1 AU. Although the early evolution of the disk is qualitatively similar among the two systems, they diverge with a Lyapunov time of order 10^2 years (Quintana et al. 2007), which ultimately leads to the formation of a substantially different planetary system: four terrestrial planets form within 1.8 AU of α Cen A. Although these N-body simulations are highly stochastic, there are clear trends in the final planetary systems (number, masses, orbits, etc.) that form in simulations with the same binary star parameters (shown further in Sections 10.4 and 10.5).

Figure 10.3 shows the results of an integration in which the disk is initially inclined by 15° relative to the binary orbital plane. The bodies in this disk are more dynamically excited than those shown in Figs. 10.1 and 10.2. In this simulation, ~25% of the initial disk mass was lost (compared with ~12% that was lost in each of the $i = 0°$ simulations shown in the previous figures), and three terrestrial planets and a single planetesimal remained by the end of the integration.

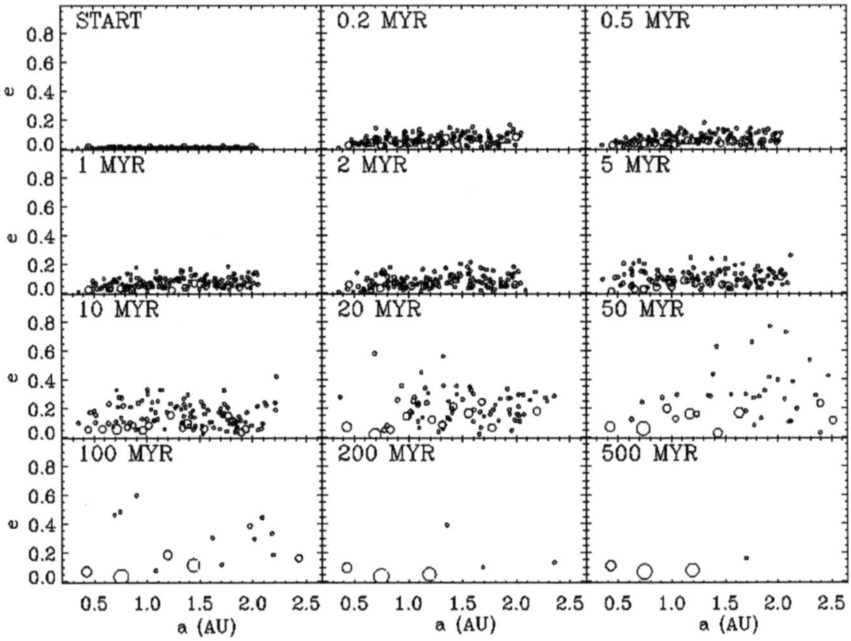

Fig. 10.3 The temporal evolution of our standard circumstellar disk centered around α Cen A and initially inclined by 15° to the binary orbital plane [simulation ACi15_2 in Quintana et al. (2002)]. The eccentricities of the embryos and planetesimals are displayed as a function of semimajor axis, and the radius of each symbol is proportional to the radius of each body that it represents. In this case, three terrestrial planets formed within ~1.2 AU, and a single planetesimal remained exterior to these planets at ~1.7 AU, all composed of approximately 75% of the initial disk mass. Similar evolution plots for all of the simulations involving α Cen A and B can be found in Quintana et al. (2002), Quintana (2003, 2004), and Lissauer et al. (2004)

We performed a total of 16 simulations of terrestrial planet growth around α Cen A in which the midplane of the disk was initially inclined by 30° or less relative to the binary orbital plane. In these simulations, when the bodies in the disk began in prograde orbits, 3–5 terrestrial planets formed around α Cen A. When $i = 180°$ relative to the binary plane, 4 or 5 planets formed. Also, 2–4 planets formed in a disk centered around α Cen B, with α Cen A perturbing the system in the same plane. The final planetary systems that formed in all of these simulations are shown in Figs. 10.8 to 10.10 of Quintana et al. (2002). The distribution of final terrestrial planet systems in the aforementioned cases (with $i \leq 30°$) is quite similar to that produced by calculations of terrestrial planet growth in the Sun–Jupiter–Saturn system.

In systems with the accreting disk initially inclined at 45° to the binary plane, from 2–5 planets formed, despite the fact that more than half of the disk mass was scattered into the central star. When the disk was inclined by 60°, the stability of the planetary embryos decreased dramatically, and almost all of the planetary embryos and planetesimals were lost from these systems. Figures of the temporal evolution of each of our α Cen simulations, and several simulations that include the Sun with neither giant planets nor a stellar companion, can be found in Quintana et al. (2002) and Quintana (2004). In contrast to planetary accretion in binary star systems, terrestrial planet growth around a star lacking both stellar and giant planet companions is slower and extends to larger semimajor axis for the same initial disk of embryos and planetesimals. Simulations by other researchers which examined various disk mass distributions around α Cen A resulted in terrestrial planet systems with semimajor axes within $\sim$1.6 AU (Barbieri et al. 2002), consistent with our results.

10.4 S-type Orbits in Other 'Wide' Binary Star Systems

In this section, we present the results from a survey ($\sim$120 numerical simulations) on the effects of a stellar companion on the final stages of terrestrial planet formation in S-type orbits around one component of a binary star system (Quintana et al. 2007). We examine binary star systems with stellar mass ratios $\mu \equiv M_C/(M_\star + M_C) = 1/3, 1/2$, or 2/3, where $M_\star$ is the mass of the star around which the protoplanetary disk is situated, and M_C is the mass of the companion. The majority of our simulations begin with equal-mass stars ($\mu = 1/2$) of either $M_\star = M_C = 0.5\text{M}_\odot$ (Set A), or $M_\star = M_C = 1\text{M}_\odot$ (Set B). Simulations were also performed with a more massive planet-hosting star (the one the disk is centered around), $M_\star = 1\text{M}_\odot$ and $M_C = 0.5\text{M}_\odot$ ($\mu = 1/3$, Set C), and also with a smaller planet hosting star of $M_\star = 0.5\text{M}_\odot$ and a more massive companion $M_C = 1\text{M}_\odot$ ($\mu = 2/3$, Set D). The binary semimajor axis, a_B, and eccentricity, e_B are chosen such that the binary periastron $q_B \equiv a_B(1 - e_B)$ takes one of the three values, $q_B = 5$ AU, 7.5 AU, or 10 AU. Binary systems with much wider periastra would have little effect on terrestrial planet formation, whereas systems with significantly smaller periastra would completely destroy (all but perhaps the very innermost part of) our initial disk of planetesimals that extends from $\sim$0.35–2 AU. Terrestrial planet accretion has been examined by Haghighipour and Raymond (2007) around one star with a stellar companion that is wide enough (tens of AU) to allow a Jupiter-like planet to remain stable, and the formation of habitable terrestrial planets was found to be feasible in some systems depending on the combined effects from the stellar/giant planet. In the simulations presented here, the binary stars are separated by $a_B = 10, 13\frac{1}{3}, 20$, or 40 AU, and the eccentricities are varied in the range $0 \leq e_B \leq 0.875$. The largest semimajor axis for which particles can be stable in any of the systems that we explore is 2.6 AU (Holman and Wiegert 1999; Quintana et al. 2007); we therefore omit giant planets analogous to those in the Solar System (which orbit beyond 5 AU) from our integrations.

For our accretion simulations, our exploration of parameter-space has two coupled goals. On one hand, we want to determine the effects of the binary orbital elements on the final terrestrial planet systems produced. On the other hand, for a given binary configuration, we want to explore the distribution of possible resulting planetary systems (where the results must be described in terms of a distribution due to the sensitive dependence on the initial conditions). We have performed from

3 to 30 integrations for each wide binary star configuration (μ, a_B, and e_B) considered, with small differences in the initial conditions: a single planetesimal near 0.5, 1, or 1.5 AU is shifted forward along its orbit by a small amount (from 0 to 9 m) prior to the integration. Ideally, of course, one would perform larger numbers of integrations to more fully sample the distributions of results, but computer resources limit our sample size. We discuss trends in the characteristics of final system properties below.

The stellar mass-ratio, μ, and the periastron distance q_B strongly influence where terrestrial planets can form in 'wide' binary star systems. The effect of q_B on the distribution of final planetary system parameters (i.e., number, masses, etc.) is demonstrated in Figs. 10.4 to 10.6. The semimajor axis of the outermost planet can be used as a measure of the size of the terrestrial planet system. Figure 10.4 shows the distribution of the semimajor axis of the outermost final planet formed in each simulation for systems with $q_B = 5$ AU (top panel), 7.5 AU, (middle panel), and 10 AU (lower panel). Note that twice as many integrations have been performed in

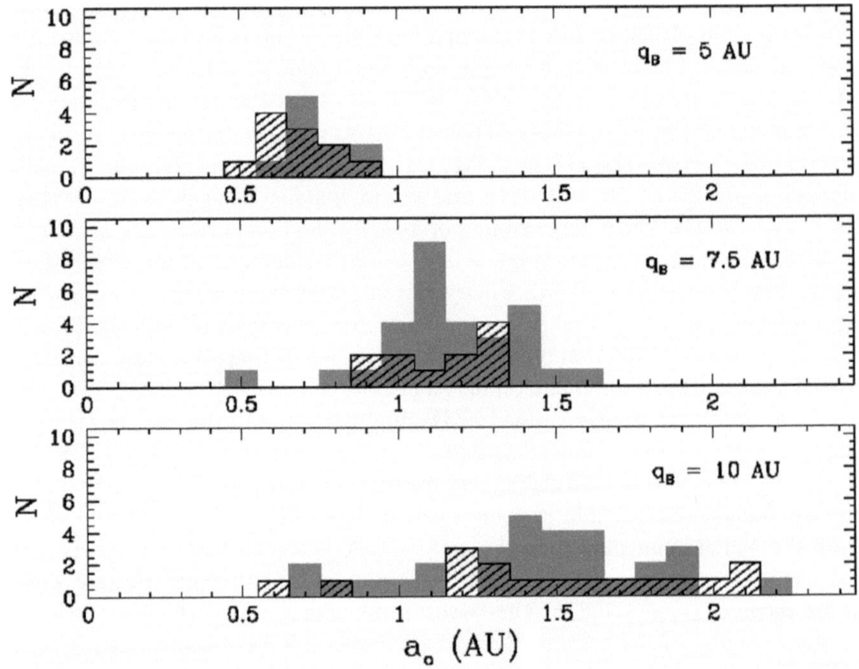

Fig. 10.4 The distribution of the semimajor axis of the outermost final planet, a_0, formed for binary star systems with $q_B = 5$ AU (*top panel*), $q_B = 7.5$ AU (*middle panel*), and $q_B = 10$ AU (*bottom panel*). The gray bars represent simulations from Set A with $M_\star = M_C = 0.5\ M_\odot$, whereas the dashed bars represent systems from Set B with $M_\star = M_C = 1.0\ M_\odot$. The effect of the binary stars is evident in systems with $q_B \leq 7.5$ AU. Although the semimajor axes extend to larger values in binary systems with larger periastra, the inner edge of the distribution is roughly determined by the inner edge of the initial disk of embryos, as in the case of a single star (i.e., the presence of different stellar companions has a minimal effect on the inner terrestrial region)

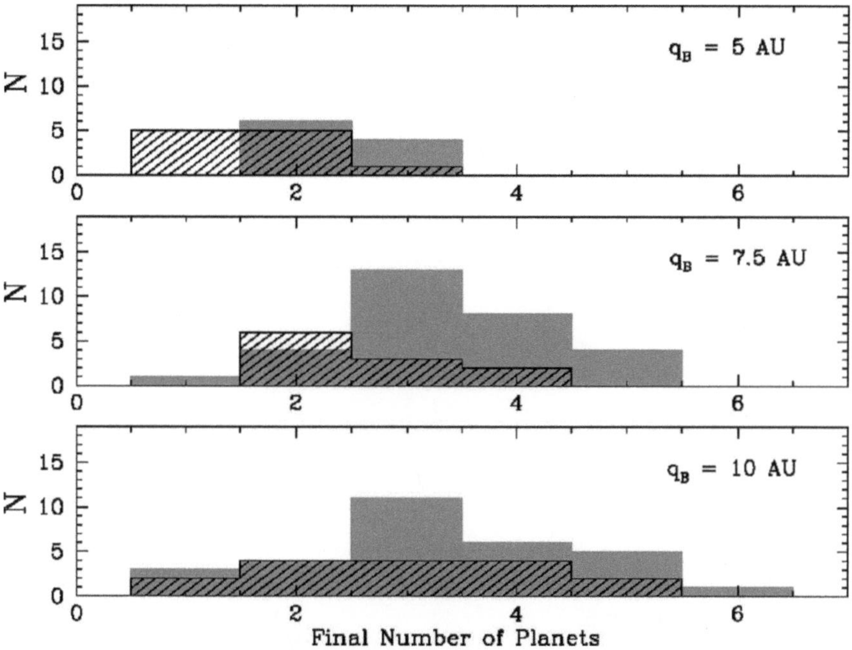

Fig. 10.5 Distributions of the number of final planets formed for binary star systems with $q_B = 5$ AU (*top panel*), $q_B = 7.5$ AU (*middle panel*), and $q_B = 10$ AU (*bottom panel*). The bar types correspond to the different sets of runs as described in Fig. 10.4. The typical number of final planets clearly increases in systems with larger stellar periastra, and also when the stars are less massive (assuming equal-mass stars)

Set B (shown with gray bars) than in Set A (dashed bars). Figure 10.4 shows a clear trend: as the binary periastron increases, the distribution of semimajor axes of the outermost planet becomes wider and its expectation value shifts to larger values. The distributions of the total number of final planets formed are shown in Fig. 10.5 for simulations with $q_B = 5$ AU, 7.5 AU, and 10 AU. In general, a smaller binary periastron results in a larger percentage of mass-loss, and a smaller number of final planets. In our ensemble of more than 100 wide binary star simulations with equal-mass stars, from 1 to 6 planets formed with semimajor axes smaller than or equal to 2.2 AU in binary systems with $q_B = 10$ AU, from 1 to 5 planets formed within 1.7 AU for systems with $q_B = 7.5$ AU, and from 1 to 3 planets formed within 0.9 AU when $q_B = 5$ AU. In the $q_B = 7.5$ AU set with equal mass stars of 1 $M_\odot$, an average of 2.8 planets formed in our largest set of 30 integrations. The distribution extends slightly farther out when the stars are more massive relative to the disk.

Figure 10.6 shows the distribution of final planetary masses (in units of the Earth's mass, $M_\oplus$) formed in systems with $q_B = 5$ AU, 7.5 AU, and 10 AU. The median mass of the final planets doesn't depend greatly on q_B. This result suggests that planet formation remains quite efficient in the stable regions, but that the size of the stable region shrinks as q_B gets smaller. This trend is consistent with the decline

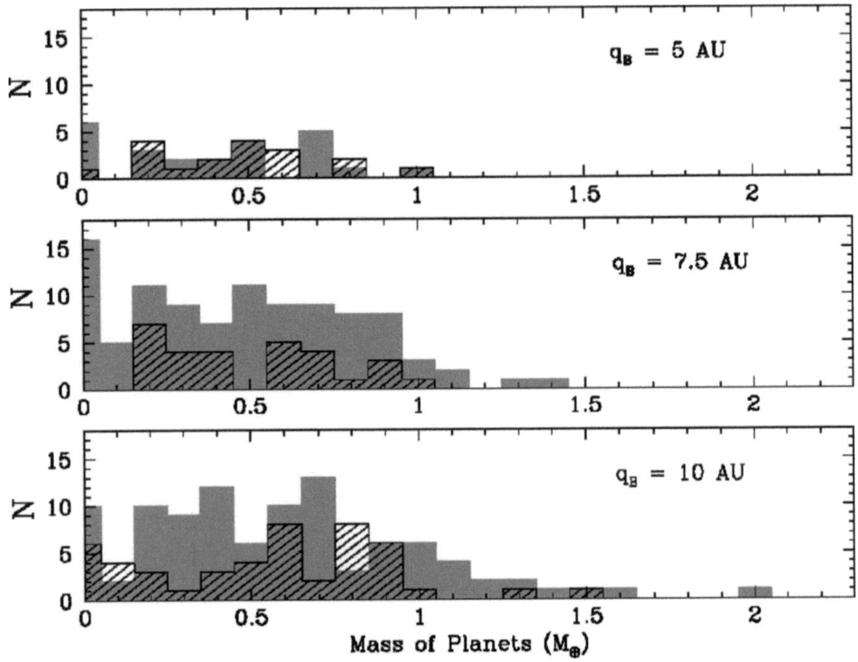

Fig. 10.6 Distributions of the final masses of planets formed around one star within binary star systems with $q_B = 5$ AU (*top panel*), $q_B = 7.5$ AU (*middle panel*), and $q_B = 10$ AU (*bottom panel*). The bar types correspond to the different sets of runs as described in Fig. 10.4. Although the size of the stable region shrinks as q_B gets smaller, the median mass of the final planets does not vary greatly for a given q_B, suggesting that planet formation remains efficient in the stable regions

in the number of planets seen in Fig. 10.5. When the periastron value becomes as small as 5 AU, planets only form within 1 AU, and the mass distribution tilts toward planets with masses $m_p < M_\oplus$, i.e., the formation of Earth-like planets is compromised.

10.5 P-type Orbits Within Close Binary Star Systems

We also performed simulations of the late stages of terrestrial planet formation within a circumbinary disk surrounding various 'close' short-period binary star systems. The combined mass of the binary stars is set equal to 1 $M_\odot$ in all of these simulations, with Jupiter-like and Saturn-like planets perturbing the system. This allows us to compare the effects from the binary stars to those from a single 1 $M_\odot$ star on the same initial disk. We examine binary systems with the stellar mass-ratio μ set equal to either 0.2 or 0.5. Binary star semimajor axes range from $a_B = 0.05$ AU – 0.4 AU, while e_B begins at 0, 1/3, 0.5, or 0.8. We examine various combinations of (a_B, e_B) such that the stellar apastron (the maximum separation of the stars) $Q_B \equiv a_B(1 + e_B)$ is 0.05 AU $\leq Q_B \leq$ 0.4 AU. For most of the simulations, the midplane of the circumbinary disk begins coplanar to the stellar orbit, but for one set of binary star parameters a relative inclination of $i = 30°$ is investigated. The initial circumbinary disk of embryos and planetesimals is the same as that used for simulating accretion within our Solar System (Chambers 2001), in the α Centauri AB system (Section 10.3), and in the wide binary star systems presented above (Section 10.4). We use a 'close-binary' algorithm which follows the accretion evolution of each body in the disk relative to the center of mass of the binary star system. To account for the stochastic nature of these simulations, each binary star system under study is simulated five or six times with slightly different initial conditions for the circumbinary disk. We statistically compared our results to a large set (more than 30) of simulations of the Sun–Jupiter–Saturn system that began with (virtually the same) initial disk mass distribution (Quintana et al. 2006).

Figure 10.7 shows the evolution of the circumbinary disk centered around two 0.5 $M_\odot$ binary stars with $a_B = 0.1$ AU and in a circular orbit. The initial disk is the same as that shown in the first panel of Fig. 10.1, but the locations of the circles show the orbital semimajor axes and eccentricities of the represented bodies relative to center of mass of the binary stars. The perturbations on the inner edge of the disk are apparent within the first million years, and in this case five terrestrial planets have formed within 100 Myr, and continue on stable orbits for the remainder of the 500 Myr integration. Figures 10.8 and 10.9 show the results from two simulations with equal (0.5 $M_\odot$) mass stars with larger apastron values. In Fig. 10.8 the stellar

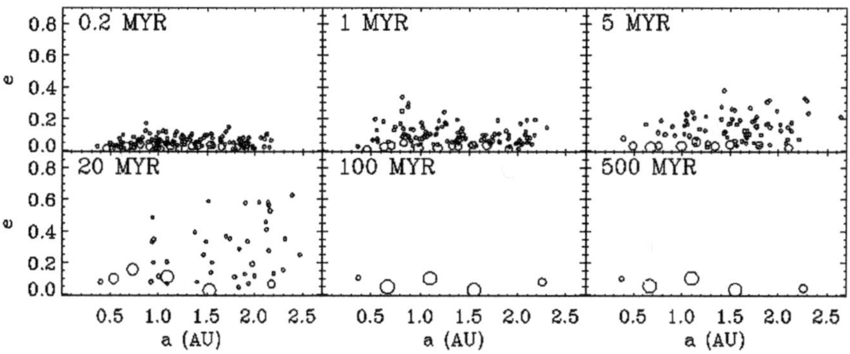

Fig. 10.7 The temporal evolution of the circumbinary disk around binary stars with $a_B = 0.1$ AU, $e_B = 0$, and equal-mass stars of $M_* = 0.5\ M_\odot$ [simulation CB_.1_0_.5_c in Quintana et al. (2006)]. Jupiter- and Saturn-like planets are also included. The initial disk is the same as the disk shown in the first panel of Fig. 10.1, and the embryos and planetesimals are represented by *circles* whose sizes are proportional to the physical sizes of the bodies. The locations of the circles show the *orbital semimajor axes* and eccentricities of the represented bodies relative to center of mass of the binary stars. The binary stars perturb, but do not truncate, the inner edge of the disk. The initially dynamically cold disk is excited during the first 10 Myr, especially in the outer region where giant planet perturbations are the greatest. Approximately 12% of the initial disk mass was lost from this system, and by 100 Myr into the simulation, five planets on low eccentricity orbits have formed and survived for the remainder of the simulation

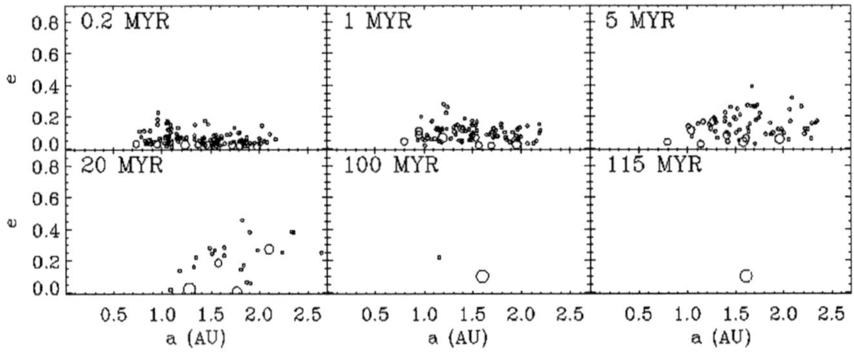

Fig. 10.8 The temporal evolution of a simulation that began with a larger stellar separation ($a_B = 0.2$ AU) and a higher binary eccentricity ($e_B = 0.5$ AU) than the system shown in Fig. 10.7 [simulation CB_.2_.5_.5_d in Quintana et al. (2006)]. The binary system (which has an apastron of $Q_B = 0.3$) is composed of equal-mass stars of $M_* = 0.5\ M_\odot$, and Jupiter-like and Saturn-like planets are included. The initial disk is the same as that shown in the *first panel* of Fig. 10.1, and the symbols are the same as those described in Fig. 10.7. The effects of the binary stars are strong early in the simulation, and the inner edge of the disk is truncated at ~0.7 AU. Approximately 79% of the initial disk mass was lost from this system, and after the last ejection at 115 Myr, only one planet remained in the system at ~1.6 AU

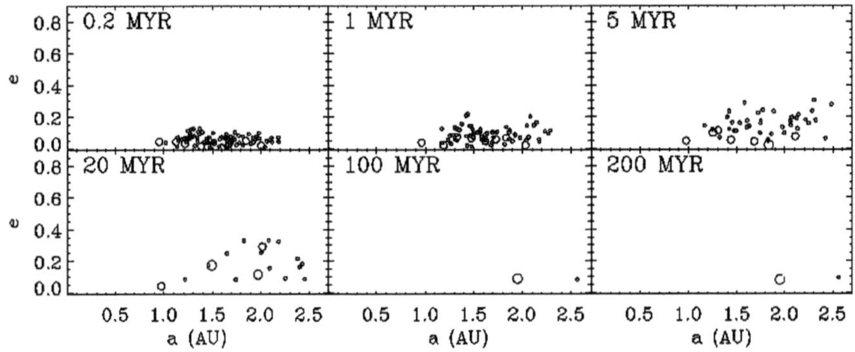

Fig. 10.9 The temporal evolution of the circumbinary disk around binary stars with $a_B = 0.3$ AU, $e_B = 1/3$, $\mu = 0.5$, and $Q_B = 0.4$ [simulation CB_.3_.33_.5_d in Quintana et al. (2006)]. The effect of the stellar companion is apparent in the first panel where the inner part of the disk is already substantially excited and truncated at ~0.8 AU. Eccentricities remain high throughout the evolution, and by 100 Myr only one planet more massive than the planet Mercury has formed in the terrestrial planet zone, with a single planetesimal remaining just beyond 2.5 AU. In this simulation, approximately 91% of the initial disk mass was lost, with 70% of the initial disk mass lost within the first 50 Myr

separation is $a_B = 0.2$ AU and $e_B = 0.5$ ($Q_B = 0.3$), and in Fig. 10.9 the stars are separated by $a_B = 0.3$ AU and have an eccentricity of $e_B = 1/3$ ($Q_B = 0.4$). In both simulations, more than 70% of the initial disk mass was lost from the system within the first 50 Myr, and each simulation resulted in a single planet more massive than the planet Mercury in our Solar System.

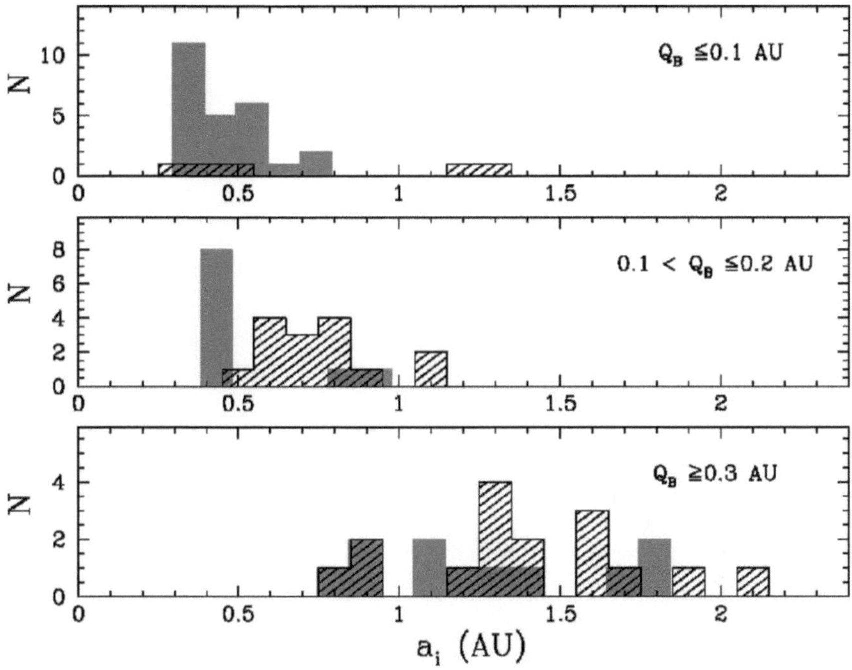

Fig. 10.10 Distributions of the semimajor axis of the innermost final planet, a_i, formed in binary star systems with $Q_B \leq 0.1$ AU (*top panel*), $0.1 < Q_B \leq 0.2$ AU (*middle panel*), and $Q_B \geq 0.3$ AU (*bottom panel*). The *gray bars* represent simulations in which the binary stars began on circular ($e_B = 0$) orbits, and the *dashed bars* represent systems with $1/3 \geq e_B \geq 0.8$. The binary stars truncate the inner edge of the initial planetesimal/embryo disk in all simulations with $Q_B \geq 0.3$ AU. Terrestrial planets (more massive than the planet Mercury) were able to form within 1 AU from the binary stars' center of mass in most simulations with $Q_B \leq 0.2$ AU, and in simulations with $a_B = 0.2$ AU, $e_B = 0.5$, and $\mu = 0.2$ ($Q_B = 0.3$), as well as in some simulations with $a_B = 0.3$ AU and $e_B = 0$ ($Q_B = 0.3$)

Figures 10.10 to 10.12 show distributions of the semimajor axis of the innermost planet, the number of final planets, and the final masses of the planets, respectively, that formed around binary stars with $Q_B \leq 0.1$ AU (top panels), 0.1 AU $< Q_B \leq$ 0.2 AU (middle planets), and $Q_B \geq 0.3$ AU (lower panels), respectively. Simulations with binary stars on initially circular orbits are shown with gray bars, and binary stars with higher eccentricities ranging from 1/3 to 0.8 are shown with dashed bars. At least one planet (more massive than the planet Mercury) formed in all of the simulations, although fewer and more diverse systems formed around binary stars with larger apastron values. Earth-mass planets within 1 AU formed in most simulations in which $Q_B \leq 0.2$ AU, and also in simulations with $Q_B = 0.3$ AU where $(a_B, e_B) = (0.3, 0)$ and $(0.2, 0.5)$ with $\mu = 0.2$ in the case of the latter. Planets did not form within 1 AU, however, in systems with $Q_B = 0.3$ AU and $(a_B, e_B) = (0.2, 0.5)$ when the binary stars had equal masses ($\mu = 0.5$). Note that all of our results are consistent with the circumbinary orbital stability criteria of Holman and Wiegert (1999).

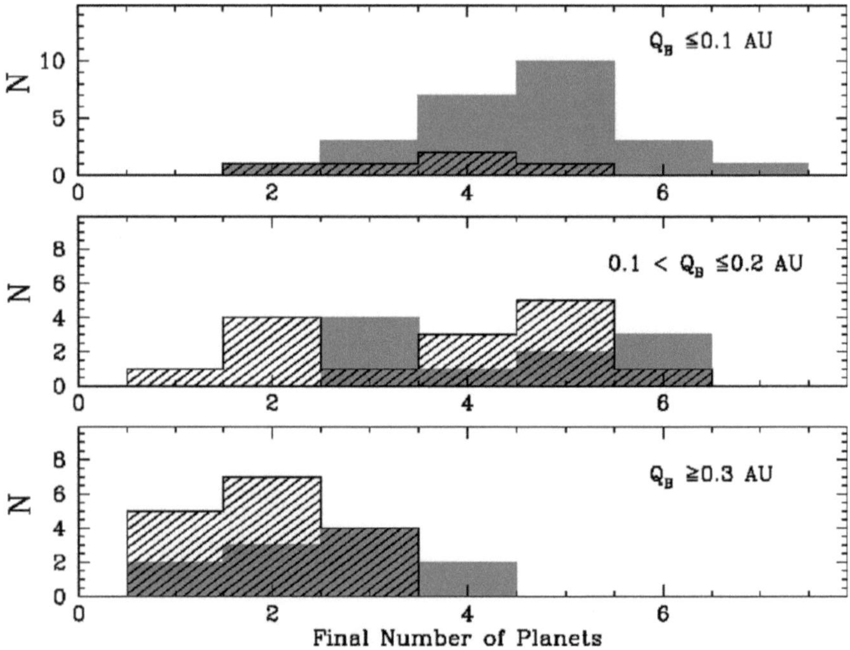

Fig. 10.11 Distributions of the number of final planets formed around close binary star systems with $Q_B \leq 0.1$ AU (*top panel*), $0.1 < Q_B \leq 0.2$ AU (*middle panel*), and $Q_B \geq 0.3$ AU (*bottom panel*). The *bar types* correspond to the different sets of runs as described in Fig. 10.10. Fewer planets tend to form around binary star systems with larger apastron values, as expected. Furthermore, in the binary systems with higher values of Q_B, fewer planets tend to form around binaries with $e_B > 0$ than around binary stars on circular orbits ($e_B = 0$)

In summary, close binary stars with maximum separations $Q_B \leq 0.2$ AU and small e_B had little effect on the accreting bodies, and in most of these simulations terrestrial planets formed over essentially the entire range of the initial disk mass distribution (and even beyond 2 AU in many cases). The stellar perturbations cause orbits to precess, thereby moving secular resonances out of the region equivalent to the solar system's inner asteroid belt, allowing terrestrial planets to form from our initially compact disk and remain in stable orbits as far as 2.98 AU from the center of mass of the binary stars. The effects of the stellar perturbations on the inner edge of the planetesimal disk became evident in systems with $e_B = 0$ and $Q_B > 0.2$ AU, and in most of the simulations with $e_B > 0$. Terrestrial-mass planets can still form around binary stars with high eccentricity values, but the planetary systems tend to be sparser and more diverse. Binary stars with $Q_B \gtrsim 0.3$ AU perturb the accreting disk such that the formation of Earth-like planets near 1 AU is unlikely. Despite these constraints, at least one terrestrial planet formed in each of our simulations, and those that formed around binary stars with $Q_B \leq 0.2$ AU were statistically consistent (in mass, final number, and orbital characteristics) to final planet systems that formed in our large set of Sun–Jupiter–Saturn simulations.

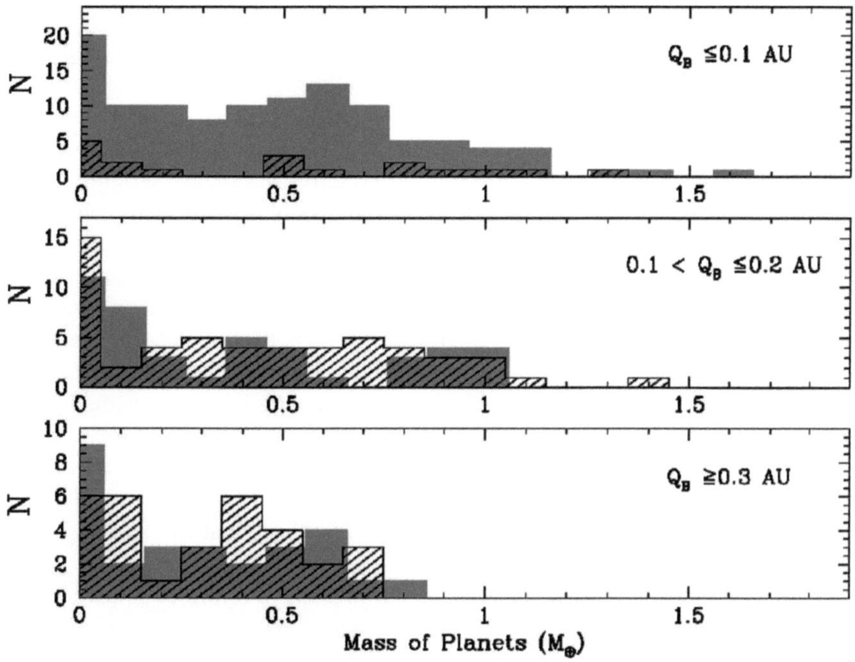

Fig. 10.12 Distributions of the final masses of planets formed around close binary star systems with $Q_B \leq 0.1$ AU (*top panel*), $0.1 < Q_B \leq 0.2$ AU (*middle panel*), and $Q_B \geq 0.3$ AU (*bottom panel*). The *bar types* correspond to the different sets of runs as described in Fig. 10.10. Although the inner edge of the disk is truncated around binary stars with larger values of Q_B, the median mass of the final planets that form does not vary greatly for a given Q_B, showing that (similar to the results of our wide binary calculations) planet formation remains efficient in the regions of stability

10.6 Conclusions

Our exploration of binary star parameter-space provides rough constraints on where and what types of terrestrial planets may form within binary star systems. Note that statistics of the final planetary systems, and also their long term dynamical stability, for nearly all of the individual simulations that we performed are presented in Quintana et al. (2002); Quintana (2003, 2004); Lissauer et al. (2004); Quintana et al. (2006); Quintana et al. (2007). We find that the presence of a binary companion of order 10 AU away acts to limit the number of terrestrial planets and the spatial extent of the terrestrial planet region around one member of a binary star system, as shown by Figs. 10.4 to 10.6. To leading order, the periastron value q_B is the most important parameter in determining binary effects on planetary outcomes in wide binary star systems (more predictive than a_B or e_B alone), whereas Q_B is the most influential parameter for accretion within circumbinary disks. For a given binary periastron q_B, fewer planets tend to form in wide binary systems with larger values of (a_B, e_B), as shown in Fig. 10.5.

Binary companions also limit the extent of the terrestrial planet region in nascent planetary systems orbiting one member of the stellar pair. As shown in Figs. 10.4 to 10.6, wider binaries allow for larger systems of terrestrial planets. Although the binary periastron is the most important variable in determining the extent of the final system of terrestrial planets (as measured by the semimajor axis of the outermost planet), for a given periastron, the sizes of the terrestrial planet systems show a wide distribution. In these simulations, the initial disk of planetesimals extends out to 2 AU, so we do not expect terrestrial planets to form much beyond this radius. For binary periastron $q_B = 10$ AU, the semimajor axis of the outermost planet typically lies near 2 AU, i.e., the system explores the entire available parameter-space for planet formation. Since these results were obtained with equal-mass stars (including those with $M = 1.0\ M_\odot$), we conclude that the constraint $q_B \gtrsim 10$ AU is sufficient for binaries to allow the formation of terrestrial planet systems like that in our solar system. With smaller binary periastron values, the resulting extent of the terrestrial planet region is diminished. When the binary periastron decreases to 5 AU, the typical planetary system extends only out to $\sim$0.75 AU and no system has a planet with semimajor axis beyond 0.9 AU (but note that we did not perform simulations with $q_B = 5$ AU and small e_B).

While the number of forming planets and their range of orbits is restricted by binary companions, the masses and eccentricities of those planets are much less affected (at least during the late stages of planetary growth, the only portion of planetary accumulation that our simulations address). The distribution of planet masses is nearly independent of binary periastron (see Fig. 10.6), although the wider binaries allow for a few slightly more massive terrestrial planets to form. Finally, we note that the timescales required for terrestrial planet formation in these systems lie in the range of approximately 20 to 200 Myr, consistent with previous findings (Chambers 2001; Quintana 2004), and are generally shorter in binary star systems with smaller q_B (see Fig. 10.13).

Whitmire et al. (1998) analyzed the effects of perturbations by a binary companion on planetesimals during the earlier stages of planetary growth. Assuming that collisions at velocities larger than 100 m/s disrupt planetesimals, they found that if two solar-mass stars have a periastron smaller than 16 AU, then planetary growth at 1 AU is inhibited. This criterion is more limiting than the results of our simulations suggest, but note that the model by Whitmire et al. (1998) does not include gas. Perturbations by a gaseous disk can align planetesimal orbits, reducing collision velocities and thereby allowing growth to proceed and produce bodies of the sizes that we use as initial conditions over a larger range of semimajor axis (Kortenkamp and Wetherill 2000).

This work has important implications regarding the question of what fraction of stars might harbor terrestrial planetary systems. The majority of solar-type stars live in binary systems, and as shown in this chapter, binary companions can disrupt both the formation of terrestrial planets and their long-term prospects for stability. Approximately half of the known binary systems are wide enough (in this context, having sufficiently large values of periastron) so that Earth-like planets can remain stable over the entire 4.6 Gyr age of our solar system

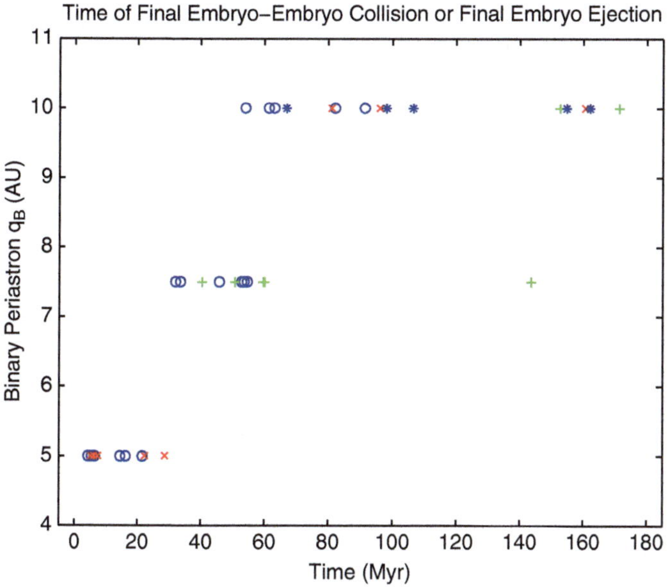

Fig. 10.13 The elapsed simulation time of the last major event in each wide binary star simulation (with equal mass stars of 0.5 $M_\odot$), where the event is either the final ejection of a planet that has grown more massive than the planet Mercury, or the final collision among two planets that are each as massive as Mercury. This final ejection/collision time (in Myr) is shown for the three values of binary periastron q_B that we investigated: $q_B = 5, 7.5$, and 10 AU. The simulations with binary eccentricity $e_B = 0$ are shown with *blue asterisk* symbols, simulations with $e_B = 0.25$ are shown with *green* + symbols, integrations performed with $e_B = 0.5$ are shown with *red x* symbols, and simulations with the largest eccentricities $e_B > 0.5$ are shown with *blue circle* symbols. The timescales required for terrestrial planet accretion around one member of a binary star system typically decrease in systems with smaller q_B, and (for a given q_B) are generally shorter in binary systems with larger values of e_B

(David et al. 2003; Fatuzzo et al. 2006). For the system to be stable out to the distance of Mars's orbit, the binary periastron must be greater than about 7 AU, and about half of the observed binaries have such periastra. Our work on the formation of terrestrial planets shows similar trends. When the periastron of the binary is larger than about 10 AU, even for the case of equal-mass stars, terrestrial planets can form over essentially the entire range of orbits allowed for single stars with a Jupiter-mass planet at 5.2 AU (out to the edge of the initial planetesimal disk at 2 AU). When periastron is smaller than 10 AU, however, the distributions of planetary orbital parameters are strongly affected by the presence of the binary companion. Specifically, the number of terrestrial planets and the spatial extent of the terrestrial planet region both decrease with decreasing binary periastron. When the periastron becomes as small as 5 AU, planets form only within 1 AU, thus the formation of Earth-like planets near 1 AU is inhibited. Terrestrial planet formation has also been numerically studied around solar-type binary star systems with larger periastra than the binary systems presented in this chapter ($a_B > 20$ AU) (Haghighipour 2006;

Haghighipour and Raymond 2007), which allow the stability of Jupiter-like planets near 5 AU. These binary star/giant planets systems can be effective in perturbing and delivering volatiles to the inner terrestrial region (Haghighipour and Raymond 2007).

Given the enormous range of orbital parameter-space sampled by known binary systems, from contact binaries to separations of nearly a parsec, the range of stellar orbits where terrestrial planet formation is affected is only a little narrower than the range where the stability of Earth-like planets is compromised. As a result, approximately 40–50% of binaries are wide enough to allow both the formation and the long-term stability of Earth-like planets in S-type orbits encircling one of the stars. Furthermore, approximately 10% of main sequence binaries are close enough to allow the formation and long-term stability of terrestrial planets in P-type circumbinary orbits (David et al. 2003; Quintana et al. 2006). Given that the galaxy contains more than 100 billion star systems, and that roughly half remain viable for the formation and maintenance of Earth-like planets, a large number of systems remain habitable based on the dynamic considerations of this research.

Acknowledgements Primary support for this research was provided by the NASA Astrobiology Institute under the NASA Ames Investigation "Linking our Origins to our Future". We are grateful for the contributions from our collaborators Fred C. Adams, John E. Chambers, and Martin J. Duncan, and we thank Michael J. Way for providing additional CPUs at NASA ARC. E.V.Q. received support in various stages of this research from NASA GSRP, NAS/NRC and NASA NPP fellowships, and the University of Michigan through the Michigan Center for Theoretical Physics (MCTP).

References

Barbieri, M., Marzari, F., & Scholl, H. 2002, A&A, 396, 219
Bodenheimer, P., Hubickyj, O., & Lissauer, J. J. 2000, Icarus, 143, 2
Chambers, J. E. 2001, Icarus, 152, 205
Chambers, J. E., Quintana, E. V., Duncan, M. J., & Lissauer, J. J. 2002, AJ, 123, 2884
David, E., Quintana, E. V., Fatuzzo, M., & Adams, F. C. 2003, PASP, 115, 825
Duquennoy, A., & Mayor, M. 1991, A&A, 248, 485
Fatuzzo, M., Adams, F. C., Gauvin, R., & Proszkow, E. M. 2006, PASP, 118, 1510
Haghighipour, N. 2006, ApJ, 644, 543
Haghighipour, N., & Raymond, S. N. 2007, ApJ, 666, 436
Holman, M. J., & Wiegert, P. A. 1999, AJ, 117, 621
Kokubo, E., & Ida, S. 1998, Icarus, 131, 171
Kortenkamp, S. J., & Wetherill, G. W. 2000, Icarus, 143, 60
Lissauer, J. J. 1993, ARA&A, 31, 129
Lissauer, J. J., Quintana, E. V., Chambers, J. E., Duncan, M. J., & Adams, F. C. 2004, RevMexAA, 22, 99
Marzari, F., & Scholl, H. 2000, ApJ, 543, 328
Mathieu, R. D., Ghez, A. M., Jensen, E. L. N., & Simon M. 2000, in Protostars and Planets IV, ed. Mannings, V., Boss, A. P., & Russell, S. S. (Tucson: Univ. of Arizona Press), 703
Quintana, E. V., Lissauer, J. J., Chambers, J. E., & Duncan, M. J. 2002, ApJ, 576, 982
Quintana, E. V. 2003, in ASP Conf. Ser. 294, Scientific Frontiers in Research on Extrasolar Planets, ed. Deming, D., & Seager, S. (San Francisco: ASP), 319

Quintana, E. V. 2004, Planet Formation in Binary Star Systems, Thesis (Ph.D.), University of Michigan, Ann Arbor (Ann Arbor: UMI Company)
Quintana, E. V., & Lissauer, J. J. 2006, Icarus, 185, 1
Quintana, E. V., Adams, F. C., Lissauer, J. J., & Chambers, J. E. 2007, ApJ, 660, 807
Safronov, V. S., 1969. Evolution of the Protoplanetary Cloud and Formation of the Earth and the Planets (Moscow: Nauka Press)
Thebault, P., Marzari, F., Scholl, H., Turrini, D., & Barbieri, M. 2004, A&A, 427, 1097
Weidenschilling, S. J. 1977, A&SS, 51, 153
Whitmire, D. P., Matese, J. J., Criswell, L., & Mikkola, S. 1998, Icarus, 132, 196
Wiegert, P. A., & Holman, M. J. 1997, AJ, 113, 1445

Chapter 11
Planetary Dynamics and Habitable Planet Formation in Binary Star Systems

Nader Haghighipour, Rudolf Dvorak, and Elke Pilat-Lohinger

11.1 Introduction

How our planet was formed, how life came about, and whether life exists elsewhere in the universe are among some of the long-standing questions in human history. The latter, which has been the main drive behind many decades of searching for planets outside the solar system, is one of the most outstanding problems in planetary science and astrobiology. Although no Earth-like planet has yet been found, the success of observational techniques in identifying now more than 350 extrasolar planets has greatly contributed to addressing this question, and has extended the concept of habitability to billions of miles beyond the boundaries of our solar system. It is now certain that our planetary system is not unique and many terrestrial-size planets may exist throughout the universe.

The orbital and physical diversity of the currently known extrasolar planets play a crucial role in their habitability. In general, whether a planet can be habitable depends on its physical and dynamical properties, and the luminosity of its host star. The notion of habitability is normally defined based on the life as we know it, and uses the physical and orbital characteristics of Earth as an example of a habitable planet. In other words, a planet is habitable if it is Earth-like so it can develop and sustain Earthly life. This definition of habitability requires that a potentially habitable planet to maintain liquid water on its surface and in its atmosphere. The planet's capability in maintaining water is determined by its size and orbital motion, the luminosity of the central star, and the distribution of water in the circumstellar material from which the planet was formed.

How extrasolar habitable planets are formed is a widely addressed question that is still unresolved. While models of planetary accretion in the inner solar system

N. Haghighipour (✉)
Institute for Astronomy and NASA Astrobiology Institute, University of Hawaii-Manoa, Honolulu 96822, Hawaii
e-mail: nader@ifa.hawaii.edu

R. Dvorak and E. Pilat-Lohinger
Institute for Astronomy, Türkenschanzstraße 17, University of Vienna, Vienna 1180, Austria
e-mail: dvorak@astro.univie.ac.at; lohinger@astro.univie.ac.at

N. Haghighipour (ed.), *Planets in Binary Star Systems*, Astrophysics and Space Science Library 366, DOI 10.1007/978-90-481-8687-7_11,

present pathways (although in some cases incomplete) toward the formation of planets such as Earth and Venus, the orbital diversity of extrasolar planets present strong challenges to the applicability of these models to other planetary environments. For instance, systems with close-in giant planets may require massive protoplanetary disks to ensure that while planetesimals and protoplanets are scattered as giant planets migrate, terrestrial bodies can form and be stable. Systems with multiple planets also present a great challenge to terrestrial planet formation since the orbital architectures of such systems may limit the regions of the stability of smaller objects.

In a system with two stars, the situation is even more complicated. The interaction between one star and the protoplanetary disk around the other may inhibit planet formation by truncating the disk and removing circumstellar material (Artymowicz and Lubow 1994). This interaction may also prevent the growth of km-size planetesimals to larger objects by increasing the relative velocities of these bodies and causing their collisions to result in fragmentation. Despite such difficulties, planets have, however, been detected in binary star systems (see Table 11.1) and observers have been able to identify three moderately close (<20 AU) binaries, namely γ Cephei (Hatzes et al. 2003), GL 86 (Eggenberger et al. 2001), and HD 41004 (Zucker et al. 2004), whose primary stars are hosts to Jupiter-like planets.

The detection of planets in binary star systems is not a surprise. There is much observational evidence that indicates the most common outcome of the star formation process is a binary system (Mathieu 1994; White and Ghez 2001). Also, as shown by Prato and Weinberger in the first chapter, there is substantial evidence for the existence of potentially planet-forming circumstellar disks in multiple star systems (Mathieu 1994; Akeson et al. 1998; Rodriguez et al. 1998; White et al. 1999; Silbert et al. 2000; Mathieu et al. 2000; Trilling et al. 2007). These all point to the fact that planet formation in binaries is robust and many of these systems may harbor additional giant planets and/or terrestrial-size objects. This chapter is devoted to study the latter.

Whether binaries can harbor potentially habitable planets depends on several factors including the physical properties and the orbital characteristics of the binary system. While the former determines the location of the habitable zone (HZ), the latter affects the dynamics of the material from which terrestrial planets are formed (i.e., planetesimals and planetary embryos), and drives the final architecture of the planets assembly. In order for a habitable planet to form in a binary star system, these two factors have to work in harmony. That is, the orbital dynamics of the two stars and their interactions with the planet-forming material have to allow terrestrial planet formation in the habitable zone, and ensure that the orbit of a potentially habitable planet will be stable for long times. We organize this chapter with the same order in mind. We begin in Section 11.2 by presenting a general discussion on the motion of planets in binary stars and their stability. Section 11.3, has to do with the stability of terrestrial planets, and in Section 11.4, we discuss habitability and the formation of potentially habitable planets in a binary-planetary system.[1]

[1] The phrase "binary-planetary system" is used to identify binary star systems in which one of the stars is host to a giant planet.

Table 11.1 Planets in double stars (Raghavan et al. 2006)

Star	a_b (AU)	a_p (AU)	$M_p \sin i\,(M_{Jup})$	e_p
HD38529	12,042	0.129	0.78	0.29
		3.68	12.7	0.36
HD40979	6,394	0.811	3.32	0.23
HD222582	4,746	1.35	5.11	0.76
HD147513	4,451	1.26	1.00	0.52
HD213240	3,909	2.03	4.5	0.45
Gl 777 A	2,846	0.128	0.057	0.1
		3.92	1.502	0.36
HD89744	2,456	0.89	7.99	0.67
GJ 893.2	2,248	0.3	2.9	–
HD80606	1,203	0.439	3.41	0.927
55 Cnc	1,050	0.038	0.045	0.174
		0.115	0.784	0.02
		0.24	0.217	0.44
		5.25	3.92	0.327
GJ 81.1	1,010	0.229	0.11	0.15
		3.167	0.7	0.3
16 Cyg B	860	1.66	1.69	0.67
HD142022	794	2.8	4.4	0.57
HD178911	789	0.32	6.292	0.124
Ups And	702	0.059	0.69	0.012
		0.83	1.89	0.28
		2.53	3.75	0.27
HD188015	684	1.19	1.26	0.15
HD178911	640	0.32	6.29	0.124
HD75289	621	0.046	0.42	0.054
GJ 429	515	0.119	0.122	0.05
HD196050	510	2.5	3.00	0.28
HD46375	314	0.041	0.249	0.04
HD114729	282	2.08	0.82	0.31
ϵ Ret	251	1.18	1.28	0.07
HD142	138	0.98	1.00	0.38
HD114762	132	0.3	11.02	0.25
HD195019	131	0.14	3.43	0.05
GJ 128	100	1.30	1.8	0.15
HD120136	240	0.05	4.13	0.01
γ Cep	20.3	2.03	1.59	0.2
GL 86	21	0.11	4.01	0.046
HD41004 AB	23	1.7	2.64	0.5
HD196885	17.23	2.63	2.96	0.462

11.2 Planetary Motion in Binary Systems and Stability

As mentioned earlier, dynamical stability is essential to the habitability of a planetary system. This issue is particularly important in binary star systems since the gravitational perturbation of the stellar companion limits stable planetary orbits around the other star to only certain regions of the phase-space. In this section, we discuss this issue in more detail.

Study of the stability of small bodies in binary stars has a long history in planetary science. Among some of the early works are the papers by Graziani and Black (1981), Black (1982), and Pendleton and Black (1983), where the authors studied the stability of a planet around a star of a binary within the framework of a general three-body system, and showed that in a binary with equal-mass stars, the orbital stability of the planet is independent of its orbital inclination (also see paper by Harrington (1977), and articles by Innanen et al. (1997) and Musielak et al. (2005) for more recent works on this subject).

The notion of stability has been discussed by different authors within different contexts. In a review article in 1984, Szebehely introduced 50 definitions for the stability of a planetary orbit in a multi-body system (Szebehely 1984). For instance, while Harrington (1977) considers an orbit stable if the semimajor axis and eccentricity of the object do not undergo secular changes, Szebehely (1980, 1981) defines orbital stability based on the integrals of motion and curves of zero velocity. Within the context of habitability, an object is stable if it has the capability of maintaining its orbital parameters (i.e., semimajor axis, eccentricity, and inclination) at all times. In other words, an object is stable if small variations in its orbital parameters do not progress exponentially, but instead vary sinusoidally. Instability occurs when a perturbative force causes drastic changes in the orbital parameters of the object so that it leaves the gravitational field of the system, or collides with other bodies.

The stability of a planetary orbit in dual-star systems depends also on the type of its orbit. From a dynamical point of view, three types of motion are recognized in double-star systems (see Figs. 11.1 and 11.2, and Dvorak 1984):

1. The S-type (or the satellite-type), where the planet moves around one stellar component
2. The P-type (or the planet-type), where the planet surrounds both stars in a distant orbit
3. The L-type (or the libration-type), where the planet moves in the same orbit as the secondary (i.e., locked in a 1:1 mean motion resonance), but 60° ahead or behind[2]

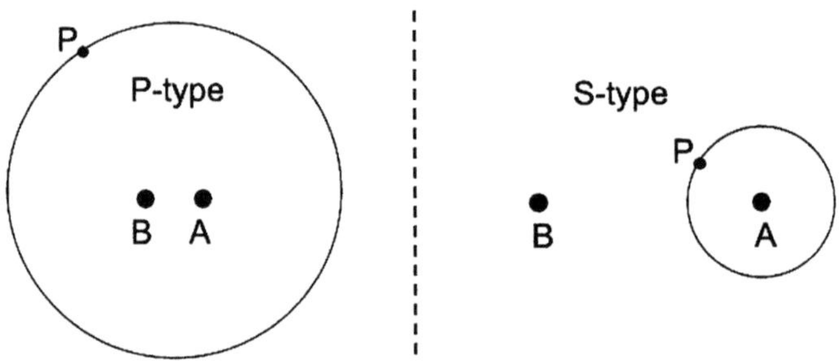

Fig. 11.1 S-type and P-type binary-planetary systems. A and B represent the stars of the binary, and P depicts the planet

[2] An earlier classification by Szebehely (1980) divides the planetary orbits in binary systems into three categories: *inner* orbit, where the planet orbits the primary star, *satellite* orbit, where it orbits the secondary star, and the *outer* orbit, where the planet orbits the entire binary system.

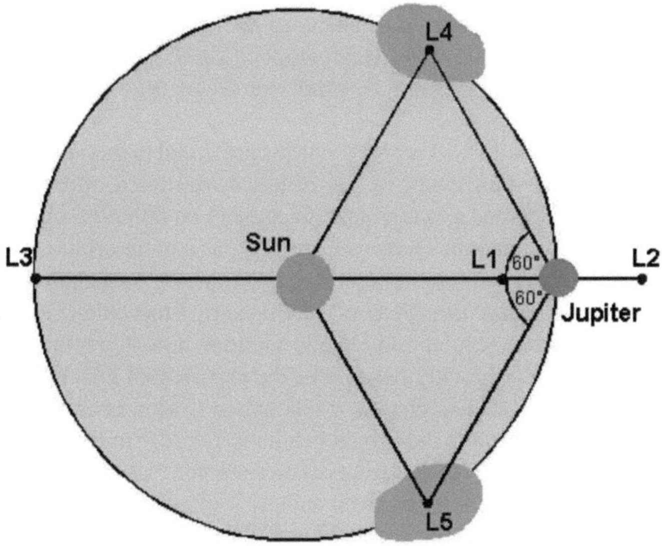

Fig. 11.2 Schematic view of the stable region around the Lagrange points L_4 and L_5 in the restricted three body problem with the Sun and Jupiter as primary bodies

Within the framework of elliptical restricted three-body problem (ER3BP),[3] many authors have studied the stability of planets in binaries for different types of above-mentioned planetary orbits (Dvorak 1984, 1986; Rabl and Dvorak 1988; Dvorak et al. 1989; Benest 1988a,b, 1989, 1993, 1996, 1998; Dvorak and Lohinger 1991; Lohinger and Dvorak 1993; Wiegert and Holman 1997; Holman and Wiegert 1999; Pilat-Lohinger and Dvorak 2002). However, because until 2003, no planet had been detected in or around a double star, the applicability of the results of these studies were only to hypothetical systems. The discovery of the first planet in a moderately close binary by Hatzes et al. (2003) changed this trend and encouraged many researcher to revisit this problem and explore the stability of planets in binaries by considering more realistic cases (see, for instance, Innanen et al. 1997; Pilat-Lohinger et al. 2003; Dvorak et al. 2003a,b, 2004b; Musielak et al. 2005; Haghighipour 2006).

In this chapter, we focus on the stability and habitability of planets in S-type orbits. As shown in Table 11.1, all the currently known planets in binary systems (regardless of the separation of the binary) are of this kind. We present the results of the studies of the general stability of S-type orbits and discuss their application to real binary systems, in particular the system of γ Cephei. Since the discovery of a giant planet around the primary of this double star (Hatzes et al. 2003), many studies have been done on the stability and habitability of this binary and the possibility of

[3] In an elliptical restricted three-body problem, the planet is considered to be a massless particle and its motion is studied in the gravitational field of two massive stars. The stars of the binary revolve around their center of mass in an unperturbed elliptical Keplerian orbit.

the formation of giant and Earth-like planets around its stellar components (Dvorak et al. 2003a; Thébault et al. 2004; Haghighipour 2005; Verrier and Evans 2006; Torres 2007). We finish this section by briefly reviewing the stability of planets in P-type and L-type orbits.

The numerical simulations of planetary orbits presented in this section are mostly carried out within the framework of the elliptical, restricted, three-body system, where the planet is regarded as a massless object with no influence on the dynamics of the binary. To determine the *character* of the motion of an orbit, we either use a chaos indicator, or carry out long-term orbital integrations. As a chaos indicator, we use the *fast Lyapunov indicator* (FLI) as developed by Froeschlé et al. (1997). FLI can distinguish between regular and chaotic motions in a short time, and chaotic orbits can be found very quickly because of the exponential growth of this vector in the chaotic region. For most chaotic orbits only a few number of primary revolutions is needed to determine the orbital behavior. In order to distinguish between stable and chaotic motions, we define a critical value for FLI which depends on the computation time. This method has been applied to the studies of many extrasolar planetary systems by Pilat-Lohinger and Dvorak (2002), Dvorak et al. (2003a,b); Pilat-Lohinger et al. (2003, 2005); Bois et al. (2003); Érdi and Pál (2003); Sandor et al. (2006).

When carrying out long-term orbital integrations, a fast and reliable characterization of the motion can be achieved by making maps of the maximum eccentricity of the orbit of the planet calculated for each integration of its orbit. The maximum eccentricity maps can be used as a useful indicator of orbital stability, especially for studies of the motion of terrestrial-size planets in the habitable zone of their host stars. Examples of such studies can be found in the works of Dvorak et al. (2003a, 2004b), Funk et al. (2004), Érdi et al. (2004), Asghari et al. (2004), and Pilat-Lohinger et al. (2006).

11.2.1 Stability of S-Type Orbits

The motion of a planet in an S-type orbit is governed by the gravitational force of its host star and the perturbative effect of the binary companion. Since the latter is a function of the distance between the planet and the secondary star, the orbit of the planet will be less perturbed if this distance is large. In other words, a planet in an S-type orbit will be able to maintain its orbit for a long time if it is sufficiently close to its parent star (Harrington 1977). By numerically integrating the motion of a massless object in an S-type orbit, Rabl and Dvorak (1988) (hereafter RD) and Holman and Wiegert (1999) (hereafter HW) have shown that the maximum value of the semimajor axis of a stable S-type orbit varies with the binary mass-ratio, semimajor axis, and eccentricity as,

$$\begin{aligned} a_c/a_b = {} & (0.464 \pm 0.006) + (-0.380 \pm 0.010)\mu + (-0.631 \pm 0.034)e_b \\ & + (0.586 \pm 0.061)\mu e_b + (0.150 \pm 0.041)e_b^2 + (-0.198 \pm 0.047)\mu e_b^2 \,. \end{aligned} \tag{11.1}$$

In this equation, a_c, the *critical* semimajor axis, is the upper limit of the semimajor axis of a stable S-type orbit, a_b and e_b are the semimajor axis and eccentricity of the binary, and $\mu = M_2/(M_1 + M_2)$, where M_1 and M_2 are the masses of the primary and secondary stars, respectively. Figure 11.3 shows the variation of a_c with the binary mass-ratio and eccentricity. As expected, S-type orbits in binaries with larger secondary stars on high eccentricities are less stable. The $\pm$ signs in Eq. (11.1) define a lower and an upper value[4] for the critical semimajor axis which correspond to a transitional region that consists of a mix of stable and unstable orbits. Such a dynamically *gray* area, in which the state of a system changes from stability to instability, is known to exist in multi-body environments, and is a characteristic of any dynamical system. Similar studies have been done by Moriwaki and Nakagawa (2004) and Fatuzzo et al. (2006) who obtained critical semimajor axes slightly larger than given by Eq. (11.1).

It is necessary to mention that in simulations of RD and HW, the initial orbit of the planet was considered to be circular. In a series of numerical integrations (Pilat-Lohinger and Dvorak 2002) (hereafter PLD) considered non-zero values for the initial eccentricity of the planet and by assuming the following initial conditions, they analyzed the influence of the planet's eccentricity on its orbital stability. For the binary, these authors assumed

- A semimajor axis of 1 AU
- An eccentricity between 0 and 0.9 in steps of 0.1
- An initial starting point for the secondary star at either its periastron or apastron

For the planet, which moves around the primary in the same plane as the orbit of the binary (i.e., coplanar orbits), they considered

- A semimajor axis between 0.1 and 0.9 AU
- An initial eccentricity between 0 and 0.5 in increments of 0.1 for all binary mass-ratios
- A starting point with different angular positions (i.e., mean anomaly = 0° or 90° or 180° or 270°)

Figure 11.4 shows a comparison of the results of the three studies by RD, HW, and PLD, in a binary with a mass-ratio of $\mu = 0.2$. In this figure, the value of the critical semimajor axis of the planet is shown for different values of its eccentricity and the eccentricity of the binary. The boundaries of the stability zone corresponding to HW simulations (calculated using Eq. 11.1) are shown in dotted lines. As shown in the top panel of this figure, stability zones of low-eccentricity orbits, as obtained by PLD, are in a good agreement with the results of HW. However, for planets with larger orbital eccentricities, as shown in the lower panel, the size of the stability zone decreases as the eccentricity of the planet increase. The plotted stability boundaries for such orbits fall outside the HW stable zone and are closer to the planet-hosting

[4] Orbits with semimajor axes smaller than the lower value or larger than the upper value are certainly unstable.

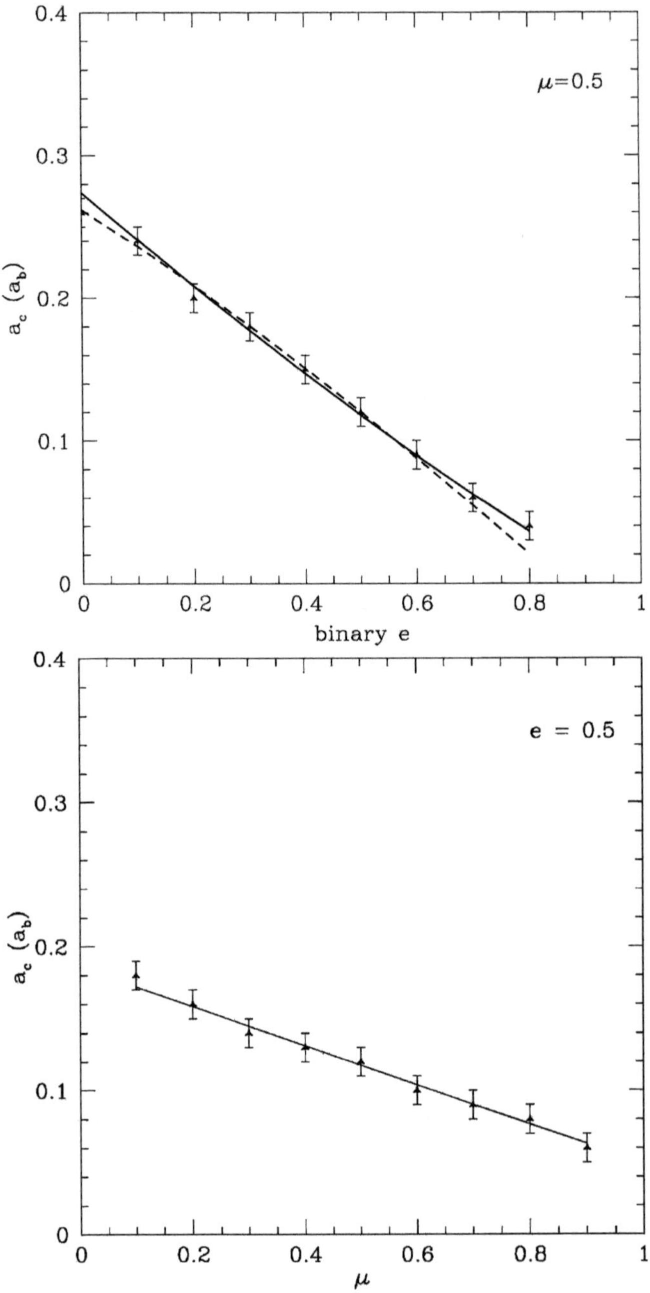

Fig. 11.3 Graphs of the critical semimajor axis (a_c) of an S-type binary-planetary system, in units of the binary semimajor axis (Holman and Wiegert 1999). The graph on the top shows a_c as a function of the binary eccentricity for an equal-mass binary. The graph on the bottom corresponds to the variations of the critical semimajor axis of a binary with an eccentricity of 0.5 in term of the binary's mass-ratio. The *solid* and *dashed line* on the *top panel* depict the empirical formulae as reported by Holman and Wiegert (1999) and Rabl and Dvorak (1988), respectively

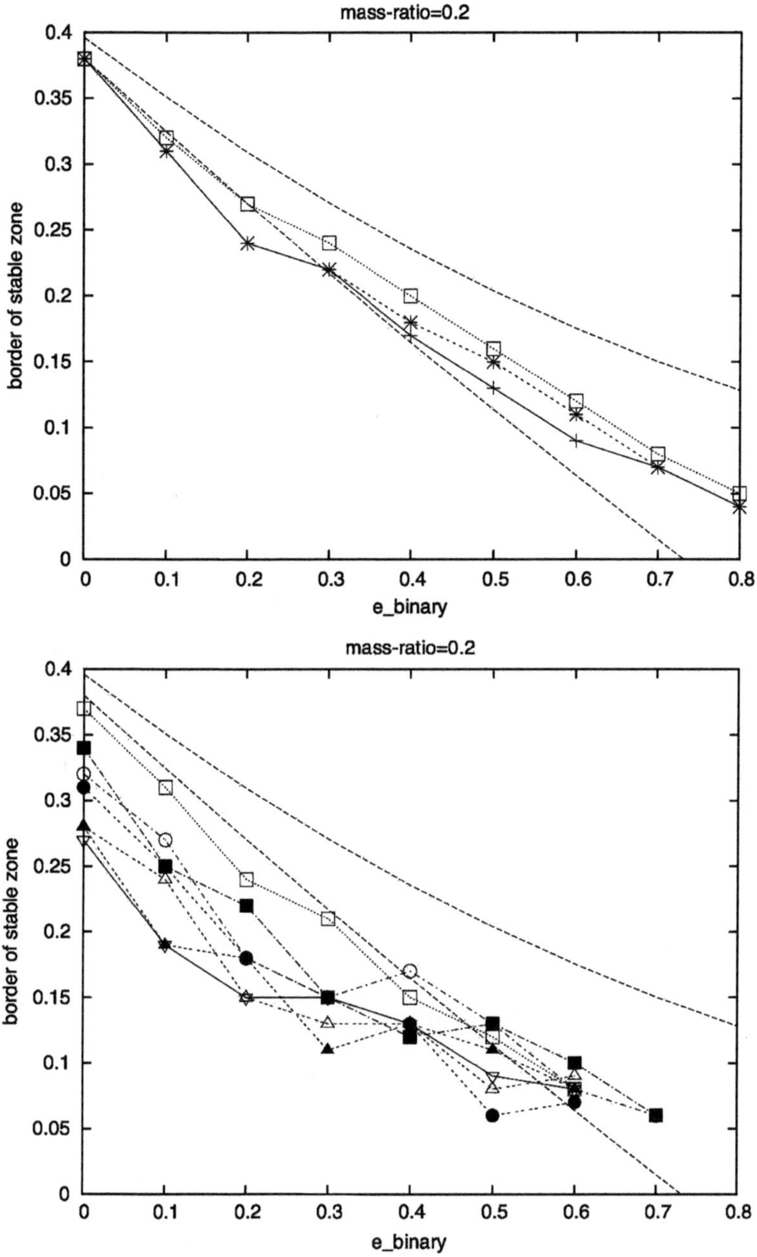

Fig. 11.4 A comparison of the results of PLD with those of HW. The *upper panel* shows the results for $e_p = 0$ (*solid line with crosses*) and $e_p = 0.1$ (*dashed line with stars*) as obtained by PLD. The *dotted line with white squares* shows the results obtained by HW. The area between the two *dashed lines* defines the boundary of the stable zone according to Eq. (11.1). The *lower panel* shows the results of PLD for e_p ranging from 0.3 to 0.9

Table 11.2 Stable zone (in units of the binary semimajor axis) of an S-type orbit for different values of the mass-ratio and eccentricity of the binary. The given size for each (μ, e_b) pair is the lesser of the values obtained by HW and PLD

	Mass-ratio (μ)								
e_b	0.1	0.2	0.3	0.4	0.5	0.6	0.7	0.8	0.9
0.0	0.45	0.38	0.37	0.30	0.26	0.23	0.20	0.16	0.13
0.1	0.37	0.32	0.29	0.27	0.24	0.20	0.18	0.15	0.11
0.2	0.32	0.27	0.25	0.22	0.19	0.18	0.16	0.13	0.10
0.3	0.28	0.24	0.21	0.18	0.16	0.15	0.13	0.11	0.09
0.4	0.21	0.20	0.18	0.16	0.15	0.12	0.11	0.10	0.07
0.5	0.17	0.16	0.13	0.12	0.12	0.09	0.09	0.07	0.06
0.6	0.13	0.12	0.11	0.10	0.08	0.08	0.07	0.06	0.045
0.7	0.09	0.08	0.07	0.07	0.05	0.05	0.05	0.045	0.035
0.8	0.05	0.05	0.04	0.04	0.03	0.035	0.03	0.025	0.02

star. This can also be seen in Table 11.2, where the lesser of the values of the inner boundary of the stable region (i.e., the semimajor axis of the last stable orbit) as obtained by HW and PLD, has been recorded. These results indicate that the stability criteria presented by HW are not applicable to eccentric S-type orbits.

It is necessary to mention that as oppose to RD and HW who determined the stable zone of a planet by identifying its escaping orbits within a certain computation time, PLD used a chaos indicator to characterize the long-term behavior of the planet's motion. Although because of the application of FLI, the computation time in PLD was much shorter than in RD and HW, their results are, however, valid for much longer times. In some cases in simulations by PLD, the application of FLI resulted in a slightly larger stable region compare to that of HW. This is due to the fact that, as oppose to the latter, in which eight starting points were used, PLD used only four starting positions. Test-computations, using a different grid for the FLI-maps, and for computation times over 10^4, 10^5 and 10^6 periods of the binary were also carried out. However, they did not change the result significantly.

Table 11.3 shows the variations of the size of the stable zone in simulations of PLD in terms of the eccentricities of the binary and planet, and for different binary mass-ratios. As shown here, as the eccentricity of the binary increases, the boundary of the stable zone varies from 0.04 (for an initially eccentric motion in a binary with an eccentricity of 0.5 and mass-ratio of $\mu = 0.9$) to 0.45 (for an initially circular motion in a circular binary with $\mu = 0.1$). Table 11.3 also shows that the size of the stable region does not have a strong dependence on the eccentricity of the planet. This dependence is not, however, negligible, especially if a planet is close to the border of the chaotic motion and moves in a highly eccentric orbit. A presentation of the 3-D stability plots for different mass-ratios with a detailed discussion can be found in PLD.

An interesting application of the analysis of HW and PLD is to the stability of terrestrial planets and smaller objects. Since in the calculations of the critical semimajor axis by these authors, a giant planet was consider to be a test particle, given that the mass of a Jovian-type planet is approximately two orders of magni-

Table 11.3 Stable zone (normalized to a_b) of an S-type orbit

Stable zone			
Mass-ratio (μ)	e_b	$e_p = 0$	$e_p = 0.5$
0.1	0	0.45	0.36
	0.5	0.18	0.13
0.2	0	0.40	0.31
	0.5	0.16	0.12
0.3	0	0.37	0.28
	0.5	0.14	0.11
0.4	0	0.30	0.25
	0.5	0.12	0.07
0.5	0	0.27	0.22
	0.5	0.12	0.07
0.6	0	0.23	0.21
	0.5	0.10	0.07
0.7	0	0.20	0.18
	0.5	0.09	0.07
0.8	0	0.16	0.16
	0.5	0.09	0.05
0.9	0	0.13	0.12
	0.5	0.06	0.04

tude larger than the mass of a terrestrial-class object, the stability criteria of HW and PLD can be readily generalized to identify regions around the stars of a binary where terrestrial-class planets can have long-term stable orbits (Quintana et al. 2002, 2006, 2007). This results can also be used to identify regions where smaller objects, such as asteroids, comets, and/or dust particles may reside. Although these analyses do not include non-gravitational forces, their applications to observational data has been successful and have identified dust bands, possibly due to the collision among planetesimals, in several wide S-type binaries (see Figs. 1.8 and 1.9 of the first chapter, and Trilling et al. 2007).

11.2.1.1 Application to the Binary γ Cephei

Gamma Cephei is one of the most interesting double star systems that host a planet. At a distance of approximately 11 pc from the Sun, and with a semimajor axis and an eccentricity of 18.5 AU and 0.36, respectively, this system present a prime example of a moderately close binary with a planet in an S-type orbit. The primary of γ Cephei, a 1.6 solar-mass K1 IV sub-giant (Fuhrmann 2004) is host to a stellar companion, an M4 V star with a mass of 0.4 solar-masses (Neuhäuser et al. 2007; Torres 2007), and a Jovian-type planet with a mass of 1.7 Jupiter-mass and an eccentricity of 0.12 at 1.95 AU (Hatzes et al. 2003).

The mass-ratio of γ Cephei binary is 0.2 making this system a suitable example for applying the stability analysis of S-type orbits as discussed in the previous section. An overview of the size of the stable region for the giant planet of this system is shown in Fig. 11.5 where the planet maintained its orbit for 1,000 time

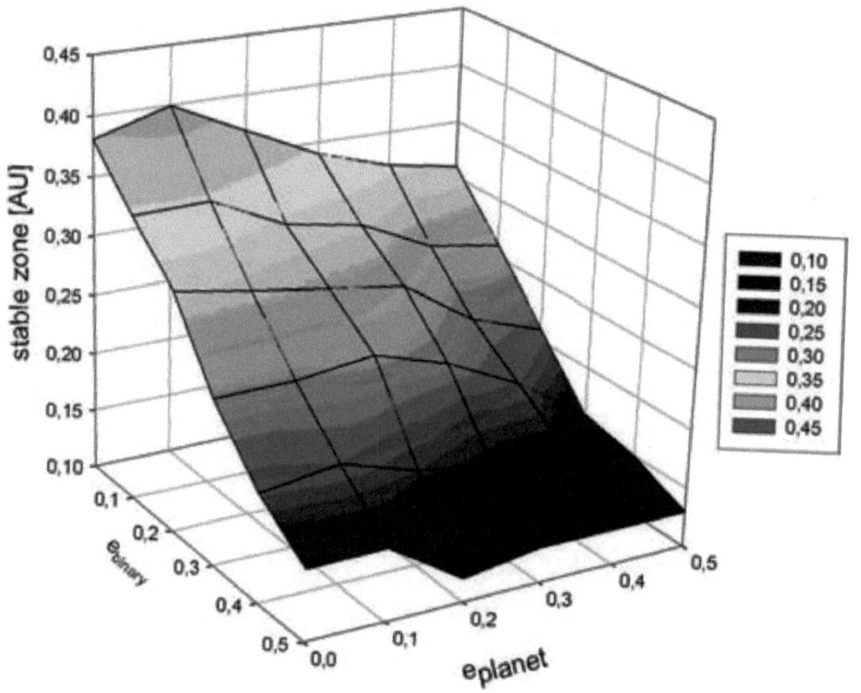

Fig. 11.5 The stable zone of an S-type orbit in a binary with mass-ratio $\mu = 0.2$ (e.g., γ Cephei). As shown here, the extent of the stable zone is more strongly affected by the eccentricity of the binary than that of the planet (Pilat-Lohinger et al. 2004)

units. As shown in this figure, the zone of stability for the giant planet extends to approximately 3.16 AU.[5] Direct integration of the binary and the planet for different values of the binary eccentricity indicate that the orbit of the planet is stable when $0.2 \leq e_b \leq 0.45$ (Haghighipour 2006). Samples of the results of these integrations are shown in Fig. 11.6. Numerical integrations of the orbit of the planet for different values of its inclination with respect to the plane of the binary (i_p) show that this object is stable for inclinations less than 40°. Figure 11.7 shows the semimajor axes and orbital eccentricities of the system for $e_b = 0.2$ and for $i_p = 5°, 10°$, and 20°.

Numerical simulations also indicated the possibility of a Kozai resonance in the γ Cephei system. Kozai resonance has been studied in binary-planetary systems by several authors (Haghighipour 2004; Haghighipour 2005a; Verrier and Evans 2006; Takeda et al. 2006, 2008; Malmberg et al. 2007; Saleh and Rasio 2009). As demonstrated by Kozai (1962), in a three-body system with two massive bodies and a small object, such as an S-type binary-planetary system, the orbital eccentricity of the planet can reach high values at large inclinations due to the exchange of

[5] Test-computations for $\mu = 0.3, 0.5$ and 0.7, up to 100,000 time units showed the same qualitative results.

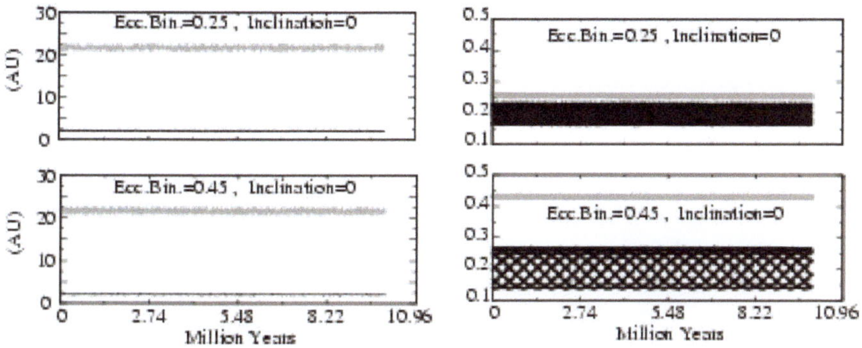

Fig. 11.6 Graphs of the semimajor axes (*left*) and eccentricities (*right*) of the giant planet (*black*) and binary (*green*) of γ Cephei for different values of the eccentricities of the binary (Haghighipour 2004). The mass-ratio of the binary is 0.2

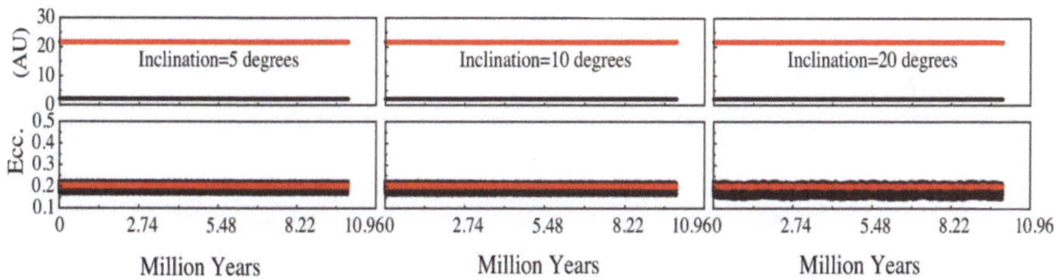

Fig. 11.7 Graphs of the semimajor axes (*top*) and eccentricities (*bottom*) of the giant planet (*black*) and binary (*red*) of γ Cephei. The initial eccentricity of the binary at the beginning of numerical integration and the value of its mass-ratio were equal to 0.20 (Haghighipour 2006)

angular momentum between the planet and the secondary star. In such cases, the longitude of the periastron of the planet, ω_p, librates around a fix value. Figure 11.8 shows this for the giant planet of γ Cephei. As shown here, ω_p librates around 90° (Haghighipour 2004, 2005a). The inclination of the planet of γ Cephei, when in a Kozai resonance, is related to its longitude of periastron and orbital eccentricity (e_p) as Innanen et al. (1997)

$$\sin^2\omega_p = 0.4\csc^2 i_p, \tag{11.2}$$

and

$$\left(e_p^2\right)_{\max} = \frac{1}{6}\left[1 - 5\cos(2i_p)\right]. \tag{11.3}$$

Equation (11.2) indicates that the Kozai resonance may occur if the orbital inclination of the small body is larger than 39.23°. For instance, as shown by Haghighipour (2004, 2005a), in the system of γ Cephei, Kozai resonance occurs at $i_p = 60°$. For the minimum value of i_p, the maximum value of the planet's orbital eccentricity is reached and, as given by Eq. (11.3), is equal to 0.764. Figures 11.8 and 11.9 show that e_p stays below this limiting value at all times.

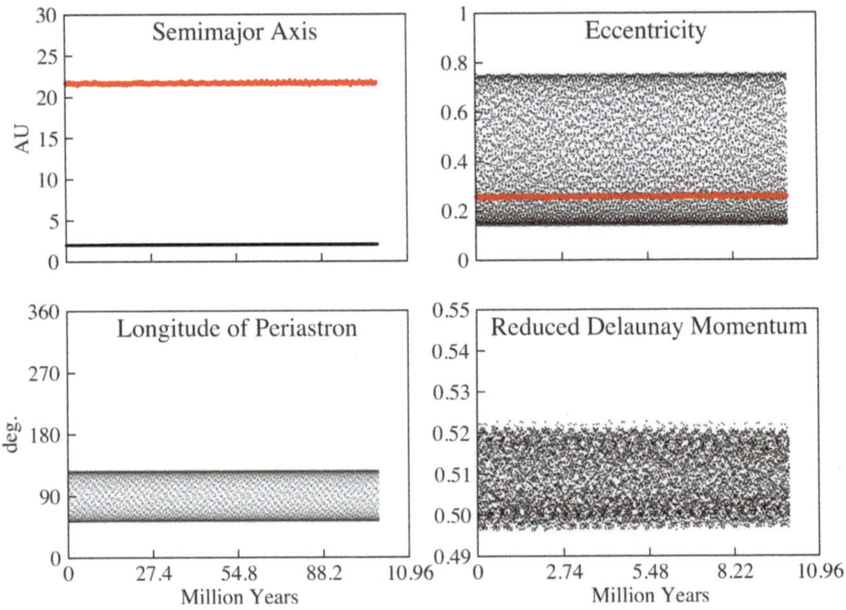

Fig. 11.8 Graphs of the semimajor axis and eccentricity of the giant planet (*black*) and binary (*red*) of γ Cephei (*top*) and its longitude of periastron and reduced Delaunay momentum (*bottom*) in a Kozai resonance (Haghighipour 2004; Haghighipour 2005a). As expected, the longitude of the periastron of the giant planet oscillates around 90° and its reduced Delaunay momentum is constant

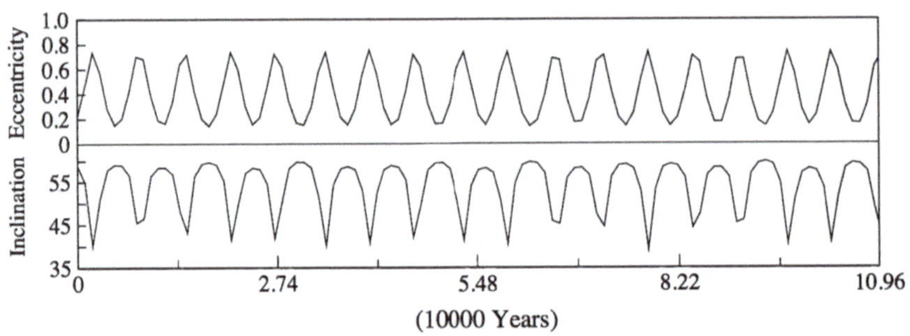

Fig. 11.9 Graphs of the eccentricity and inclination of the giant planet of γ Cephei in a Kozai Resonance (Haghighipour 2004; Haghighipour 2005a). As expected, these quantities have similar periodicity and are 180° out of phase

11.2.1.2 Application to Binaries Gliese 86 and HD41004

The application of the stability analysis of Section 11.2.1 to the planet of the binary Gliese 86 indicates that the orbit of this planet is stable. This is not surprising since with a semimajor axis of 0.11 AU, this planet is close enough to the primary star to be immune from the perturbation of the other stellar companion.

In the case of HD41004, the application of the stability analysis of Section 11.2.1 is not straightforward; the orbital parameters of the planet in this binary has not been uniquely determined. The value of the semimajor axis of this planet varies between 1.31 and 1.7 AU, and its orbital eccentricity seems to be quite high (between 0.39 and 0.74) (Zucker et al. 2004). Since the eccentricity of the binary HD41004 is unknown, to determine the stable zone of this system, simulations were carried out for different sets of orbital parameters as a function of the binary eccentricity. The results indicate that the stability of the planet is strongly correlated with its orbital eccentricity and the eccentricity of the binary. Simulations show that in all cases, in order to obtain stability, binary eccentricity has to be smaller than 0.6. For high values of the planet's eccentricity (e.g., 0.74), the value of the binary eccentricity has to become even smaller (less than 0.15) to ensure that the orbit of the planet will stay stable.

11.2.2 Stability of P-Type Orbits

Although, no circumbinary planet has yet been discovered, stability of P-type orbits has been a subject of research for many years (Ziglin 1975; Szebehely 1981; Dvorak 1984, 1986, 1989; Kubala et al. 1993; Holman and Wiegert 1999; Broucke 2001; Pilat-Lohinger et al. 2003; Musielak et al. 2005). In general, a planet in a P-type orbit is stable if its distance from the binary is so large that the perturbations of the binary stars cannot disturb its motion. Such a stable planet cannot, however, orbit the binary too far from its center of mass since galactic perturbations and the effects of passing stars may make the orbit of the planet unstable. As shown by Dvorak (1984), for circular binaries, this distance is approximately twice the separation of the binary, and for eccentric binaries (with eccentricities up to 0.7) the stable region extends to four time the binary separation. Subsequent studies by Dvorak (1986, 1989) and Holman and Wiegert (1999) have shown that Dvorak's 1984 results can be formulated by introducing a critical semimajor axis below which the orbit of the planet will be unstable;

$$\begin{aligned} a_c/a_b = {} & (1.60 \pm 0.04) + (5.10 \pm 0.05)e_b + (4.12 \pm 0.09)\mu \\ & + (-2.22 \pm 0.11)e_b^2 + (-4.27 \pm 0.17)e_b\mu + (-5.09 \pm 0.11)\mu^2 \\ & + (4.61 \pm 0.36)e_b^2\mu^2 \,. \end{aligned} \tag{11.4}$$

Figure 11.10 shows the value of a_c for different values of the binary eccentricity. Similar to S-type orbits, the $\pm$ signs in Eq. (11.4) define a lower and an upper value for the critical semimajor axis a_c, and set a transitional region that consists of a mix of stable and unstable orbits. We refer the reader to Holman and Wiegert (1999), Pilat-Lohinger et al. (2003) and Pilat-Lohinger and Funk (2006) for more details. Using this stability criteria in analysis of their observational data (Trilling et al. 2007) have been able to detect circumbinary dust bands, possibly resulted from the collision of planetesimal, around several close binary stars (first chapter, Figs. 1.8 and 1.9).

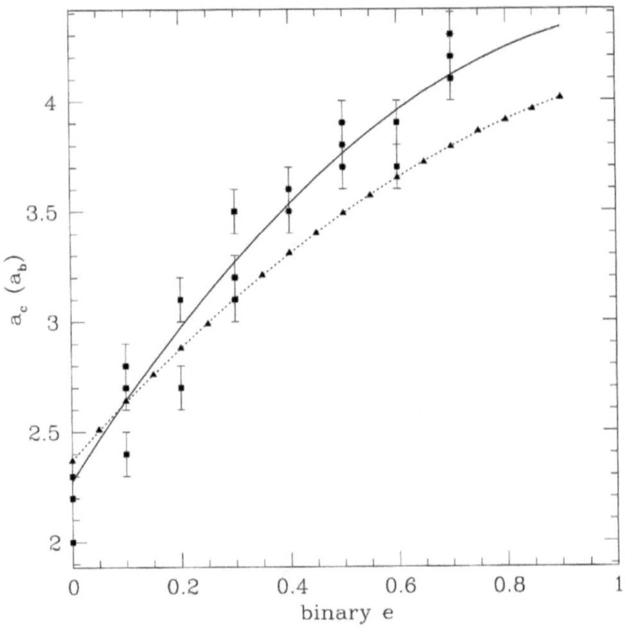

Fig. 11.10 Critical semimajor axis as a function of the binary eccentricity in a P-type system. The *squares* correspond to the result of stability simulations by Holman and Wiegert (1999) and the *triangles* represent those of Dvorak et al. (1989). The *solid line* corresponds to Eq. (11.4). As indicated by Holman and Wiegert (1999), the figure shows that at outer regions, the stability of the system fades away

A dynamically interesting feature of a circumbinary stable region is the appearance of islands of instability. As shown by Holman and Wiegert (1999), islands of instability may develop beyond the inner boundary of the *mixed zone*, which correspond to the locations of $(n : 1)$ mean-motion resonances. The appearance of these unstable regions have been reported by several authors under various circumstances (Hénon and Guyot 1970; Dvorak 1984, 1989; Rabl and Dvorak 1988). Extensive numerical simulations would be necessary to determine how the overlapping of these resonances would affect the stability of P-type binary-planetary orbits.

11.2.3 Stability of L-Type Orbits

The L-type orbit, in which an object librates around one of the binary's Lagrangian triangular points (Fig. 11.2), may not be entirely relevant to planetary motions in double star systems. The reason is that such an orbital configuration requires $M_2/(M_1 + M_2) \leq 1/26$, which is better fulfilled in systems consisting of a star and a giant planet. Recent simulation by Haghighipour et al. (2008) have shown that in systems with a close-in giant planet, L-type planetary orbits with low eccentricities can be stable for long times. We refer the reader to Section 11.3.2 and the paper by Pilat-Lohinger et al. (2003) for more details on the stability of these orbits.

11.3 Terrestrial Planets in Binaries

In the previous section, a general analysis of the dynamics of a planet in a binary star system was presented. However, within the context of habitability, the interest falls on the motion and long-term stability of Earth-like planets. It would be interesting to extend studies of the habitability, similar to those by Jones et al. (2001) and Menou and Tabachnik (2003), to binary star system, in particular those in which a giant planet already exists, and analyze the dynamics of fictitious Earth-like planets in such complex environments. In this section, we focus on this issue.

In general, four different types of orbits are possible for a terrestrial planet in a binary system that hosts a giant planet:

- **TP-i:** the terrestrial planet is inside the orbit of the giant planet.
- **TP-o:** the terrestrial planet is outside the orbit of the giant planet.
- **TP-t:** the terrestrial planet is a Trojan of the primary (or secondary) or the giant planet.
- **TP-s:** the terrestrial planet is a satellite of the giant planet.

In principle, the study of the stability of these orbits requires the analysis of the dynamics of a complicated N-body system consisting of two stars, a giant planet, and a terrestrial-class object. Except for a few special cases, the complexities of these systems do not allow for an analytical treatment of their dynamics, and require extensive numerical integrations. Those special cases are:

- Binaries with semimajor axes larger then 100 AU in which the secondary star is so far away from the primary (the planet-hosting star) that its perturbative effect can be neglected (Norwood and Haghighipour 2002).
- Binaries in which the giant planet has an orbit with a very small eccentricity (almost circular).
- Binaries in which, compared to the masses of the other bodies, the mass of the terrestrial planet is negligible. In these systems, within the framework of ER3BP, one can define curves of zero-velocity, the barriers of the motion of the fictitious terrestrial planet, using the Jacobi constant (Dvorak and Freistetter 2003).

When numerically studying the dynamics of a terrestrial planet in a binary-planetary system, integrations have to be carried out for a vast parameter-space. These parameters include the semimajor axes, eccentricities, and inclinations of the binary and the two planets, the mass-ratio of the binary, and the ratio of the mass of the giant planet to that of its host star. The angular variables of the orbits of the two planets also add to these parameters. Although such a large parameter-space makes the numerical analysis of the dynamics of the system complicated, numerical integrations are routinely carried out to study the dynamics of terrestrial planets in binary-planetary systems. The reason is that such numerical computations allows for the investigation of the stability of many terrestrial planets and for a grid of their initial conditions in one or only a few simulations. Given the small size of these objects compared to that of a giant planet (e.g., 1/300 in case of Earth and Jupiter), to the zeroth order of approximations, the effect of a terrestrial planet on the motion of a giant planet can be ignored, and the terrestrial planet can be considered as a test particle. This simplification makes it possible to study the stability of thousands of possible orbits of a terrestrial-class body in one integration. Several studies of the dynamical evolution of a terrestrial planet in a binary system have used this simplification and have shown that the final results are quite similar to the results of the numerical integrations of an actual four-body system (Dvorak et al. 2004b; Érdi and Sándor 2005).

In the rest of this section, we present the results of the studies of the dynamics of a terrestrial planet, focusing primarily on TP-i and TP-o orbits.[6] Since no extrasolar terrestrial planet has yet been discovered, the only possible approach for a detailed dynamical analysis of the orbit of such an object in a binary system is to consider a specific extrasolar planetary system in a double star, and study the dynamics of a fictitious terrestrial planet for different values of its orbital elements, and those of the binary and its giant planets. As an example, we will consider the binary-planetary system of γ Cephei. For the orbital elements of this system, we use the values given by Haghighipour (2006) and Neuhäuser et al. (2007).

11.3.1 Stability of TP-i *and* TP-o *Orbits*

A study of the dynamics of a *full* four-body system consisting of a terrestrial planet in a TP-i or TP-o orbit in γ Cephei indicates that the orbit of this planet can only be stable in close neighborhood of the primary star and outside the influence zone[7] of the giant planet (Fig. 11.11, Haghighipour 2006). Integrations also show that, while the habitable zone of γ Cephei is unstable (Dvorak et al. 2003a; Haghighipour 2006), it is possible for an Earth-like planet to have a stable TP-i orbit in a region between 0.3 and 0.8 AU from the primary star, and when its orbit is coplanar with that of the giant planet with an inclination less than 10°. In the region outside the orbit of the giant planet, i.e., when the terrestrial planet is in a TP-o orbit, the perturbations from the giant planet and the secondary star affect the stability of this object. For instance, for the values of the binary eccentricity equal to $e_b = 0.25$, 0.35, and 0.45, the periastron of the secondary star will be as close as 13.9, 12.0,

[6] The stability of TP-t and TP-s orbits has recently been studies in a few articles by Schwarz et al. (2007a,b), Nauenberg (2002) and Domingos et al. (2006).

[7] The influence zone of a planetary object with a mass m_p around a star with a mass M is defined as the region between $3R_H - a_p(1-e_p)$ and $3R_H + a_p(1+e_p)$, where a_p and e_p are the semimajor axis and eccentricity of the planet, and $R_H = a_p(1-e_p)(m_p/3M)^{1/3}$ is its Hill radius.

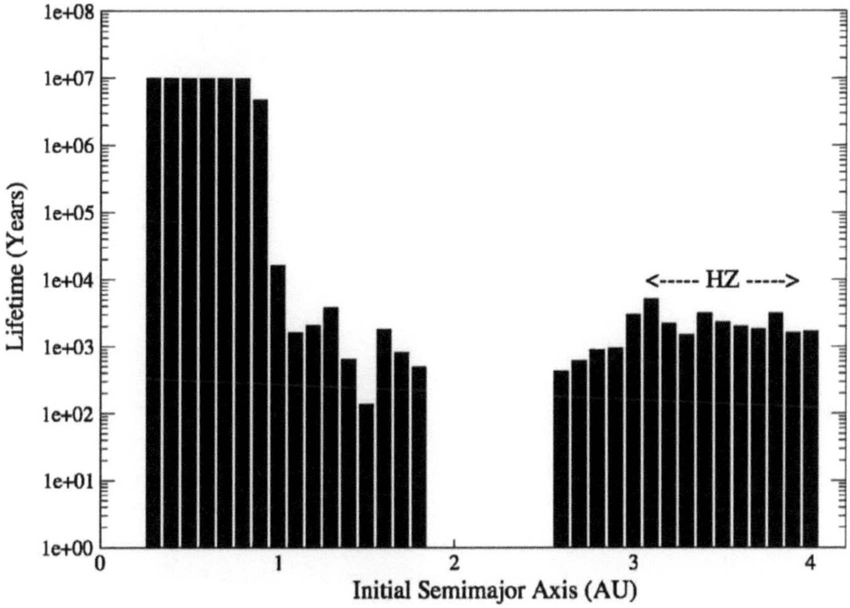

Fig. 11.11 Graph of the lifetime of an Earth-size object in a circular orbit around the primary of γ Cephei. The habitable zone of the primary has been indicated by HZ. No planet was placed in the region between the aphelion and perihelion distances of the giant planet of the system. As shown here, only Earth-size planets close to the primary star maintain their orbits for long times (Haghighipour 2006)

and 10.2 AU, respectively. At these distances, the secondary will have strong effects on the stability of the orbit of a terrestrial planet with a semimajor axis between 2.5 and 5.8 AU. Simulations show that no orbit survives in this region longer than approximately 10^5 years. For TP-o orbits inside 2.5 AU, the perturbation of the giant planet is the main factor in the instability of the orbit of the terrestrial planet.

To study the effect of orbital inclination on the stability of a terrestrial planet in γ Cephei, the region between the host-star and the giant planet of this system was examined for different values of the inclination of a fictitious terrestrial-size object, with and without the secondary star (Pilat-Lohinger et al. 2004). Figure 11.12 shows the results. While dynamical models using two massive bodies (i.e., primary star and the giant planet) show a vast region of stability for a massless terrestrial planet (gray area in the lower panel of Fig. 11.12), models with three massive bodies (i.e., primary, secondary, and the giant planet), show a decrease in the stable region (see upper panel of Fig. 11.12). They also show an arc-like chaotic path with an island of stability around 1 AU, which corresponds to the 3:1 mean motion resonance with the giant planet.

The instability of the orbit of a terrestrial planet in γ Cephei system (in particular in its habitable zone) has been studied only for prograde orbits. Recently in an article by Gayon and Bois (2008), the authors investigated the stability of retrograde

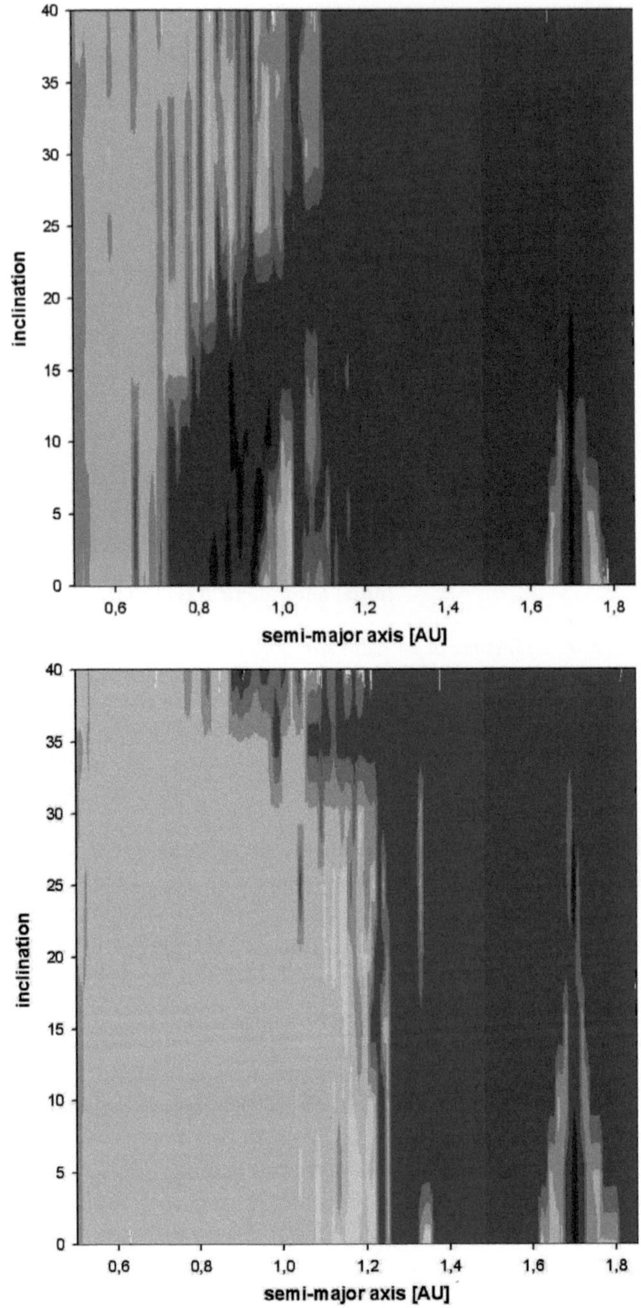

Fig. 11.12 Stability maps for a fictitious terrestrial planet in the γ Cephei system. *Dark regions* represent chaotic zones and *gray regions* correspond to stability. The *upper graph* shows the stability of a terrestrial planet in a restricted four-body problem (R4BP) [i.e., γ Cephei binary + detected planet + fictitious (massless) planet]. The *lower panel* shows the results in a restricted three-body problem (R3BP) [i.e., primary + detected planet + fictitious planet] (Pilat-Lohinger et al. 2004)

orbits in extrasolar planetary systems with multiple planets and showed that in systems were prograde orbits are unstable (e.g., HD 73256), retrograde orbits may survive for long times. The stability of retrograde orbits in planetary systems has been known for many years (Harrington 1972, 1975, 1977; Donnison and Mikulskis 1994). Such long-term stable orbits have also been observed among Jupiter's retrograde irregular satellites (Jewitt and Haghighipour 2007). To investigate whether retrograde TP-i and TP-o orbits can survive in a binary-planetary system, the motion of a massless terrestrial planet in these orbits was simulated for 1 Myr in the binary of γ Cephei. Figure 11.13 shows the results. From this figure one can see that for semimajor axes smaller than 1.8 AU, a TP-i orbit is stable. However, at close distances to the giant planet, this orbit suffers from strong perturbations from this object and becomes unstable (straight long lines around 2 AU where the eccentricity of terrestrial planet reaches unity). Figure 11.13 also shows that for a TP-o orbit, a stability region exists for initial semimajor axes ranging from 2.5 to 7.2 AU, with the minimum perturbation received at the semimajor axis of 3.2 AU. Beyond 7.2 AU, terrestrial planet becomes unstable due to the perturbation from the secondary star. The three small peaks in the stable region of Fig. 11.13 correspond to mean-motion resonances between the terrestrial planet and the giant planet. A comparison between this figure and Fig. 11.11, in which the lifetime of a terrestrial planet in a

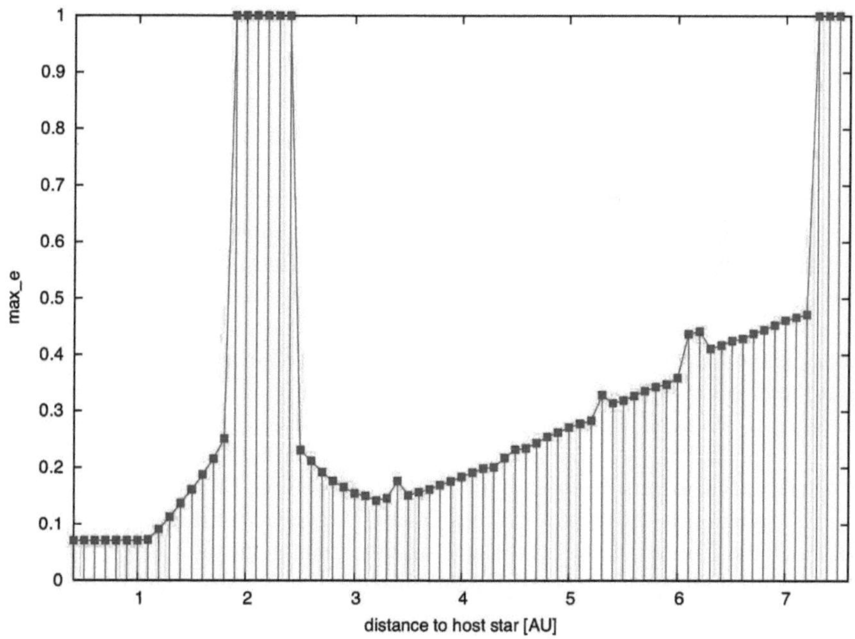

Fig. 11.13 Graph of the eccentricity of a massless terrestrial planet in a retrograde orbit in the γ Cephei system. Integrations were carried out for 1 Myr within a restricted four-body system. The three peaks at 3.4, 5.3, and 6.1 AU correspond to 2:1, 4:1, and 5:1 mean-motion resonances between the terrestrial and the giant planets

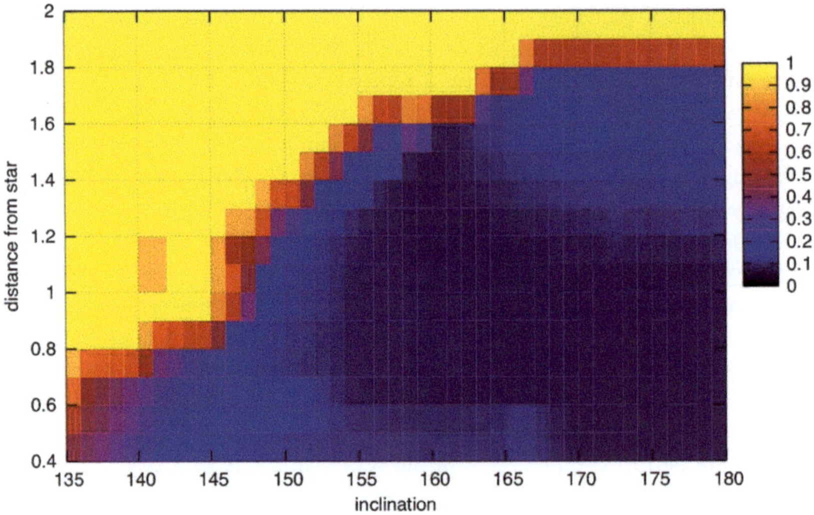

Fig. 11.14 Stability diagram for a retrograde TP-i orbit for different values of the orbital inclination. The binary eccentricity is 0.35. Stability in shown in *blue* whereas *yellow* corresponds to chaotic orbits

prograde circular orbit is shown, clearly indicates that γ Cephei has a large stable region for retrograde orbits, in particular in the habitable zone of its primary star.

Figure 11.14 shows the results of the integrations of a terrestrial planet in a retrograde TP-i orbit in the γ Cephei system, in terms of the initial orbital inclination of this object. The initial semimajor axis of the terrestrial planet was varied between 0.4 and 2.0 AU, and its initial orbital inclination was chosen to be between 135° and 180°. The eccentricity of the binary was $e_b = 0.35$. As shown in this figure, the upper stability limit for a terrestrial planet in retrograde orbits is 1.8 AU corresponding the inclinations between 165° and 180°, and the lower limit is 0.8 AU for an inclination of 145°. For more inclined orbits of a fictitious terrestrial planet, this lower limit drops to 0.4 AU. Simulations for different values of the binary eccentricity indicate that the direct effect of the binary orbit on the stability of a terrestrial planet in a retrograde TP-i orbit is negligible, and it only affects the eccentricity of the orbit of the giant planet ($\delta e_{GP} = 0.08, 0.1, 0.11$ for $e_b = 0.25, 0.35, 0.45$, respectively). It is the latter that affects the orbit of the fictitious planet in a TP-i orbit.

The situation is different for a retrograde TP-o orbit. As shown in Fig. 11.15, although the effect of the 2:1 resonance with the giant planet at 3.1–3.6 AU makes the orbit of a retrograde terrestrial planet unstable, large stable regions, especially

→

Fig. 11.15 Stability diagram for a retrograde TP-o orbit for different values of the orbital inclination. The binary eccentricity is 0.25, 0.35, and 0.45 from *top* to *bottom*. Stability in shown in *blue* whereas *yellow* corresponds to chaotic orbits. As shown here, instability extends to large distances as the binary eccentricity increases

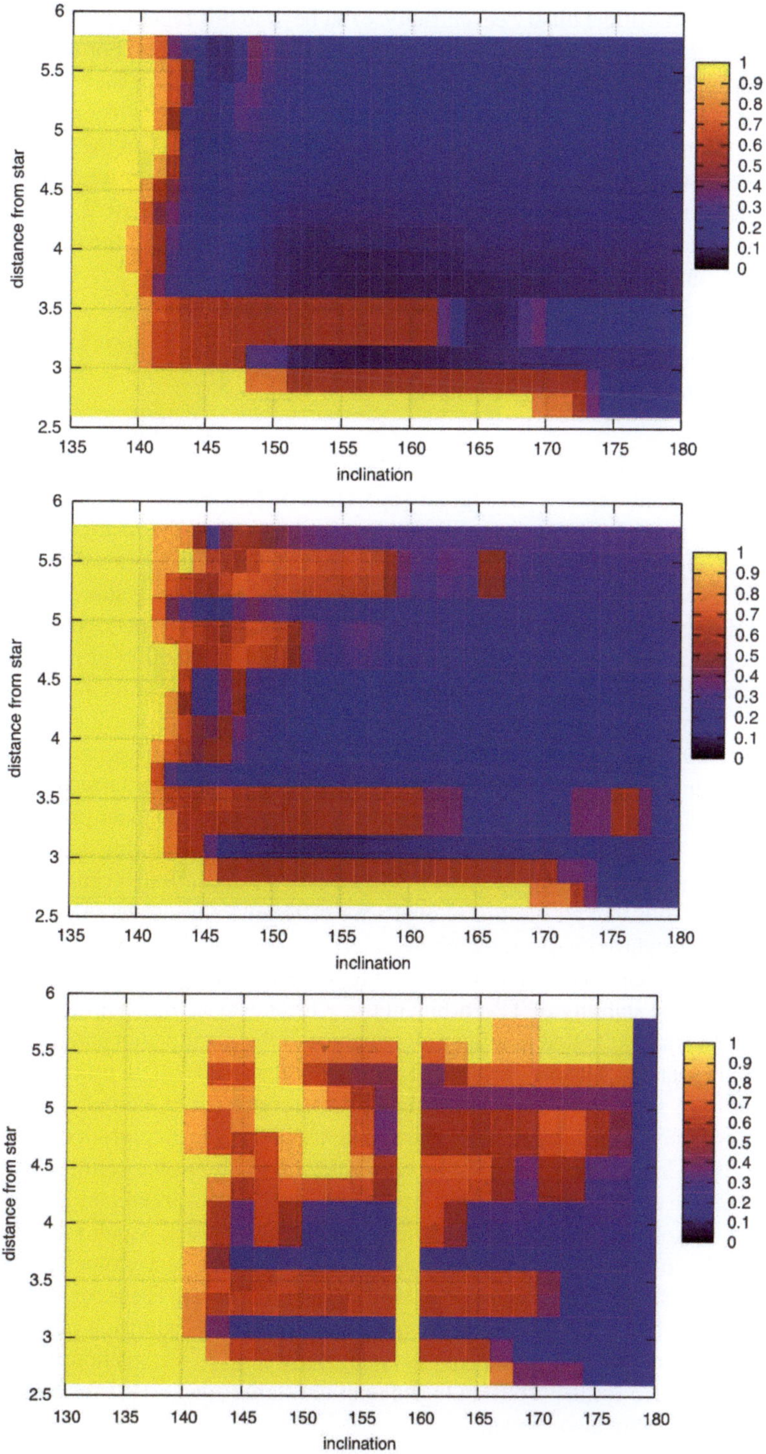
6
5.5
5
4.5
4
3.5
3
2.5
distance from star
135 140 145 150 155 160 165 170 175 180
inclination
1
0.9
0.8
0.7
0.6
0.5
0.4
0.3
0.2
0.1
0
6
5.5
5
4.5
4
3.5
3
2.5
distance from star
135 140 145 150 155 160 165 170 175 180
inclination
1
0.9
0.8
0.7
0.6
0.5
0.4
0.3
0.2
0.1
0
6
5.5
5
4.5
4
3.5
3
2.5
distance from star
130 135 140 145 150 155 160 165 170 175 180
inclination
1
0.9
0.8
0.7
0.6
0.5
0.4
0.3
0.2
0.1
0

for lower values of the binary eccentricity, exist beyond this region and for different values of the initial inclination of the orbit of the terrestrial planet. For instance, for $e_b = 0.25$, the region of stability extends from 3.5 to 5.8 AU for the values of the inclinations ranging from 145° to 180°. At the 2:1 mean-motion resonance also a stable region exists for inclinations between 160° and 180°, when the eccentricity of the terrestrial planet is smaller than 0.2. In the case of a binary with $e_b = 0.35$ (Fig. 11.15, middle graph), the unstable region corresponding to the 2:1 resonance extends to higher inclinations. For the values of the eccentricity of the binary larger than 0.45, instability extends to almost all inclinations.

11.3.2 Stability of TP-t *Orbits*

An interesting case of a stable orbit is when the terrestrial planet is in a Lagrange equilibrium point either at an angular separation of 60° ahead of a giant planet or behind it (Fig. 11.2). In the simplified dynamical model of the restricted (circular and elliptic) three-body system, many investigations exist concerning the stability of such a planet in terms of the mass-ratio of the planet-hosting star and its giant planet (Rabe 1961), and the eccentricity of the giant planet's orbit (Deprit and Rom 1970). Within the context of extrasolar planetary systems, stability of the orbit of a terrestrial planet in a Lagrangian point has been studied by Laughlin and Chambers (2002), Menou and Tabachnik (2003), Sándor and Érdi (2003), Érdi et al. (2004), Schwarz (2005), Schwarz et al. (2007a,b), Schwarz et al. (2009). Recent simulations by Haghighipour et al. (2008) and Capen et al. (2009) show that terrestrial planets as Trojans of giant planets can also exist in systems where the giant planet transits its host star. Figure 11.16 shows an example of such systems. In this figure, a solar-type star is host to a Jupiter-size transiting planet in a 3-day circular orbit. The graph shows the stability of an Earth-size object for different values of its semimajor axis and orbital eccentricity. As shown here, a terrestrial planet in a 1:1 resonance with the giant planet can maintain a stable orbit for eccentricities ranging from 0.2 to 0.5.

To study the stability of TP-t orbits in the γ Cephei system, the orbit of a terrestrial planet was integrated in a Lagrangian point of the giant planet and in a general four-body system. Because as shown by Schwarz et al. (2007a), in order for a small body to have a stable Trojan orbit in an elliptical, restricted, three-body system, the orbital eccentricity of the giant planet cannot exceed 0.3, the eccentricity of the giant planet of γ Cephei was set to this value. Figure 11.17 shows the results. As shown here, a region of stability exists for a Trojan terrestrial planet around the giant planet. Small variations in the orbital eccentricity of the giant planet, which are due to the perturbation of the secondary companion, cause the apparent asymmetry in the location of the stable orbits around the two Lagrangian points. These regions contain very stable orbits with eccentricities smaller than 0.2. The extension of semimajor axis associated with this asymmetry is small (only 2.5%). A transition from an eccentricity of 0.35 for the orbits on the edge of the stable zone, to 0.5 for orbits in the unstable region are also shown.

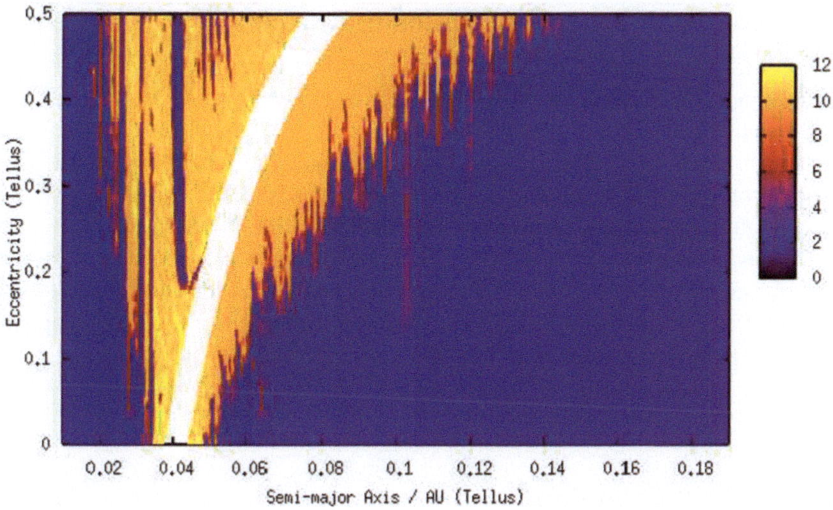

Fig. 11.16 Graph of the stability of an Earth-sized planet in a system consisting of a Sun-like star and a transiting Jupiter-mass object in a 3 days orbit. *Blue* shows stability whereas *red* corresponds to chaos. An island of stability corresponding to the 1:1 mean-motion resonance is shown (Haghighipour et al. 2008; Capen et al. 2009)

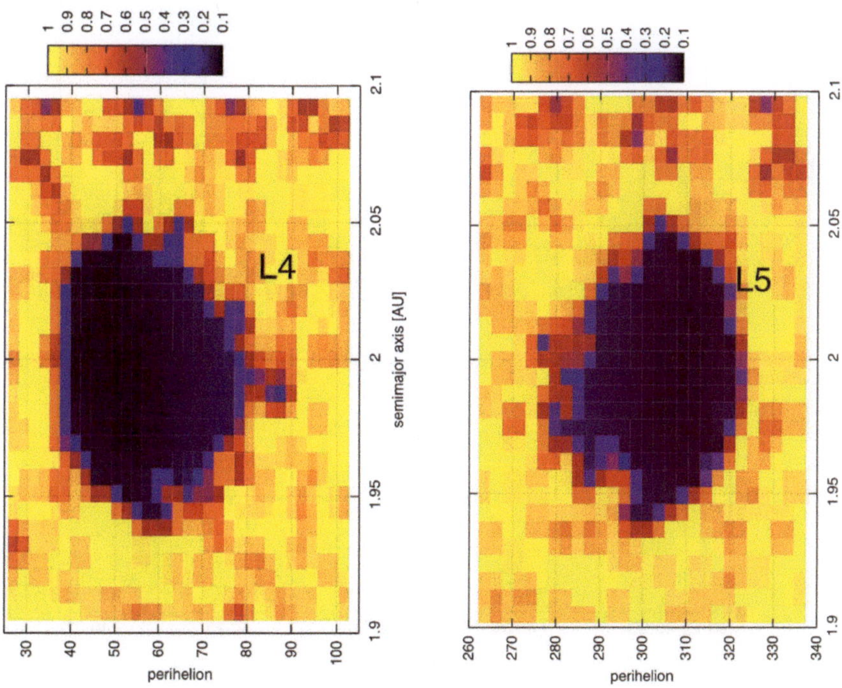

Fig. 11.17 Stability diagram for a TP-t orbit for the region around the Lagrange points L_4 and L_5 of the giant planet in the γ Cephei system. The angular distance to the Lagrange point are shown on the horizontal axes and the distance in AU is in the direction of the line connecting the host-star to the Lagrange point. The dark region shows the stable region

11.4 Habitable Planet Formation in Binaries

As seen in previous chapters, planet formation in close and moderately close binary star systems is an active topic of research. Whether models of giant and terrestrial planet formation around single stars can be extended to binary systems depends strongly on the orbital elements of the binary, its mass-ratio, and the types of its stars. While the detection of systems such as L1551 (Fig. 11.18) implies that planet formation in binaries may proceed in the same fashion as around single stars, simulations such as those by Heppenheimer (1978), Artymowicz and Lubow (1994), Whitmire et al. (1998) and Pichardo et al. (2005) indicate that a stellar component in an eccentric orbit can considerably affect planet formation by

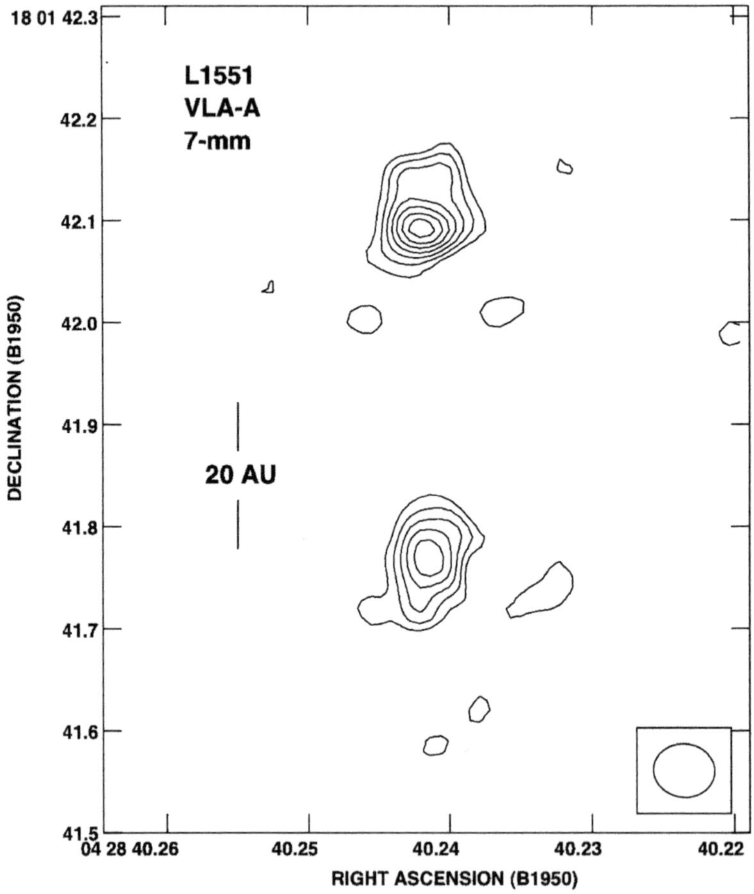

Fig. 11.18 Interferometric observation of the binary system L1551 (Rodriguez et al. 1998). Two compact sources are evident in the map. The separation of the binary is 45 AU and the disk around each core extends to approximately 10 AU. The mass of each disk is approximately 0.06 solar-masses

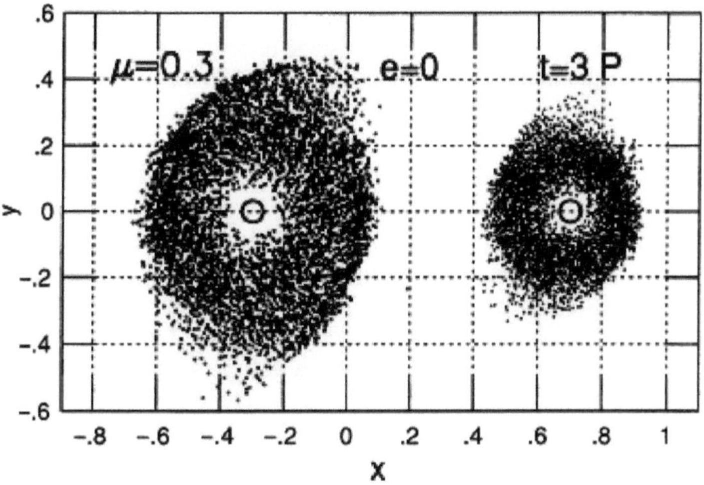

Fig. 11.19 Disk truncation in and around binary systems (Artymowicz and Lubow 1994). The *top graphs* show circumstellar disk in a binary with a mass-ratio of 0.3. Note the disk truncation when the eccentricity of the binary is increased from 0 to 0.3. The *bottom graphs* show similar effect in a circumbinary disk. The mass-ratio is 0.3 and the binary eccentricity is 0.1. The numbers inside each graph represent the time in units of the binary period. The axes are in units of the binary semimajor axis

- Increasing the relative velocities of planetesimals, which may cause their collisions to result in breakage and fragmentation
- Truncating the circumprimary disk of embryos to smaller radii, which causes the removal of material that may be used in the formation of terrestrial planets (Fig. 11.19)
- Destabilizing the regions where the building blocks of these objects may exist

Prior to the detection of planets in binary stars, studies of planet formation in these systems were limited to only some specific or hypothetical cases. For instance, Heppenheimer (1974, 1978), Drobyshevski (1978), Diakov and Reznikov (1980), Whitmire et al. (1998), Kortenkamp et al. (2001) studied planet formation in binaries where the system consisted of Sun and Jupiter, and the focus was on the effect of Jupiter on the formation of inner planets of our solar system. Whitmire et al. (1998) studied planet formation in binaries, in particular those resembling some of extrasolar planets, in which the secondary star has a mass in the brown dwarf regime. Barbieri et al. (2002), Quintana et al. (2002), Lissauer et al. (2004) also studied the late stage of terrestrial planet formation (i.e., growth of planetary embryos to terrestrial-size objects) in the α Centauri system.

The detection of the giant planet of γ Cephei changed this trend. By providing a real example of a planetary system in a binary star, this discovery made theorist take a deeper look at the models of planet formation and focus their efforts on explaining how this planet was formed and whether such systems could harbor smaller planetary objects. The results of their works, however, have made the matter quite complicated. For instance, while simulations as those presented in the chapter by Quintana and Lissauer imply that the late stage of terrestrial planet formation may proceed successfully in binary star systems and result in the formation of terrestrial-class objects, simulations of earlier stages have not been able to model the accretion of planetesimals to planetary embryos. On the other hand, as indicated by Marzari and co-authors in the seventh chapter, despite the destructive role of the binary companion, i.e., increasing the relative velocities of planetesimals, which causes their collisions to result in erosion, growth of these objects to larger sizes may still be efficient as the effect of the binary companion can be counterbalanced by dissipative forces such as gas-drag and dynamical friction. As shown by these authors, for planetesimals of comparable sizes, the combined effect of gas-drag and the gravitational force of the secondary star may result in the alignment of the periastra of small objects and increase the efficiency of their accretion by reducing their relative velocities (Marzari et al. 1997; Marzari and Scholl 2000). However, the efficiency of this mechanism depends on the size of the planetesimals,[8] the eccentricity of the planetesimals disk, and the orbital elements of the binary system. As shown by Paardekooper et al. (2008), depending on the perturbation of the secondary star, the eccentricity of the disk may reach a limiting value below which the encounter velocities of planetesimals are within a factor 2 of their corresponding values in a circular disk, and above that, the encounter velocities become so high that planetesimal accretion is inhibited. The application of these simulations to the α Centauri system has shown that the growth of planetesimals to planetary embryos may be impossible within 0.5 AU of the primary star of this system (Thébault et al. 2008). However, simulations by Thébault et al. (2009) and Marzari et al. (2009) indicate that this process can be efficient around α Centauri B, and planetary embryos can form within the terrestrial/habitable region of this star. Similar results have been obtained by Xie and Zhou (2008, 2009) when numerically integrating a slightly inclined disk of planetesimals around the primary of γ Cephei. As shown by these authors, the gas-drag causes the sorting of inclined planetesimals according to their sizes, and increases the efficiency of their accretion by decreasing their relative velocities. For larger values of planetesimals inclinations, accretion is more efficient in wide (e.g., >70 AU) binaries (Marzari et al. 2009).

As one can notice, a common starting point in all these simulations is after km-sized object or larger bodies have already formed. The reason is that among the four stages of planet formation, that is,

- Coagulation of dust particles and their growth to centimeter-sized objects
- Growth of centimeter-sized particles to kilometer-sized bodies (planetesimals)
- Formation of Moon- to Mars-sized protoplanets (also known as planetary embryos) through the collision and accretion of planetesimals
- Collisional Growth of planetary embryos to terrestrial-sized objects

[8] For colliding bodies with different sizes, depending on the size distribution of small objects, and the radius of each individual planetesimal, the process of the alignment of periastra may instead increase the relative velocities of the two objects, and cause their collisions to become eroding (Thébault et al. 2006).

the last two can be more readily studied in a system of double stars. At these stages, the dominant force in driving the dynamics of objects is their mutual interactions through their gravitational forces, and the simulations can be done using N-body integrations. In tenth chapter of this volume, Quintana and Lissauer have presented the results of a series of such simulations, and investigated terrestrial planet formation in binary systems such as α Centauri. Given that the late stage of terrestrial planet formation is a slow process, which may take a few hundred million years, it is possible that during the first few million years of this process, giant planets are also formed at large distances from the planet-hosting star. Similar to terrestrial planet formation in our solar system, these objects will play a vital role in the formation, distribution, and water-content of terrestrial-class objects in binary systems. Within the context of habitable planet formation, this implies that the formation of terrestrial planets has to be simulated while the effects of the secondary and the giant planet(s) of the system are also taken into account.

11.4.1 Habitable Zone

Life, as we know it, requires liquid water. A potentially habitable planet has to be able to maintain liquid water on its surface and in its atmosphere. The capability of a planet in maintaining water depends on many factors such as its size, interior dynamics, atmospheric circulation, and orbital parameters (semimajor axis and orbital eccentricity). It also depends on the brightness of the central star at the location of the planet. These properties, although at the surface unrelated, have strong intrinsic correlations, and combined with the luminosity of the star, determine the system's *habitable zone*. For instance, planet's interior dynamics and atmospheric circulation generate a CO_2 cycle, which subsequently results in greenhouse effect. The latter helps the planet to maintain a uniform temperature. This process can, however, be disrupted if the planet is too close or too far from the central star. In other words, the distance of the planet from the central star must be such that the amount of the radiation received by the planet allow liquid water to exist on its surface and in its atmosphere. The orbital elements of the planet, on the other hand, have to ensure that this object will maintain a stable orbit at all times.

The width of a habitable zone and the location of its inner and outer boundaries vary with the luminosity of the central star and the planet's atmospheric circulation models (Menou and Tabachnik 2003; Jones et al. 2005, 2006). Conservatively, the inner edge of a habitable zone can be considered as the distance closer than which water on the surface of the planet evaporates due to a runaway greenhouse effect. In the same manner, the outer edge of the habitable zone is placed at a distance where, in the absence of CO_2 clouds, runaway glaciation will freeze the water and creates permanent ice on the surface of the planet. Using these definitions of the inner and outer boundaries of a habitable zone, Kasting et al. (1993) have shown that a conservative range for the habitable zone of the Sun would be between 0.95

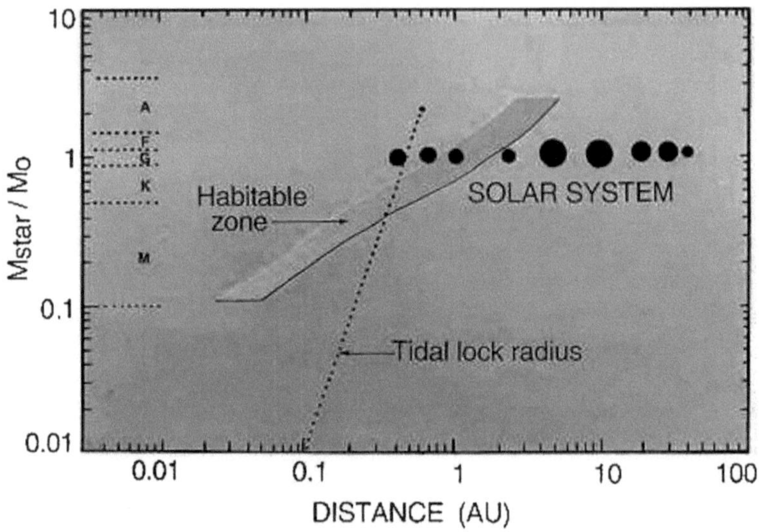

Fig. 11.20 Habitable zone (Kasting et al. 1993)

and 1.15 AU from this star (Fig. 11.20). As noted by Jones et al. (2005), however, the outer edge of this region may extend to farther distances (Forget and Pierrehumbert 1997; Williams and Kasting 1997; Mischna et al. 2000) close to 4 AU.

Since the notion of habitability is based on life on Earth, one can calculate the boundaries of the habitable zone of a star by comparing its luminosity with that of the Sun. For a star with a surface temperature T and radius R, the luminosity L and its brightness $F(r)$ at a distance r are given by

$$F(r) = \frac{1}{4\pi} L(R,T) r^{-2} = \sigma T^4 R^2 r^{-2}\,, \tag{11.5}$$

where σ is the Boltzmann constant. Using Eq. (11.5) and the fact that Earth is in the habitable zone of the Sun, the radial distances of the inner and outer edges of the habitable zone of a star can be obtained from

$$r = \left(\frac{T}{T_S}\right)^2 \left(\frac{R}{R_S}\right) r_E\,. \tag{11.6}$$

In this equation, T_S and R_S are the surface temperature and radius of the Sun, respectively, and r_E represents the distance of Earth from the Sun (i.e., the inner and outer edges of Sun's habitable zone). Using Eq. (11.6), the habitable zone of a star can be defined as a region where an Earth-like planet receives the same amount of radiation as Earth receives from the Sun, so that it can develop and maintain similar habitable conditions as those on Earth.

11.4.2 Formation of Habitable Planets in S-Type Binaries

As mentioned earlier, a potentially habitable planet has to have a stable orbit in the habitable zone of its host star. Simulations of the stability of an Earth-size planet in the γ Cephei system (Fig. 11.11) indicate that in an S-type binary, the region of the stability of this object is close to the primary, where the terrestrial planet is safe from the perturbations of the giant planet and the secondary star. This implies that in order for the primary to host a habitable planet, its habitable zone has to also fall within those distances where the orbit of an Earth-like planet is stable. Within this framework, Haghighipour and Raymond (2007) considered a binary star with a giant planet in an S-type orbit and simulated the late stage of the formation of Earth-like planets in the habitable zone of its primary star. In their simulations, these authors assumed that

- The primary star is Sun-like with a habitable zone extending from 0.9 AU to 1.5 AU (Kasting et al. 1993).
- A Jupiter-mass planet has already formed in a circular orbit at 5 AU from the primary star.
- The collisional growth of planetesimals has been efficient and has formed a disk of planetary embryos (e.g., via oligarchic growth; Kokubo and Ida 1998).
- The water-mass fraction of embryos is similar to the current distribution of water in primitive asteroids of the asteroid belt (Abe et al. 2000). That is, embryos inside 2 AU are dry, the ones between 2 and 2.5 AU contain 1% water, and those beyond 2.5 AU have a water-mass fraction of 5% (Raymond et al. 2004, 2005a,b; Raymond 2006a,b).
- The initial iron content for each embryo is obtained by interpolating between the values of the iron contents of the terrestrial planets (Lodders and Fegley 1998; Raymond et al. 2005a,b), with a dummy value of 40% in place of Mercury because of its anomalously high iron content.

The model of Haghighipour and Raymond (2007) also includes a circumprimary disk of 115 Moon-to Mars-sized bodies, with masses ranging from 0.01 to 0.1 Earth-masses. These objects were randomly distributed between 0.5 and 4 AU by 3 to 6 mutual Hill radii. The masses of embryos were increased with their semimajor axes (a) and the number of their mutual Hill radii (Δ) as $a^{3/4}\Delta^{3/2}$ (Raymond et al. 2004). The surface density of the disk was assumed to vary as $r^{-1.5}$, where r is the radial distance from the primary star, and was normalized to a density of 8.2 g/cm^2 at 1 AU. Figure 11.21 shows the graph of one of such disks where the total mass is approximately four Earth-masses.

The late stage of terrestrial planet formation (Wetherill 1996) was simulated by numerically integrating the orbits of the planetary embryos for different values of the mass (0.5, 1.0, 1.5 solar-masses), semimajor axis (20, 30, 40 AU), and orbital eccentricity (0, 0.2, 0.4) of the secondary star. The collisions among planetary embryos (which are the consequence of the increase of their eccentricities due to their interactions with the secondary star (Charnoz et al. 2001) and the giant planet) were considered to be perfectly inelastic, with no debris generated, and no changes in the

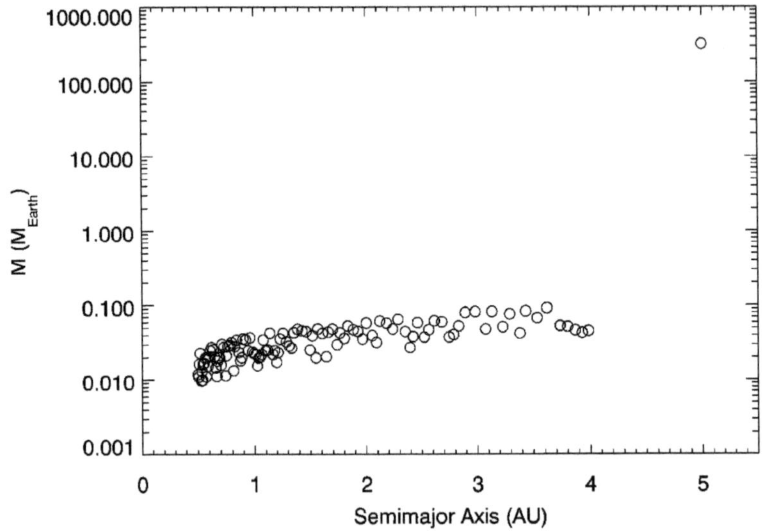

Fig. 11.21 Radial distribution of original protoplanetary objects (Haghighipour and Raymond 2007)

morphology and structures of the impacting bodies. Similar to the current models of the formation of terrestrial planets in our solar system, Haghighipour and Raymond (2007) adopted the model of Morbidelli et al. (2000) in which water-rich bodies originating in the solar system's asteroid belt were the primary source of Earth's water.[9] The delivery of water to a terrestrial planet was then facilitated by allowing transfer of water from one embryo to another during their collision. Figure 11.22 shows the results of several of this simulations for a binary with a mass-ratio of 0.5. The inner planets of the solar system are also shown for a comparison. As shown here, several Earth-size planets, some with substantial amount of water, are formed in the habitable zone of the primary star.

An interesting result shown in Fig. 11.22 is the relation between the orbital eccentricity of the stellar companion and the water content of the final bodies. As shown here, in systems where the secondary star has larger orbital eccentricity, the amount of water in final planets is smaller. This can be seen more clearly in Fig. 11.23, where the final assembly and water contents of planets are shown for a circular and an eccentric binary. As shown here, for identical initial distributions of planetary embryos (i.e., simulations on the same rows), the total water content of the system on the left, where the secondary star is in a circular orbit, is higher than that of the system on the right, where the orbit of the secondary is eccentric. This can be at-

[9] It is important to emphasize that the delivery of water to the inner part of the solar system might not have been entirely due to the radial mixing of planetary embryos. Smaller objects such as planetesimals in the outer region of the asteroid belt, and comets originating in the outer solar system, might have also contributed Raymond et al. (2007).

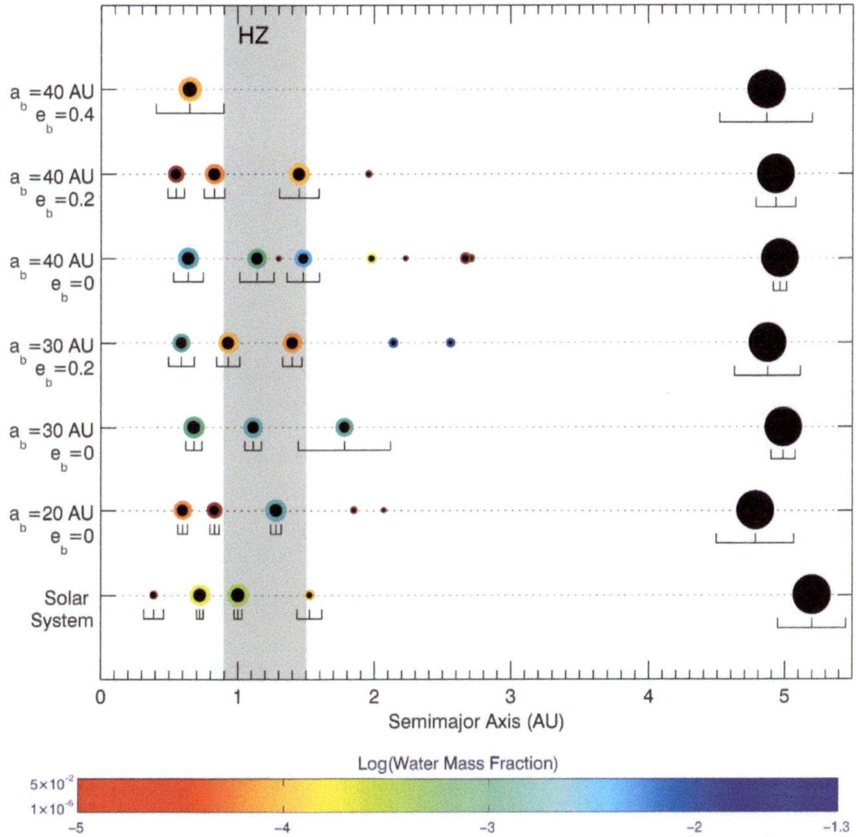

Fig. 11.22 Results of the simulations of habitable planet formation in a binary-planetary system with $\mu = 0.5$, for different values of the eccentricity (e_b) and semimajor axis (a_b) of the stellar companion. The inner planets of the solar system are shown for a comparison. As seen from this figure, several Earth-like planets with substantial amount of water are formed in the habitable zone of the star (Haghighipour and Raymond 2007)

tributed to the fact that in an eccentric binary, because of the close approach of the secondary star to the disk of planetary embryos, most of the water-carrying objects at the outer regions of the disk leave the system prior to the formation of terrestrial planets (Artymowicz and Lubow 1994; David et al. 2003). Simulations indicate that on average 90% of embryos in these systems were ejected during the integration (i.e., their semimajor axes exceeded 100 AU) and among them, 60% collided with other protoplanetary bodies prior to their ejection from the system. A small fraction of embryos (~5%) also collided with the primary or secondary star, or with the Jupiter-like planet of the system (Haghighipour and Raymond 2007).

In a binary-planetary system, the destabilizing effect of the secondary star is enhanced by the presence of the giant planet. Similar to our solar system, in these binaries, the Jovian-type planet perturbs the motion of embryos and enhances their radial mixing and the rate of their collisions by transferring angular momentum

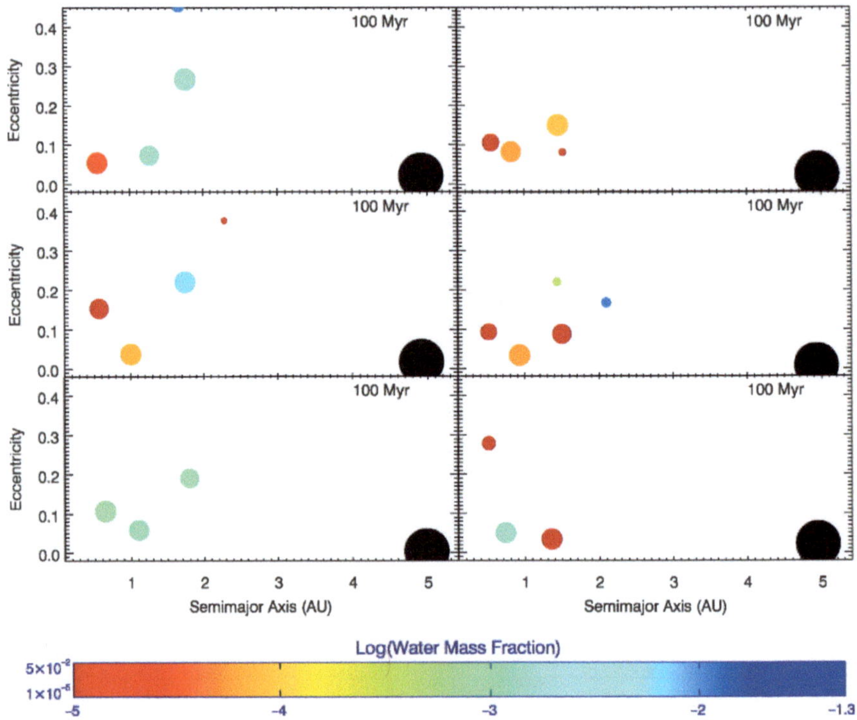

Fig. 11.23 The *left column* shows the results of three simulations for different distribution of planetary embryos in a binary with equal-mass Sun-like stars. The orbit of the secondary star is circular with a radius of 30 AU. The *right column* shows the results of simulations for the same binary stars and similar distributions of planetary embryos where the secondary is in an orbit with a semimajor axis of 40 AU and eccentricity of 0.2 (Haghighipour and Raymond 2007)

from the secondary star to these objects (Chambers and Cassen 2002; Levison and Agnor 2003; Raymond et al. 2004; Raymond 2006a). Figure 11.24 shows this in more details. The binary systems in these simulations have mass-ratios of 0.5, and their secondary stars are at 30 AU. The binary eccentricity in these systems is equal to 0, 0.2 and 0.4, from top to bottom. As shown here, as the eccentricity of the binary increases, the interaction of the secondary star with the giant planet of the system becomes stronger (see the final eccentricity of the giant planet), which causes closer approaches of this object to the disk of planetary embryos and enhancing collisions and mixing among these bodies. The eccentricities of embryos, at distances close to the outer edge of the protoplanetary disk, rise to higher values until these bodies are ejected from the system. In binaries with smaller perihelia, the process of transferring angular momentum by means of the giant planet is stronger and the ejection of protoplanets occurs at earlier times. As a result, the total water budget of such systems is small. A comparison between Figs. 11.24 and 11.25, where simulations were carried out for a binary without a giant planet, illustrates the significance of the intermediate effect of the giant planet in a better way. As shown by Fig. 11.25,

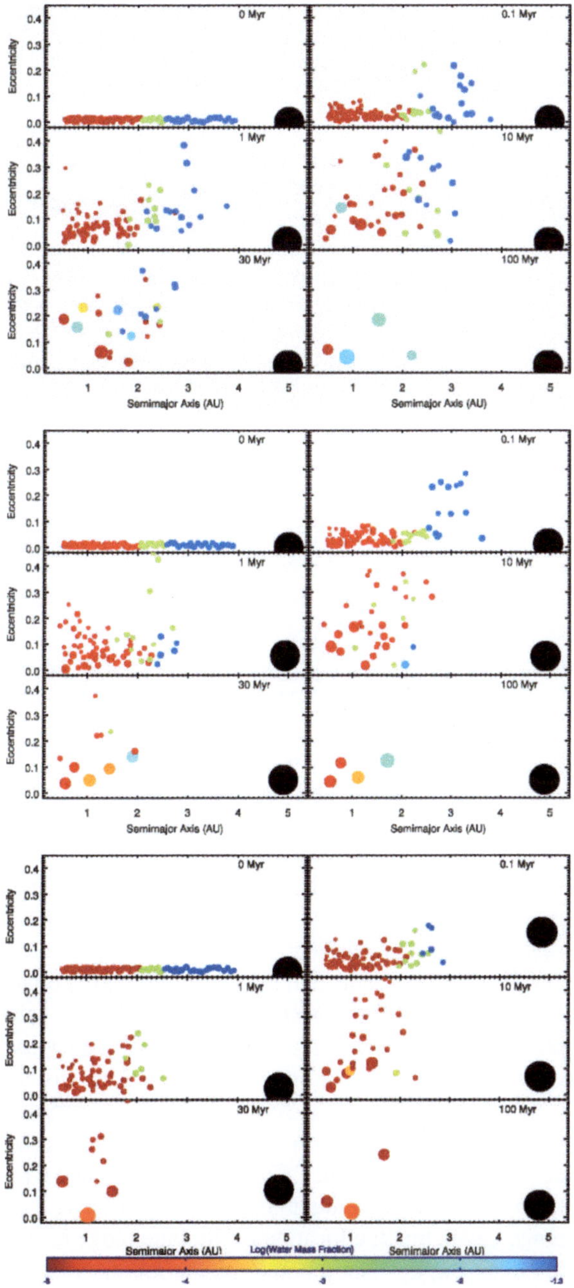

Fig. 11.24 Variation of water contents of the final planets with the eccentricity of the stellar companion. In these simulations, the primary star has a mass of 0.5 solar-masses, the semimajor axis of the binary is 30 AU, and its eccentricity is equal to 0,0.2, and 0.4, from *top to bottom* (Haghighipour and Raymond 2007)

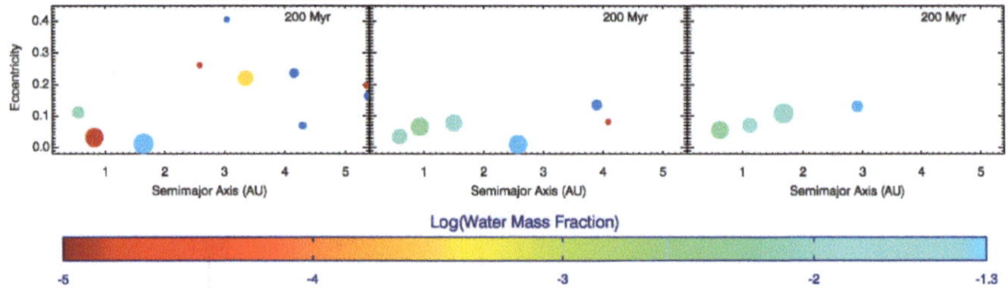

Fig. 11.25 Habitable planet formation in binary systems with no Jupiter-like planet. The stars of each binary are Sun-like and their separations are 30 AU. The orbital eccentricity of the secondary star is 0, 0.2, and 0.4, for the systems on the *left*, *middle*, and *right*, respectively. Note that compared with previous simulations, the time of integration has to increase to 200 Myr in order to form comparable terrestrial-class planets (Haghighipour and Raymond 2007)

it is still possible to form terrestrial-class planets, with significant amount of water, in the habitable zone of the primary star. However, because of the lack of the transfer of angular momentum through the Jovian-type planet, the radial mixing of these objects is slower and terrestrial planet formation takes longer.

Another interesting result depicted by Fig. 11.25 is the decrease in the number of the final terrestrial planets and increase in their sizes and accumulative water contents with increasing the eccentricity of the secondary star. As shown here, from left panel to the right, as the binary eccentricity increases, the close approach of the secondary star to the protoplanetary disk increases the rate of the interaction of these objects and enhances their collisions and radial mixing. As a result, more of the water-carrying embryos participate in the formation of the final terrestrial planets. It is important to emphasize that this process is efficient only in moderately eccentric binaries. In binary systems with high eccentricities (small perihelia), embryos may be ejected from the system (David et al. 2003), and terrestrial planet formation may become inefficient.

The results of the simulations without a giant planet imply a trend between the location of the outer terrestrial planet and the perihelion of the binary. In Fig. 11.26 this has been shown for a set of different simulations. The top panel in this figure represents the semimajor axis of the outermost terrestrial planet, $a_{\rm out}$, as a function of the binary eccentricity, e_b. The bottom panel shows the ratio of this quantity to the perihelion distance of the binary, q_b. As shown here, terrestrial planet formation in binaries without a giant planet seems to favor the region interior to approximately 0.19 times the binary perihelion distance. This has also been noted by Quintana et al. (2007) (see their Fig. 9) in their simulations of terrestrial planet formation in close binary star systems. Given that the location of the inner edge of the habitable zone is at 0.9 AU, this trend implies that a binary perihelion distance of approximately $0.9/0.19 = 4.7$ AU or larger may be necessary to allow habitable planet formation. According to these results, habitable planet formation may not succeed in binaries with Sun-like primaries that have stellar companions with perihelion distances smaller than $\sim$5 AU. This, of course, is not a stringent condition, and is neither surprising since a moderately close binary with a perihelion distance of $\sim$5 AU would

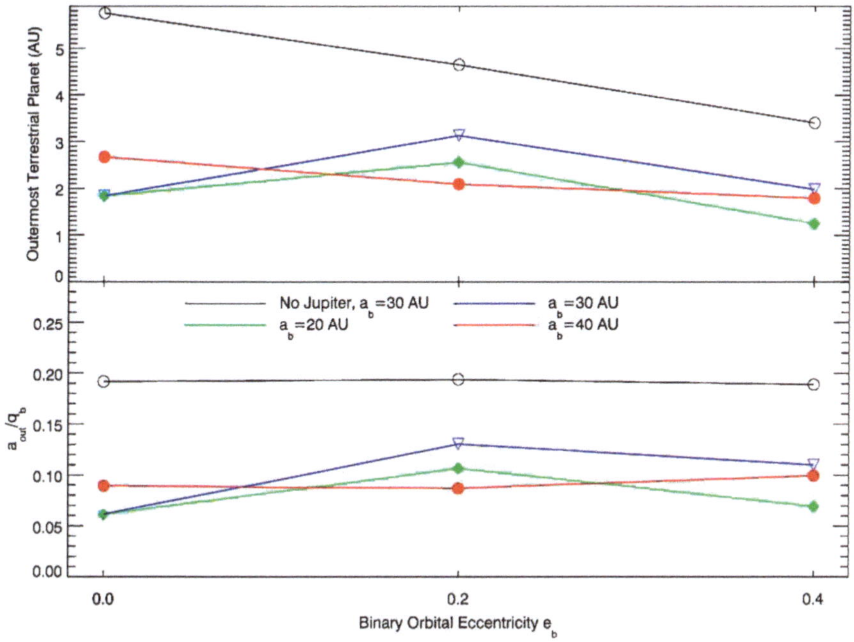

Fig. 11.26 *Top panel*: Semimajor axis of the outermost terrestrial planet. *Bottom panel*: The ratio of the semimajor axis of this object to the perihelion distance of the binary. The secondary star is solar mass (Haghighipour and Raymond 2007)

be quite eccentric, and as indicated by Holman and Wiegert (1999) the orbit of a terrestrial-class object in the region around 1 AU from the primary of such a system will be unstable [see Eq. (11.1)].[10]

In binary systems where a giant planet exist, Fig. 11.26 indicates that terrestrial planets form closer-in. The ratio a_{out}/q_b in these systems varies between approximately 0.06 and 0.13, depending on the orbital separation of the two stars. The accretion process in such systems is more complicated since the giant planet's eccentricity and its ability to transfer angular momentum are largely regulated by the binary companion.

As shown by Haghighipour and Raymond (2007), despite the stochasticity of the simulations, and the large size of the parameter-space, many simulations resulted in the formation of Earth-like planet, with substantial amount of water, in the habitable zone of the primary star. Figures 11.27 and 11.28 show the result. The orbital elements of the final objects are given in Table 11.4. It is necessary to mention that because in these simulations, all collisions have been considered to be perfectly

[10] It is, also, important to note that, because the stellar luminosity, and therefore the location of the habitable zone, are sensitive to stellar mass (Kasting et al. 1993; Raymond et al. 2007), the minimum binary separation necessary to ensure habitable planet formation will vary significantly with the mass of the primary star.

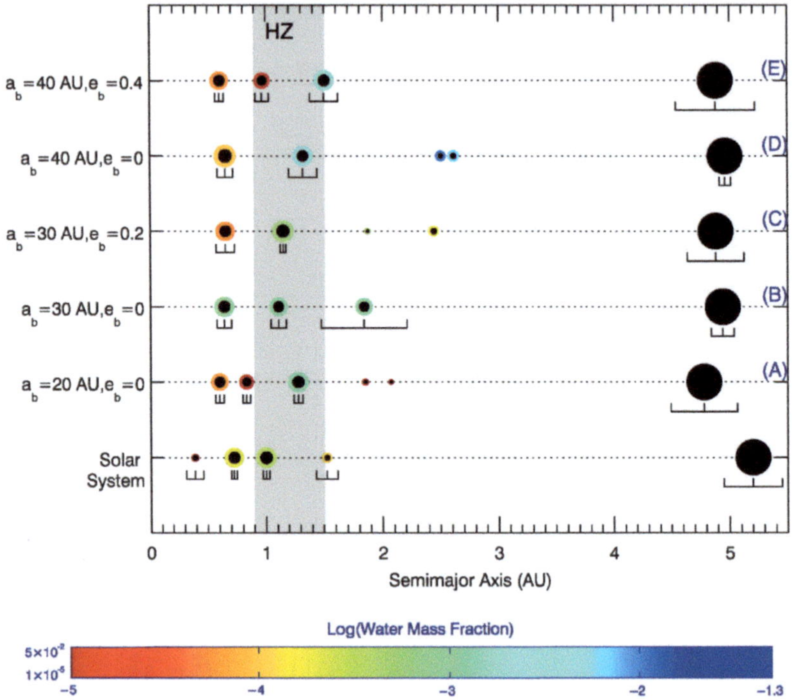

Fig. 11.27 Variation of number and water content of Earth-like objects with the perihelion of the secondary star. The mass of the secondary star in all simulations is 1 solar-mass (Haghighipour and Raymond 2007)

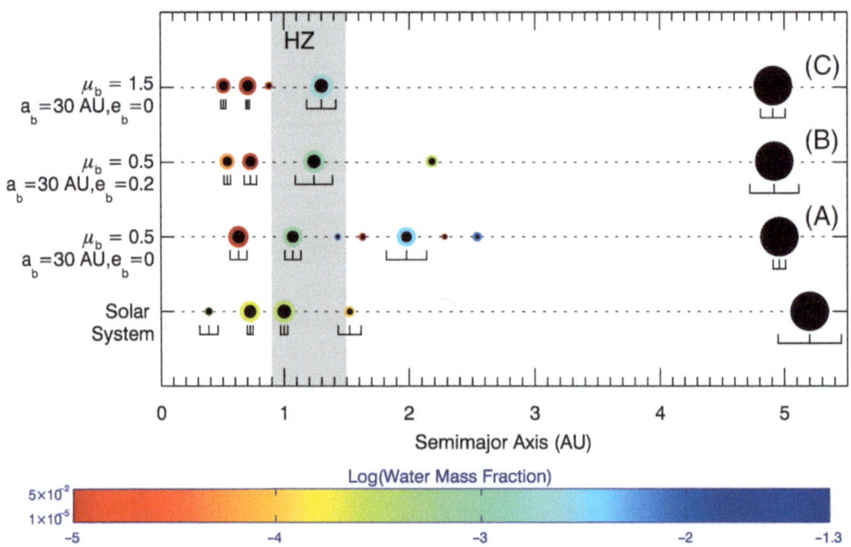

Fig. 11.28 Variation of number and water content of Earth-like objects with the mass-ratio of the binary (Haghighipour and Raymond 2007)

Table 11.4 Sun's habitable zone: 0.9–1.50 AU

Simulation	m_p ($M_\oplus$)	a_p (AU)	e_p	Water fraction
9-A	0.95	1.28	0.03	0.00421
9-B	0.75	1.11	0.06	0.00415
9-C	1.17	1.16	0.03	0.00164
9-D	0.86	1.33	0.09	0.01070
9-E	0.95	1.50	0.08	0.00868
10-A	0.74	1.07	0.06	0.00349
10-B	0.99	1.26	0.12	0.00366
10-C	1.23	1.30	0.09	0.00103

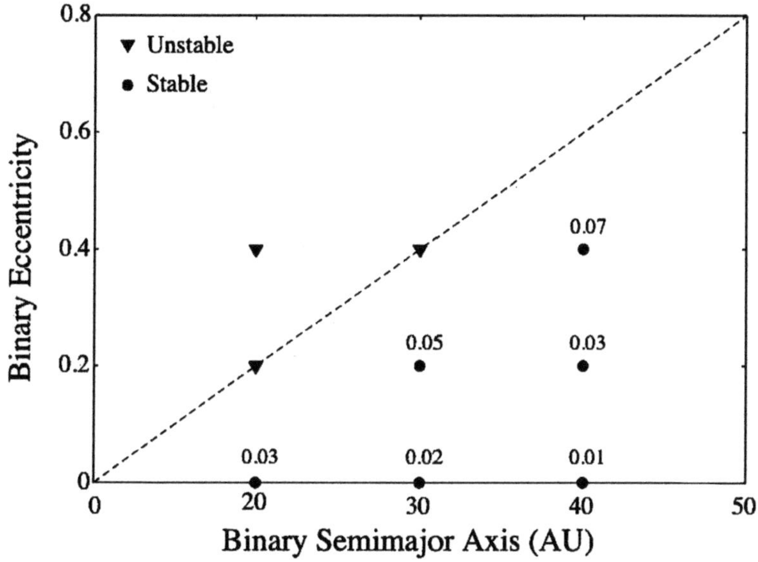

Fig. 11.29 Habitable planet formation in the (e_b, a_b) space of an equal-mass binary-planetary system. Circles correspond to binaries in which habitable planets are formed. *Triangles* represent systems in which the giant planet is unstable. The number associated with each *circle* represents the average eccentricity of the giant planet of the system during the simulation (Haghighipour and Raymond 2007)

inelastic (i.e., the water contents of the resulted planets would be equal to the sum of the water contents of the impacting bodies, and the loss of water due to the impact and the motion of the ground of an impacted body (Genda and Abe 2005; Canup and Pierazzo 2006) has been ignored), the numbers given in Table 11.4 set an upper limit for the water budget of final planets. The total water budget of these objects may in fact be 5–10 times smaller than those reported here (Raymond et al. 2004).

A study of the systems of Figs. 11.27 and 11.28 indicates that these binaries have relatively large perihelia. Figure 11.29 shows this for simulations in a binary with a mass-ratio of 0.5 in terms of the semimajor axis and eccentricity of the stellar companion. The circles in this figure represent those systems in which the giant planet maintained a stable orbit and also simulations resulted in the formation of habitable bodies. The number associated with each circle corresponds to the average eccentricity of the giant planet during the simulation. The triangles correspond to systems in which the giant planet became unstable. Given that at the beginning of each simulation, the orbit of the giant planet was circular, a non-zero value for its average eccentricity is indicative of its interaction with the secondary star. The fact that Earth-like objects were formed in systems where the average eccentricity of the giant planet is small implies that this interaction has been weak. In other words, binaries with moderate to large perihelia and with giant planets on low eccentricity orbits are most favorable for habitable planet formation. Similar to the formation of habitable planets around single stars, where giant planets, in general, play destructive roles, a strong interaction between the secondary star and the giant planet in a binary-planetary system (i.e., a small binary perihelion) increases the orbital eccentricity of this object, and results in the removal of the terrestrial planet-forming materials from the system. For more details we refer the reader to Haghighipour and Raymond (2007).

References

Abe, Y., Ohtani, E., Okuchi, T., et al.: in Origin of the Earth and the Moon, Eds. Righter K. & Canup R. (Tucson: Univ. Arizona Press, 2000), 413

Akeson, R. L., Koerner, D. W. & Jensen, E. L. N.: AJ, **505**, 358 (1998)

Artymowicz, P. & Lubow, S. H.: AJ, **421**, 651 (1994)

Asghari, N., Broeg, C., Carone L., et al.: A&A, **426**, 31 (2004)

Barbieri, M., Marzari, F. & Scholl, H.: A&A, **396**, 219 (2002)

Benest, D.: A&A, **206**, 143 (1988a)

Benest, D.: CMDA, **43**, 47 (1988b)

Benest, D.: A&A, **223**, 361 (1989)

Benest, D.: CMDA, **56**, 45 (1993)

Benest, D.: A&A, **314**, 983 (1996)

Benest, D.: A&A, **332**, 1147 (1998)

Black, D.C.: AJ, **87**, 1333 (1982)

Bois, E., Kiseleva-Eggleton, L., Rambaux, N. & Pilat-Lohinger, E.: ApJ, **598**,1312 (2003)

Broucke, R. A.: CMDA, **81**, 321 (2001)

Canup, R. M. & Pierazzo, E.: 37th Annual Lunar and Planetary Science Conference, 2146 (2006)

Chambers, J. E. & Cassen, P.: M&PS, **37**, 1523 (2002)

Charnoz, S., Thébault, P. & Brahic, A.: A&A, **373**, 683 (2001)

Capen, S., Haghighipour, N. & Kirste, S.: to appear in BAAS (2009)

David, E., Quintana, E. V., Fatuzzo, M. & Adams, F. C.: PASP, **115**, 825 (2003)

Deprit, A. & Rom, A.: A&A, **5** 416 (1970)

Diakov, B. B. & Reznikov, B. I. : Moon. Planet., **23**, 429 (1980)

Domingos, R. C., Winter, O. C. & Yokoyama, T.: MNRAS, **373**, 1227 (2006)

Donnison, J.R. & Mikulskis, D.F.: MNRAS, **266**, 25 (1994)

Drobyshevski, E. M.: Moon. Planet., **18**, 145 (1978)

Dvorak, R.: Celest. Mech., **34**, 369 (1984)

Dvorak, R.: A&A, **167**, 379 (1986)

Dvorak, R., Froeschlé, Ch. & Froeschlé, C.: A&A, **226**, 335 (1989)

Dvorak, R. & Lohinger, E.: in Proceedings of the NATO ASI Series, Ed. Roy A.E. (New York and London: Plenum Press, 1991, 439)

Dvorak, R. & Freistetter, F.: in Chaos and Stability in Planetary Systems, Eds. Dvorak, R., Freistetter, F. & Kurths, J. (Springer, Berlin Heidelberg, 2003, **LNP 683,** 50)
Dvorak, R., Pilat-Lohinger, E., Funk B. & Freistetter, F.: A&A, **398**, L1 (2003a)
Dvorak, R., Pilat-Lohinger E., Funk B. & Freistetter F.: A&A, **410**, L13-L16 (2003b)
Dvorak, R., Pilat-Lohinger, E., Schwarz, R. & Freistetter, F.: A&A, **426**, 37 (2004b)
Eggenberger, A., Udry, S. & Mayor, M.: in Scientific Frontiers in Research on Extrasolar Planets, ASP Conference Series, Eds. Deming, D. & Seager, S. (San Francisco, 2001), **294**, 43
Érdi, B. & A. Pál, A.: Dynamics of resonant exoplanetary systems, in: Proceedings of the 3rd Austrian-Hungarian Workshop on Trojans and Related Topic, Eds. F. Freistetter, R. Dvorak. & B. Érdi (Eötvös University Press), pp. 3 (2003)
Érdi, E., Dvorak, R., Sándor, Zs. & Pilat-Lohinger, E.: MNRAS, **351**, 1943 (2004)
Érdi, B. & Sándor, Zs.: CMDA, **92**, 113 (2005)
Fatuzzo, M., Adams, F. C., Gauvin, R. & Proszkow, E. M.: PASP, **118**, 1510 (2006)
Forget, F. & Pierrehumbert, R. T.: Science, **278**, 1273 (1997)
Froeschlé, C., Lega, E. & Gonczi, R.: CMDA, **67**, 41 (1997)
Fuhrmann, K.: Astron. Nachr., **325**, 3 (2004)
Funk, B., Pilat-Lohinger, E., Dvorak, R., Freistetter, F. & Erdi, B.: CMDA, **90**, 43 (2004)
Genda, H. & Abe, Y.: 36th Annual Lunar and Planetary Science Conference, 2265 (2005)
Gayon, J. & Bois, E.: A&A, **482**, 665
Graziani, F. & Black, D.C.: ApJ, **251**, 337 (1981)
Haghighipour, N.: in AIP Conference Proceedings 713: The Search for Other Worlds, Eds. S. S. Holt, & D. Deming, American Institute of Physics (New York, 2004) pp. 269
Haghighipour, N.: BAAS, **37**, 526 (2005)
Haghighipour, N.: ApJ, **644**, 543 (2006)
Haghighipour, N. & Raymond, S. N.: ApJ, **666**, 436 (2007)
Haghighipour, N., Steffen, J., Hinse, T. & Agol, A.: Am. Geophys. U., P14B-04 (2008)
Harrington, R.S.: Celest. Mech., **6**, 322 (1972)
Harrington, R.S.: AJ, **80**, 1081 (1975)
Harrington, R.S.: AJ, **82**, 753 (1977)
Hatzes, A. P., Cochran, W. D., Endl, M., et al.: ApJ, **599**, 1383 (2003)
Hénon, M. & Guyot, M.: in Periodic Orbits, Stability and Resonances, Ed. Giacaglia, G.E.O., D. Reidel Publishing Company (Dordrecht, Netherlands, 1970) pp. 349
Heppenheimer, T. A.: Icarus, **22**, 436 (1974)
Heppenheimer, T. A.: A&A, **65**, 421 (1978)
Holman M. J. & Wiegert P. A.: AJ, **117**, 621 (1999)
Innanen, K. A., Zheng, J. Q., Mikkola, S. & Valtonen, M. J.: AJ, **113**, 1915 (1997)
Jewitt D., & Haghighipour, N.: ARA&A, **45**, 261 (2007)
Jones, B. W., Sleep, P. N., Chamber, J. R.: A&A, **366**, 254 (2001)
Jones, B. W., Underwood, D. R. & Sleep, P. N.: ApJ, **622**, 1091 (2005)
Jones, B. W., Sleep, P. N. & Underwood, D. R.: 2006, ApJ, **649**, 1010 (2006)
Kasting, J. F., Whitmire, D. P. & Reynolds, R. T.: Icarus, **101**, 108 (1993)
Kokubo, E. & Ida, S.: Icarus, **131**, 171 (1998)
Kortenkamp, S. J., Wetherill, G. W. & Inaba, S.: Science, **293**, 1127 (2001)
Kozai, Y.: AJ, **67**, 591 (1962)
Kubala, A., Black, D. C. & Szebehely, V.: CMDA, **56**, 51 (1993)
Laughlin, G. & Chambers, J. E.: AJ, **124**, 592 (2002)
Levison, H. F. & Agnor, C.: AJ, **125**, 2692 (2003)
Lissauer, J. J., Quintana, E. V., Chambers, J. E., et al.: RevMexAA (Series de Conferencias), **22**, 99 (2004)
Lodders, K. & Fegley, B., Jr.: M & PS, **33**, 871 (1998)
Lohinger, E. & Dvorak, R.: A&A, **280**, 683 (1993)
Malmberg, D., Davies, M. B. & Chambers, J. E.: MNRAS, **377**, L1 (2007)
Marzari, F., Scholl, H., Tomasella, L. & Vanzani, V.: Planet. & Sp. Sci., **45**, 337 (1997)
Marzari, F. & Scholl, H.: ApJ, **543**, 328 (2000)

Marzari, F., Thébault, P., & Scholl, H.: Submitted to A&A (arXiv: 0908.0803v2)
Mathieu, R. D.: ARA& A, **32**, 465 (1994)
Mathieu, R. D., Ghez, A. M., Jensen, E. L. & Simon, M.: in Protostars and Planets IV, Eds. Mannings V., Boss A. P. & Russell S. S., (Tucson: Univ. Arizona Press, 2000), pp. 703
Menou, K. & Tabachnik, S.: ApJ, **583**, 473 (2003)
Mischna, M. A., Kasting, J. F., Pavlov, A. & Freedman, R.: Icarus, **145**, 546 (2000)
Morbidelli, A., Chambers, J., Lunine, J. I., et al.: MAPS, **35**, 1309 (2000)
Moriwaki, K. & Nakagawa, Y.: ApJ, **609**, 1065 (2004)
Musielak, Z. E., Cuntz, M., Marshall, E. A. & Stuit, T. D.: A&A, **434**, 355 (2005)
Nauenberg, M.: AJ, **124**, 2332 (2002)
Neuhäuser, R., Mugrauer, M., Fukagawa, M., et al.: A&A, **462**, 777 (2007)
Norwood, J. W. & Haghighipour, N.: BAAS, **34**, 892 (2002)
Paardekooper, S. J., Thébault, P. & Mellema, G.: MNRAS, **386**, 973 (2008)
Pendleton, Y. J. & Black, D. C.: AJ, **88**, 1415 (1983)
Pichardo, B., Sparke, L. S. & Aguilar, L. A. : MNRAS, **359**, 521 (2005)
Pilat-Lohinger, E., Suli, A., Freistetter, F., et al.: 2006, in Epsc. Conf., European Planetary Science Congress (Berlin, Germany, 2006), pp. 717
Pilat-Lohinger, E. & Dvorak, R.: CMDA, **82**, 143 (2002)
Pilat-Lohinger, E., Funk, B., Dvorak, R.: A&A, **400**, 1085 (2003)
Pilat-Lohinger, E., Dvorak, R., Bois, E. & Funk, B.: in APS Conference Series 321: Extrasolar Planets: Today and Tomorrow, Eds. Beaulieu J. P., Lecavelier des Etangs A. & Terquem C., (2004), pp. 410
Pilat-Lohinger, E.: 2005, in Dynamics of Populations of Planetary Systems, Proceedings of IAU Coll. 197, Eds. Knezevic Z. & Milani A. (Cambridge, UK, Cambridge University Press, 2005), pp. 71
Pilat-Lohinger E. & Funk, B.: in Proceedings of the 4th Austro-Hungarian Workshop, Department of Astronomy of the Eštvšs University (Budapest: 2006)
Quintana, E. V., Lissauer, J. J., Chambers, J. E. & Duncan, M. J.: ApJ, **576**, 982 (2002)
Quintana, E. V. & Lissauer, J. J.: Icarus, **185**, 1 (2006)
Quintana, E. V., Adams, F. C., Lissauer, J. J. & Chambers, J. E.: ApJ, **660**, 807 (2007)
Rabe, E.: AJ, **66**, 500 (1961)
Rabl, G. & Dvorak, R.: A&A, **191**, 385 (1988)
Raghavan, D., Henry, T. J., Mason, B. D., et al.: ApJ, **646**, 523 (2006)
Raymond, S. N., Quinn, T. & Lunine, J., I.: Icarus, **168**, 1 (2004)
Raymond, S. N., Quinn, T. & Lunine, J., I.: ApJ, **632**, 670 (2005a)
Raymond, S. N., Quinn, T. & Lunine, J., I.: Icarus, **177**, 256 (2005b)
Raymond, S. N.: ApJL, **643**, L131 (2006a)
Raymond, S. N., Barnes, R. & Kaib, N. A.: ApJ, **644**, 1223 (2006b)
Raymond, S. N., Quinn, T. & Lunine, J.I.: Astrobiology, **7**, 66 (2007)
Rodriguez, L. F., D'Alessio, P., Wilner, D. J., et al.: Nature, **395**, 355 (1998)
Saleh, L. A. & Rasio, F. A.: ApJ, **694**, 1566 (2009)
Sándor, Zs. & Érdi, B.: CMDA, **86**, 301 (2003)
Sandor, Zs., Suli., A., Erdi, B., et al.: MNRAS, **374**, 1495 (2006)
Schwarz, R.: Ph.D. thesis, University of Vienna, online database: http://media.obvsg.at/dissdb (2005)
Schwarz, R., Dvorak, R., Pilat Lohinger, E., et al.: A&A, **462**, 1165 (2007a)
Schwarz, R., Dvorak, R., Süli, Á. & Érdi, B.: A&A, **474** 1023 (2007b)
Schwarz, R., Süli, Á. & Dvorak, R.: MNRAS, **398** 2085 (2009)
Silbert, J., Gledhill, T., Duchéne, G. & Ménard, F.: ApJ, **536**, L89 (2000)
Szebehely, V.: Celest. Mech., **22**, 7 (1980)
Szebehely, V. & McKenzie, R.: Celest. Mech., **23**, 3 (1981)
Szebehely, V.: Celest. Mech., **34**, 49
Takeda, G. & Rasio, F. A.: Astrophy. Sp. Sci., **304**, 239 (2006)
Takeda, G., Kita, R. & Rasio, F. A.: ApJ, **683**, 1063 (2008)

Thébault, P., Marzari, F., Scholl, H., et al.: A&A, **427**, 1097 (2004)
Thébault, P., Marzari, F. & Scholl, H.: Icarus, **183**, 193 (2006)
Thébault, P., Marzari, F. & Scholl, H.: MNRAS, **388**, 1528 (2008)
Thébault, P., Marzari, F. & Scholl, H.: MNRAS, **393**, L21 (2009)
Torres, G: ApJ, **654**, 1095 (2007)
Trilling, D. E., Stansberry, J. A., Stapelfeldt, K. R., et al.: ApJ, **658**, 1264 (2007)
Verrier, P. E. & Evans, N. W.: MNRAS, **368**, 1599 (2006)
White, R. J., Ghez, A. M., Reid, I. N. & Schultz, G.: ApJ, **520**, 811 (1999)
White, R. J. & Ghez, A. M.: ApJ, **556**, 265 (2001)
Whitmire, D. P., Matese, J. L., Criswell, L. & Mikkola, S.: Icarus, **132**, 196 (1998)
Wiegert, P. A. & Holman, M. J.: ApJ, **113**, 1445 (1997)
Wetherill, G. W.: Icarus, **119**, 219 (1996)
Williams, D. M. & Kasting, J. F.: Icarus, **129**, 254 (1997)
Xie, J. W. & Zhou, J. L.: ApJ, **686**, 570 (2008)
Xie, J. W. & Zhou, J. L.: ApJ, **698**, 2066 (2009)
Ziglin, S. L.: Sov. Astron. Let., **1**, 194 (1975)
Zucker, S., Mazeh, T., Santos, N. C., et al.: A&A, 426, **695** (2004)

Index

MIX
Papier aus verantwortungsvollen Quellen
Paper from responsible sources
FSC® C105338

FSC
www.fsc.org

If you have any concerns about our products,
you can contact us on
ProductSafety@springernature.com

In case Publisher is established outside the EU,
the EU authorized representative is:
Springer Nature Customer Service Center GmbH
Europaplatz 3, 69115 Heidelberg, Germany

Printed by Libri Plureos GmbH
in Hamburg, Germany